Energy Storage

This updated edition explores the critical field of energy storage technologies in today's evolving energy landscape. It serves as an essential reference describing various methods of storing energy in the context of renewable energy integration and efficient resource utilization.

Energy Storage: Systems and Components, Second Edition, provides a thorough examination of current storage technologies and analytical frameworks. The book explores systems using electrochemical transformations, classical battery storage elements, flow batteries, fuel cells and hydrogen storage while also investigating storage systems based on physical principles such as gravitational potential forces, air compression and rotational kinetic energy. The book begins with generalities on energy storage systems, establishing fundamental parameter sets for analyzing efficiency and performance characteristics, with special attention to usable energy density across different technologies. It then methodically reviews major storage technologies with detailed technical descriptions and models, covering everything from physics-based solutions like mechanical gravitational systems and pumped hydro to compressed air storage, various electrochemical solutions and chemical transformations. This new edition includes expanded sections on emerging topics like e-methanol production, synfuels for seasonal storage and dry gravitational technology. Readers will benefit from the inclusion of Ragone plot theory, which enables evaluation of real energy release potential relative to power dependency during discharge processes.

This book is ideally suited for practicing engineers and researchers in the energy domain who need comprehensive technical information on storage applications and design principles. It also serves engineering students studying energy systems, renewable integration and storage technologies, providing them with fundamental concepts and practical exercises to enhance understanding.

Energy Storage

Systems and Components

Second Edition

Alfred Rufer

CRC Press
Taylor & Francis Group
Boca Raton London New York

CRC Press is an imprint of the
Taylor & Francis Group, an **informa** business

Designed cover image: Shutterstock

Second edition published 2026
by CRC Press
2385 NW Executive Center Drive, Suite 320, Boca Raton FL 33431

and by CRC Press
4 Park Square, Milton Park, Abingdon, Oxon, OX14 4RN

CRC Press is an imprint of Taylor & Francis Group, LLC

First edition published by CRC Press 2017

ISBN: 978-1-041-23319-0 (hbk)
ISBN: 978-1-041-23321-3 (pbk)
ISBN: 978-1-003-73648-6 (ebk)

DOI: 10.1201/9781003736486

Typeset in Times
by Apex CoVantage, LLC

Contents

About the Author

Alfred Rufer earned an MS at the Swiss Federal Institute of Technology Lausanne (EPFL), Lausanne, Switzerland, in 1976. In 1978, he joined ABB, where he was involved in the fields of high power electronics and control. In 1993, he became an assistant professor at EPFL. Since 1996, he has been a full professor and head of the Industrial Electronics Laboratory, EPFL. He has authored or coauthored more than 200 publications on power electronics and applications, and he holds several patents. Rufer's research activities focus on power converters as modular multilevel converters in asymmetric or hybrid technologies. Another important field initiated by Rufer is dedicated to energy storage, where many applications have been studied. Other new developments have recently been presented, for example, a low-aging, easy-to recycle hybrid energy storage device based on compressed air.

From his retirement in 2016 Alfred Rufer has lead several research projects dedicated to different energy-related topics like renewable energy sources and storage, energy efficiency in general and energy-efficient pneumatic conversion and actuators.

In 2006, Rufer was elected to the IEEE fellow grade.

Symbols

C	Capacity
DoD	Depth of charge
E	Energy (general)
E_b	Energy of the battery (in Wh)
E_{sto}	Storage capacity
e_b	Dimensionless energy of the battery (in p.u.)
e_{fw}	Specific energy of a flywheel
e_h	Specific hydraulic energy
e_m	Specific energy (in Wh/kg)
e_v	Volume energy-density (in Wh/dm^3)
J	Moment of inertia [kgm^2]
K	Form factor of a flywheel
M_e	Acceleration torque [Nm]
M_r	Load torque [Nm]
MPPT	Maximum power point tracking
MRR	Modified Ragone representation
N	Rotational speed (min^{-1})
N	Rotational speed (rad/s)
N_s	Unitless specific speed
P	Power (general) (in W)
PAF	Power amplification factor
p_m	Specific power (in W/kg)
p_v	Volume power density (in W/dm^3)
Q	Electrical charge (in Coulomb)
Q	Heat
Q	Flowrate
R_s	Series resistor
R_p	Parallel resistor
SoC	State of charge
SoE	State of energy
SoH	State of health
W	Mechanical work
η_{cn}	Normal round-trip energy efficiency
η_{cn0}	Normal ideal round-trip energy efficiency
η_0	Self-discharge energy factor
ρ	Density of the material
σ	Tensile strength
Ω	Angular velocity [rad/s]

UNITS

A Ampere
F Farad
h Hour
J Joule
m Meter
s Second
V Volt
W Watt

1 Generalities on Energy Storage

1.1 HISTORY AND CONTEXT OF USE

Large facilities for electrical energy storage have been built in the second half of the 20th century in the context of matching the variable power demand (daily cycles) with the installation of nuclear power plants, known for their mostly constant power production. Figure 1.1a shows the typical power profile of the weekly consumption, and Figure 1.1b illustrates one of the most important pump-storage facilities built in Luxemburg in this context. Many other facilities have been carried out with the same motivations, for example, the Hongrin-Leman facility in Switzerland; the Raccoon Mountain Pumped-Storage Plant west of Chattanooga, Tennessee, United States; or the Yagisawa Power Station in Japan.

From the end of the 20th century, another trend emerges in the context of the development of renewable energy sources. From the classical centralized utility of today, there is a clear movement in the direction of distributed utility of tomorrow, together with the appearance of so-called smart grids (Figure 1.2).

Renewable energy sources are known by their variation in time, or available power related to meteorology conditions. This is clearly a new motivation for the development and the realization of new energy storage systems. An additional reason can be found in the principle of decentralizing the energy production itself. This concerns the power matching between decentralized generators and their loads nearby, which can generate significant and fast variations of the local power demand. Figure 1.3 illustrates the principle of "loosing" the averaging effect of the power due to decentralization of production.

A general tendency towards an increasing use of energy storage can be observed. Two different aspects are considered:

- First, the use of storage technology in order to solve the problem of availability of sources (day-to-night shift for photovoltaic plants, as a first example, or the bridging of lack of production of fluctuating sources)
- Second, the use of energy storage technology in order to assist some problematic consumers, when the local generation cannot follow the strong and fast demand. In this context, two examples can illustrate the problem. First is the case of a micro-turbine, where the fast increase of load must respect some minimum time constant (minutes) due to thermal constraints. Second, in the case of the use of fuel cells, the design of these systems for the maximum peak power can lead to unacceptable costs. The design for the mean value is more realistic, and the highest power demand can be taken out of a storage device.

DOI: 10.1201/9781003736486-1

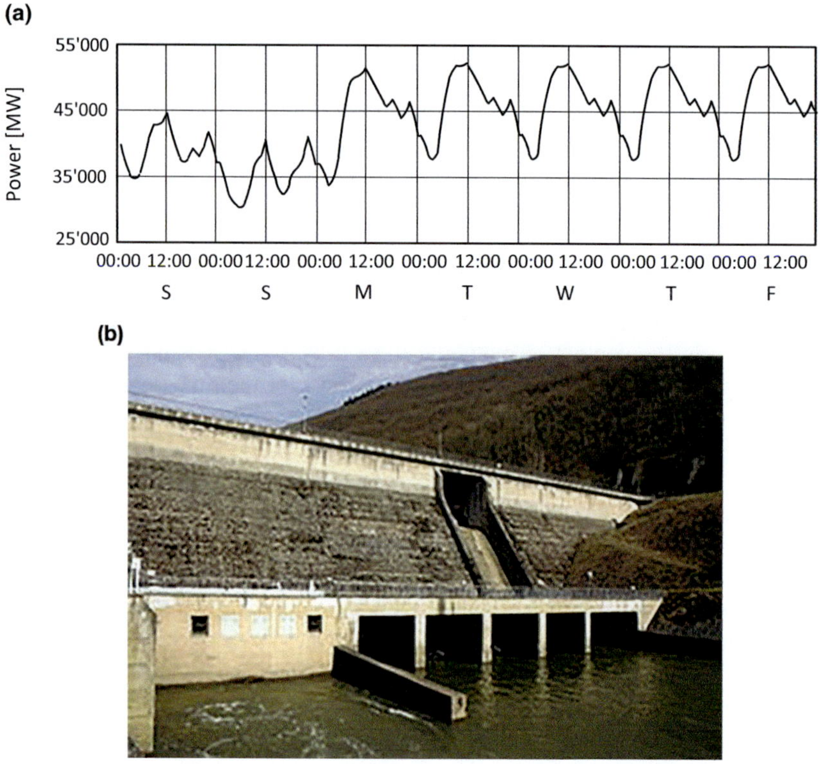

FIGURE 1.1 Daily variations of the power demand: (a) profile of the demand over a week and (b) pumped-storage plant (Vianden, Luxemburg).

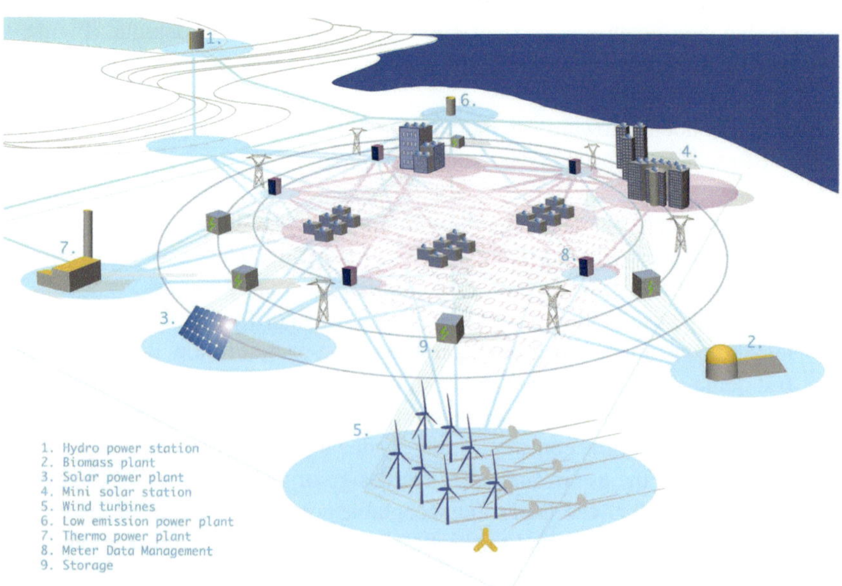

FIGURE 1.2 From centralized utility of today to distributed utility of tomorrow.

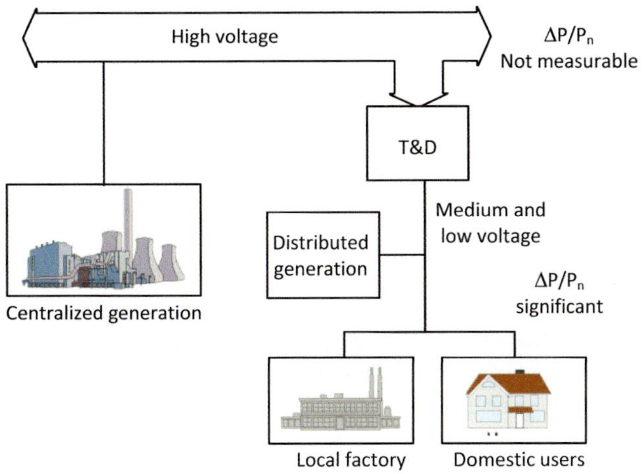

FIGURE 1.3 Power fluctuations ($\Delta P/P_n$) in centralized and decentralized power generation systems.

1.2 GENERAL DEFINITIONS

1.2.1 Definitions of Energy

The most convenient way to define energy is to use its relationship to the integral of the exchanged power.

$$E = \int_{t_1}^{t_2} P(t) \cdot dt \tag{1.1}$$

Another way is to consider the general definitions of a thermodynamic system and to combine them with the macroscopic forms of energy of the system. The sum of all energy forms of a system is called the *total energy.*

The *macroscopic energy* of a system is related to its movement and to the external effects such as gravity, magnetism or electricity.

The *microscopic energy* is related to the molecular activity of a system and is often called the *internal energy.*

A simple definition of the total energy is given by the sum of the internal energy U, the kinetic energy KE and the potential energy PE, leading to the expression of Equation 1.2: [1].

$$E = U + KE + PE = U + m\frac{V^2}{2} + mgz \tag{1.2}$$

where

m is the mass

V is the velocity of the mass from a given reference point

g is the gravitational acceleration
z is the height of the mass center from the reference point.

For a system with a rotating mass, the term for the kinetic energy becomes

$$KE_{rot} = J\frac{\omega^2}{2} \tag{1.3}$$

J is the moment of inertia
ω is the angular velocity.

Equation 1.2 is valid for closed systems. For so-called open systems, an additional term related to the flow of material must be introduced. This related energy is characterized through the mass flowrate:

$$\dot{m} = \rho\dot{v} = \rho A V_{mean} \tag{1.4}$$

where
$\dot{v}$ is the volumetric flowrate
ρ is the density
A is the section of the flow
V_{mean} is the average velocity of the flow.

The energy flowrate associated with the flow of material is given by

$$\dot{E} = \dot{m}e \tag{1.5}$$

where e is the specific energy of the material.

Magnetic and electric effects can play a major role in the energy of given systems like inductors or capacitors. These amounts are considered macroscopic energy and can be calculated as (see Appendix B):

$$E = E_{mag} + E_{el} = \frac{1}{2}LI^2 + \frac{1}{2}CU^2 \tag{1.6}$$

In Equation 1.2, U is the internal energy of a system and includes all forms of microscopic energy at the level of the molecular and atomic scale. It corresponds to kinetic and potential energy of molecules and atoms, of nucleons and electrons and to the energy that binds them together. In the case of energy storage devices, the variation of the internal energy is generally considered, and it can be described through Equation 1.7 [2].

$$dU = -pdV + TdS + \sum_i \mu_i dn_i \tag{1.7}$$

The first term on the right-hand side of the equation is the work done on the system, where p is the pressure and V the volume. The second term on the right corresponds to the heat transferred into the system in terms of temperature T and entropy S.

The third term represents the chemical energy and is the form of the internal energy related to the cohesion between the positive charged nucleus of the atoms and their negative electrons. The chemical energy binds also the atoms in the molecule. μ_i is the chemical potential of a species i and n corresponds to the number of considered moles.

In the case of an electrochemical battery, the internal energy is expressed through the Gibbs energy ΔG. This energy is related to the standard redox potential by the known Nernst law (Equation 1.8).

$$E_{SHE} = -\frac{\Delta G}{nF} = \left[E^0_{ox/red}\right]_{SHE} + \frac{RT}{nF}\ln\left(\frac{Ox}{Red}\right) \tag{1.8}$$

where

(Ox) and (Red) designate the concentrations of the oxidized and reduced forms of the redox couple

R = 8.32 J/K/mol

T is the absolute temperature

n is the number of electrons implicated in the reaction

F is the constant of Faraday, as will be explained later in Chapter 4.

In Equation 1.8, the subscript SHE indicates that the potential is referred to the standard hydrogen electrode.

1.2.2 Energy and Power

The power can be defined as the "energy flowrate" to or from a given system

$$P(t) = \frac{dE}{dt} \tag{1.9}$$

where

P (power) is expressed in Watt [W], [kW], [MW]

E (energy) is expressed in Joules [J], [kWh]

t (time) in seconds [s], hours [h].

$$1J \triangleq 1W \cdot s \triangleq 1N \cdot m$$

Example: Transformation of Electric Power into Heat

- **Energy:** Heating up 100 L of water from 20°C to 60°C needs 4.6 kWh (4.18 J/°C and per gram gives 16.7 MJ or 4.6 kWh).
- **Power:** To do this heating up within 4 hours, a 1.15 kW heater is needed; for 1 hour, a 4.6 kW heater is needed.

1.2.3 About the Distribution of Energy

Energy systems tend to become “electrically controllable” systems or to integrate the “electric vector” somewhere because the most convenient and most efficient way to transport, convert or control the power is in its electric form. Figure 1.4 shows the evident advantages of having one electric drive for each application and using an electric system for the distribution of energy. In the first line, a former mechanical

FIGURE 1.4 From “mechanical” distribution and control to the electric/electronic systems (Courtesy Musee des Arts et Sciences, Sainte-Croix): (a) Mechanical power distribution in a former workshop. (b) Electrical distribution to individual users. (c) The Watt regulator. (d) Modern electronically controlled drive. (Courtesy of Musee des Arts et Sciences, Sainte-Croix, Switzerland.)

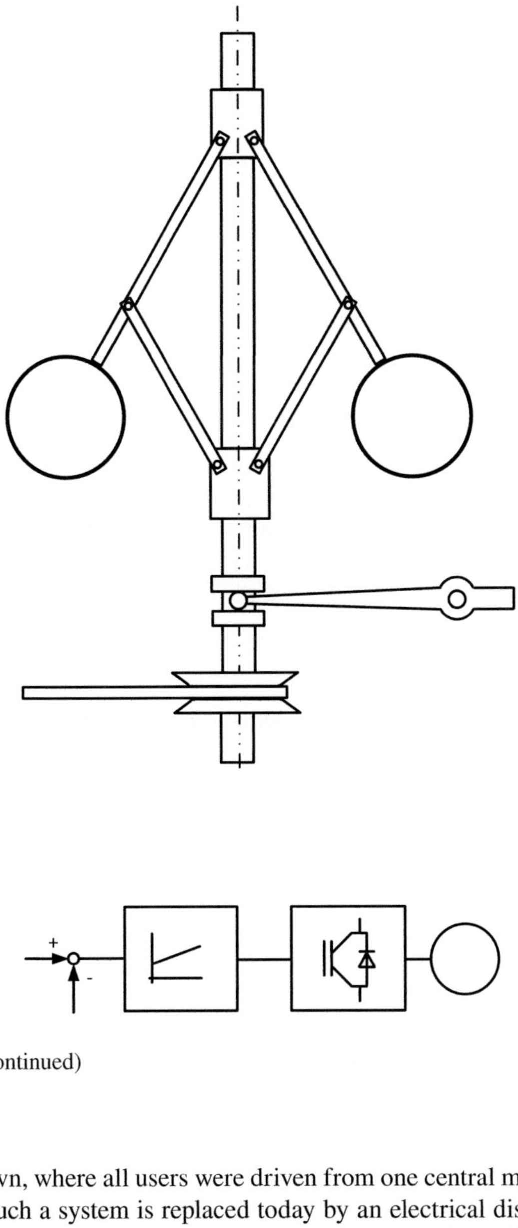

FIGURE 1.4 (Continued)

workshop is shown, where all users were driven from one central motor through pulleys and belts. Such a system is replaced today by an electrical distribution system, allowing the individual engagement and protection of each application. From the control point of view, mechanical Watt controllers have been long replaced by flexible and fast power electronic circuits.

Considering the integration of renewable energy sources as photovoltaics and wind, and taking into account that such types of sources have an unpredictable character, energy storage systems will be in most of the cases interfaced with an electricity transport and distribution system.

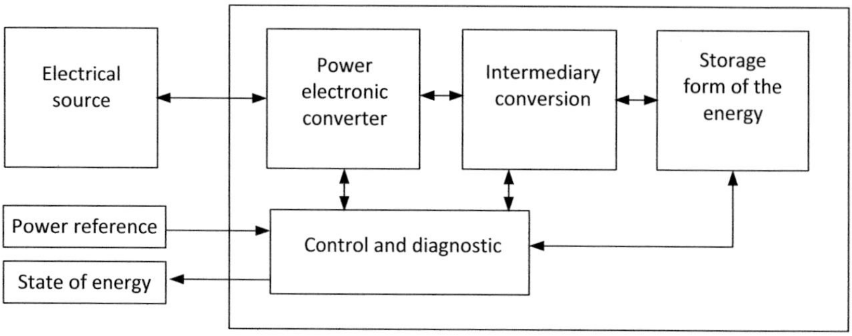

FIGURE 1.5 Electrical energy storage system (ESS). (Adapted from [3].)

On such a base, the scheme represented in Figure 1.5 will be valid for most energy storage systems [3]. The input source on the left side is interfaced with a power electronic converter circuit that allows control of the power flow to the storage means (right). In the middle, one or more additional conversions can be involved as electromechanical, electrothermal or electrochemical conversions.

1.2.4 Storage Form of Energy

In Figure 1.5, the real storage element shown in the diagram indicates the storage form of energy, representing in fact the energy reservoir. This block corresponds to one unit where the change of the internal state is directly related to the change of the energy content with a high reversibility.

The storage forms of energy can be various:

- The potential energy associated with the earth's gravity when one mass is moved from one altitude to another (for example, water in a pumped storage plant)
- The kinetic energy of a mechanical system in rotation (flywheels)
- The pressure of a compressible fluid (compressed air)
- Covalent bonds of given molecules (electrochemicals)
- Electric or magnetic state variables (electrical field, magnetic field, capacitors, superconducting inductors)

1.2.5 Intermediary Conversion

The interface between the storage form unit and the electrical level (output of the power electronic converter) may need a so-called intermediary conversion, shown in Figure 1.5. The intermediary conversion can vary.

For example, it includes the electric motor/generator and the hydraulic pump/turbine in a pump storage plant or is simply the electric driving machine of a flywheel. In the case of compressed air energy storage, the intermediary conversion can involve several successive conversions such as a water pump in a liquid piston system or a simple motor and compressor.

1.2.6 Control and Diagnostic

The control and diagnostic device is essential for numerous different conversion functions such as controlling the speed of the driving machine or simply to control the current or voltage. The control quantities are generally related to the exchanged power level during charge and discharge of the storage device.

The value of the instantaneous charge and discharge power level has to be related to the typical power density (kW/dm^3) or specific power (kW/kg) of the energy storage device itself. These parameters will be defined in more detail in Sections 3.2 and 3.3.

The control and diagnostic unit has to provide the information on the state of charge (SoC) or on the state of energy (SoE). The parameter of depth of discharge (DoD) is also used.

1.3 TECHNICAL DEFINITIONS

To define correctly the application of a given storage device and in order to have clear parameters for the design, technical definitions are needed [4].

1.3.1 C: Capacity

C is the quantity of electricity to be recovered from an accumulator. It is generally expressed in Ah and corresponds to the integral of the current.

$$C = \int_0^t i(t) \cdot dt \tag{1.10}$$

1.3.2 DoD: Depth of Discharge

DoD is the indication of the quantity of electricity already extracted from an accumulator related to its maximum capacity.

$$DoD = \frac{\int_0^t i_{dis}(t) \cdot dt}{Capacity} \tag{1.11}$$

1.3.3 SoC: State of Charge

SoC is an indication of the remaining quantity of electricity available from an accumulator, related to its maximum available capacity.

$$SoC = \frac{\text{Amount of remaining charge}}{\text{Practical capacity of an acumulator}}$$

$$SoC = \frac{Capacity - \int_0^t i_{dis}(t) \cdot dt}{Capacity} \tag{1.12}$$

State of charge and depth of discharge:

$$\text{SoC} = 1 - \text{DoD} \quad \text{or} \quad \text{DoD} = 1 - \text{SoC} \tag{1.13}$$

The SoC and DoD are related to electric charges as current integrals. This is usually applied to electrical and electrochemical accumulators. By analogy, the SoC can be replaced by the SoE, corresponding to the ratio between the remaining amount of energy available (E) and the total amount of stored energy $\left(E_{sto}\right)$. SoE is a more universal parameter and can be used for any storage device.

$$SoE = \frac{E}{E_{sto}} \tag{1.14}$$

1.3.4 "Round-Trip" Efficiency under *Normal, Ideal* and *Real* Conditions

Energy efficiency values are in a very broad context highly sensitive parameters. It is therefore of great importance to specify in a pragmatic way the exact conditions of their evaluation.

The considered energy storage cycles correspond to a series of charges, discharges and idle modes. To define the round trip efficiency properly, it is necessary to first have identical SoE at the end and at the beginning of the cycle. Even if the storage systems are of huge complexity and diverse technologies, the energy loss can be divided into two main categories, the loss due to the power transfer P_{ch} (charge and discharge) and the loss due to self-discharge P_0.

For round-trip efficiency (index *c*), one new definition and notation is now defined that concerns the ratio between the charging and discharging power. In many references, such as [3], the assumption is made that the charge and discharge powers are identical. Such conditions are called "normal", and this condition will be indicated through the index *n*.

A dedicated symbolic is now proposed:

The normal round-trip efficiency: η_{cn}

as a general definition.

Further, one should know through the definition of the used symbol for the round-trip efficiency if it includes the self-discharge loss P_0 or not.

As a consequence, the second proposal is to add the index 0 to the general symbol if it describes:

The ideal normal round-trip efficiency: η_{cn0}

Finally, if the self-discharge losses are included, the symbol is completed by the letter t, leading to

The real normal round-trip efficiency: η_{cnt}

where the additional index t is an indication that the storage time is influencing the final value of the energy balance due to self-discharge.

1.3.5 Charge and Discharge Losses

One can suppose the charge and discharge losses are of the form

$$P_{ch} \cong \alpha \cdot P^2 \tag{1.15}$$

with P being the charge or discharge power.

In the case of an electrochemical battery, the parameter α, which is related to the internal resistor, is dependent on SoE and also on the direction of the energy transfer.

In the electrochemical battery, it can be assumed that the power is proportional to the current (voltage approximatively constant). This is not the case for other storage elements like capacitors (supercapacitors) where the voltage varies strongly with the SoE.

1.3.6 Losses Due to Self-Discharge

The self-discharge losses are generally an increasing function of the SoE. They are noted as:

$$P_0(\mathrm{SoE})$$

1.3.7 Total Losses

Considering one operating cycle of the storage device that is characterized by one specific power profile P(t) and by one state of energy SoE(t), the total losses P_{loss} are equal to the sum of the charge/discharge losses added to the self-discharge losses.

$$P_{loss} = \alpha P^2(t) + P_0\left(SoE(t)\right) \tag{1.16}$$

Further, the dissipated energy E_{loss} over one complete cycle can be calculated as

$$E_{loss\_cyc} = \int_0^{t_{cyc}} [\alpha P^2(t) + P_0(SoE(t))]dt \tag{1.17}$$

By convention, the energy storage device is defined as an *energy sink*, and consequently, *the power P(t) is positive during the charge phase* and *negative during the discharge phase*. The profile of the state of energy SoE is given by

$$SoE(t) = SoE(t_0) + \frac{\int [P(t) - P_{loss}(t)]dt}{E_{sto}} \tag{1.18}$$

E_{sto} is the energetic capacity of the storage device.

As a standard example, Figure 1.6 shows a specific power profile consisting of three successive phases. First a charging phase at constant power P_M during one charging time of t_{ch}, then an idling phase characterized by a low self-discharge and finally a discharging phase at an identical power as during the charging phase but of a slightly reduced duration t_d due to the fact that the recovered energy amount during discharging can only be smaller than the amount transferred during the charging phase.

The schematic power profile of Figure 1.6 illustrates the importance of the specific characteristics of the charge/discharge cycle and intensity of the charging and discharging power as the ratio of the different phase durations, especially the duration of idling mode.

The evolution of the energy losses is indicated. Finally, the state of energy [SoE(t)] is also represented.

1.3.8 Round-Trip Efficiency

The round-trip efficiency η_c is the ratio of the recovered energy E_d during the discharge to the spent energy E_{ch} for the charging, calculated as a mean value over one charging/discharging cycle. Such a cycle can be chosen arbitrarily, but the SoE after discharge must be identical to the SoE before the charging process.

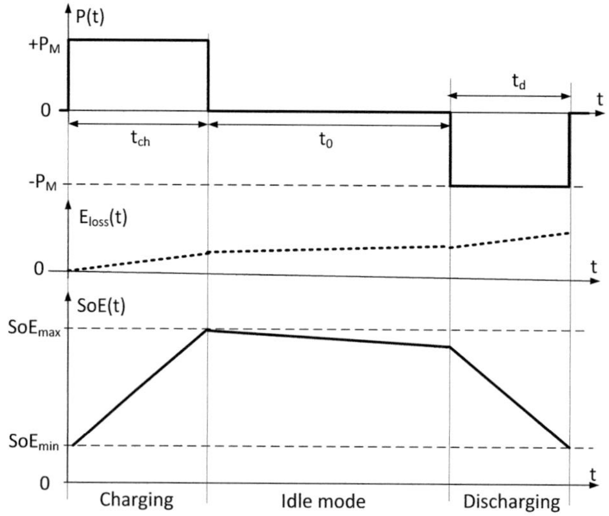

FIGURE 1.6 Typical charging and discharging cycle with power profile. (Adapted from [3].)

In the example of Figure 1.6, the ideal normal round-trip efficiency where the self-discharge loss is neglected is calculated according the definitions in Section 1.3.4 through

$$\eta_{cn0} = \frac{E_d}{E_{ch}} = \frac{P_M \cdot t_d}{P_M \cdot t_{ch}} = \frac{t_d}{t_{ch}} \tag{1.19}$$

The round-trip efficiency can also be expressed by:

$$\eta_{cnt} = \frac{E_{ch} - E_{loss}}{E_{ch}} = \frac{P_M \cdot t_{ch} - \alpha P_M^2 (t_{ch} + t_d) - \int P_0(SoE(t))dt}{P_M \cdot t_{ch}} \tag{1.20}$$

Replacing t_d with $t_d = \eta_{cnt} \cdot t_{ch}$ leads to

$$\eta_{cnt} = \frac{1 - \alpha P_M}{1 + \alpha P_M} - \frac{\int P_0(SoE)(t))dt}{P_M \cdot t_{ch}(1 + \alpha P_M)} \tag{1.21}$$

The round-trip efficiency can be expressed depending on each phase of the cycle, more precisely through consideration of the "instantaneous power efficiencies" during charging η_{ch} and discharging η_d:

$$\eta_{ch} = \frac{P_M - \alpha P_M^2}{P_M} = 1 - \alpha P_M \tag{1.22}$$

$$\eta_d = \frac{P_M}{P_M + \alpha P_M^2} = \frac{1}{1 + \alpha P_M} \tag{1.23}$$

Finally, considering the *self-discharge energy factor:*

$$\eta_0 = \frac{\int P_0(SoE(t))dt}{P_M t_{ch}}$$

we get $\eta_{cnt} = (\eta_{ch} - \eta_0) \cdot \eta_d$ (1.24)

This expression highlights very well the influence of the individual efficiencies of charging, discharging and idling (self-discharge). It must be noted that the same expression is obtained for any profile of the charging power (P_{ch}) and discharging power (P_d).

REFERENCES

[1] Cengel, Y. A., Boles, M. A., Lacroix, M., *Thermodynamique: une approche pragmatique (Cengel and Boles, Thermodynamics—an Engineering Approach)*, 6th edition, The McGraw-Hill Montreal, Canada, 2008.

[2] Girault, H., *Analytical and Physical Electrochemistry*, EPFL Press/CRC Press/Taylor and Francis Group, Lausanne, Switzerland, 2004.
[3] Multon, B., Aubry, J., Haessig, P., Ben Ahmed, H., Systèmes de stockage d'énergie électrique Techniques de l'ingénieur, *BE*, 8, 100, April 2013.
[4] Glaize, C., Geniès, S., *Les accumulateurs électrochimiques au lithium, haute température et à circulation d'électrolyte*, Hermes, Lavoisier, Paris, France, 2013.

2 Energy Storage Systems

2.1 PHYSICAL SYSTEMS

Storing energy is usually a repetitive task with a very high number of expected cycles. In contrast with the classical electrochemical batteries where the expected number of cycles is limited due to degradation phenomena, physical systems based on reversible mechanisms of various natures can be used with much longer life cycles. One good example can be found in the hydraulic sector with the classical pumped hydro-power plants. Such systems can in addition be realized for very high power levels in the range of hundreds of megawatts [1, 2].

Another principle is based on gas compression and expansion and is known as compressed air energy storage (CAES) [3, 4].

Further mechanical systems are also used based on rotational kinetic energy and are known as flywheels [5, 6].

Electrical components like capacitors or inductors are also candidates for energy storage, even if their specific energy capacity is limited. Such systems are generally used for their ability to provide high level of instantaneous power [7–9].

2.1.1 Gravitational Hydro-Pumped Storage

Hydro-pumped storage facilities comprise water reservoirs placed at different altitudes which are interfaced through a reversible pump/turbine set. The stored energy E can be described through the gravitation law

$$E = m \cdot g \cdot h \tag{2.1}$$

where

m is the displaced mass
g is the gravitational acceleration
h is the difference of the levels of both reservoirs.

A simplified diagram of a hydro-pumped storage facility is given in Figure 2.1. The main conversion components are hydraulic pumps and turbines, coupled directly to synchronous machines used as motors in accumulation mode and as generators in restitution mode. Modern pumped hydro plants use variable speed generators/motors, offering many advantages like variable power in pump mode.

Gravitational hydraulic pumped storage will be described more in detail in Chapter 7.

The gravitation force and variation of the altitude of a given mass can be used with means other than water, as proposed in Reference [10]. In such a system, blocks of concrete are transported on rails from one altitude to another.

DOI: 10.1201/9781003736486-2

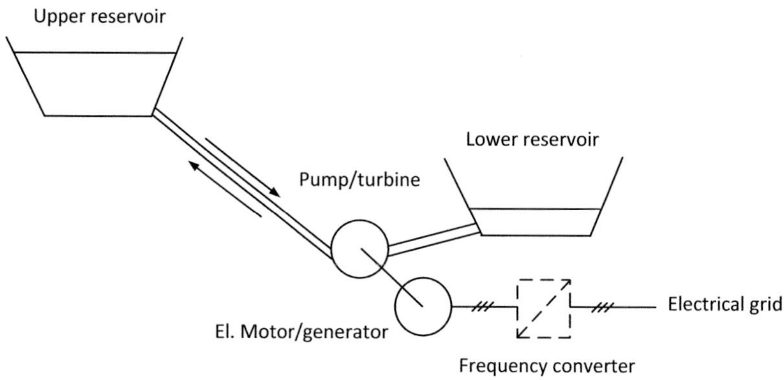

FIGURE 2.1 Principles of a hydro-pumped storage plant.

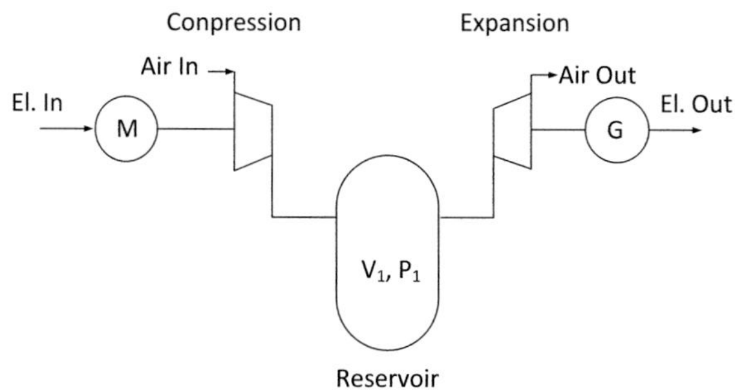

FIGURE 2.2 General scheme of CAES.

2.1.2 Compressed Air Energy Storage

CAES is based on the compression and expansion of air. Such systems generally use a compression stage comprising an electric motor driving a compression machine. Then the compressed air is stored in a reservoir. For the recovery of the stored energy, an expansion system is also provided. This expansion machine is composed of a volumetric expander driving an electric generator. The general scheme of CAES is represented in Figure 2.2.

The amount of energy stored in a reservoir of volume V_1, pressurized at a pressure level P_1 and stabilized at the same temperature as its surroundings, can be calculated through Equation 2.2 [3].

$$E = P_1 \cdot V_1 \left[\ln\left(\frac{P_1}{P_a}\right) - 1 + \frac{P_a}{P_1} \right] \tag{2.2}$$

P_a is the pressure of the surrounding (atmosphere). The expression gives the value of the maximum amount of energy that can be recovered from the reservoir that corresponds to a full expansion under isothermal conditions.

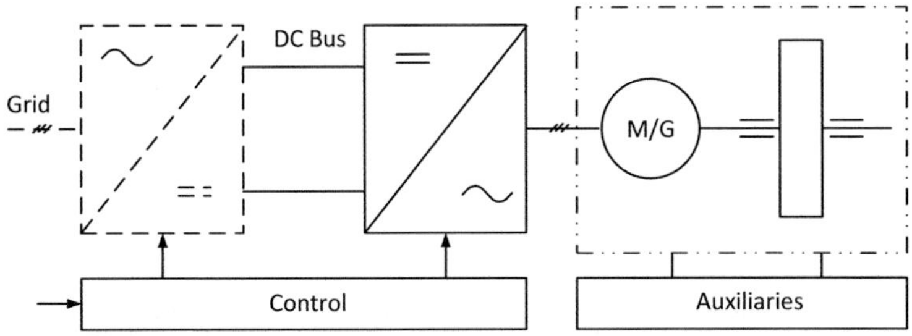

FIGURE 2.3 Flywheel system.

System performances and the description of components dedicated to the principle of CAES will be presented in Chapter 6.

2.1.3 Rotational Kinetic Energy (Flywheels)

Another possibility to store energy is given through rotational kinetic energy. Such systems also belong to the category of reversible physics phenomena. A flywheel system is generally composed of an electric machine coupled to a rotating mass. With the help of a power electronic converter, the driving torque of the variable speed machine (positive for charging and negative for discharging) can be imposed precisely and makes possible the control of the exchanged power level. Flywheels have been realized for a long time based on normal steel and limited speed, but modern equipment benefits from advanced materials like carbon composites and fast-running permanent magnet motors that can rotate at several hundred thousand revolutions per minute. Partially evacuated encapsulations reduce the aerodynamic losses of the flywheel and motor. A schematic representation of a flywheel system is given in Figure 2.3.

The amount of energy stored in a rotating mass running at a speed Ω is given by

$$E = \frac{1}{2} J \Omega^2 \tag{2.3}$$

where J is the moment of inertia.

The auxiliaries of a flywheel system provide the partial vacuum needed to reduce aerodynamic losses, and they often include the control of special bearings (magnetic, air).

The principles of flywheel storage, together with related performances, will be described more in detail in Chapter 8.

2.2 ELECTRIC SYSTEMS

2.2.1 Superconductive Magnetic Energy Storage Systems

On the basis of a purely electric/magnetic component like the inductor, energy can also be stored. The energy stored in an inductor can generally be expressed by the following law:

$$E = \frac{1}{2} L \cdot I^2 \tag{2.4}$$

L is the value of the inductance
I is the circulating current.

The maximum power P_{max} during charging and discharging is given by the values of maximum current and voltage delivered by the power electronic converter.

$$P_{max} = U_{max} I_{max} \tag{2.5}$$

A superconductive magnetic energy storage (SMES) system uses superconductivity in order to increase the energy density (higher current in the same support) and to reduce the ohmic losses to a minimum value [8].

A general scheme of a SMES system is represented in Figure 2.4.

During the charged state of a SMES system, note that the inductor current must always flow. Even if there are no variations of this current, it causes conduction losses in the inductor itself and in the interconnections but also in the freewheeling paths of the converter. For long-term idling mode, but also for security reasons, a by-pass switch is generally provided.

2.2.2 Capacitive (and Supercapacitive) Systems

The dual element of the inductor is the electric capacitor. It can be used for energy storage according the relation giving its energy content:

$$E = \frac{1}{2} C \cdot U^2 \tag{2.6}$$

C is the value of the capacitance
U is the value of the voltage across the capacitor.

The charging and discharging of a capacitor is realized by using a power electronic converter. The general scheme of a capacitive energy storage device is represented

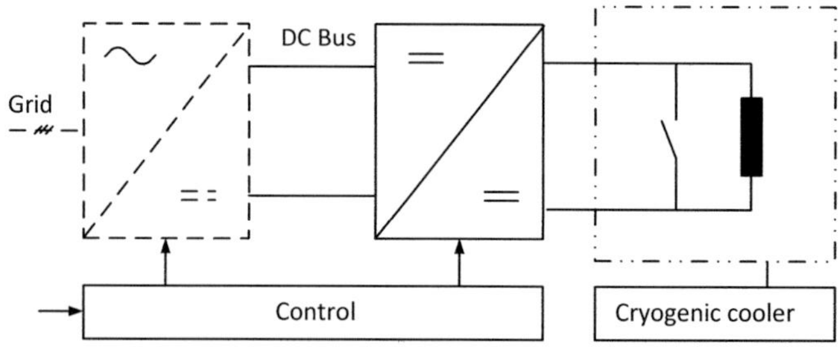

FIGURE 2.4 Superconductive magnetic energy storage system.

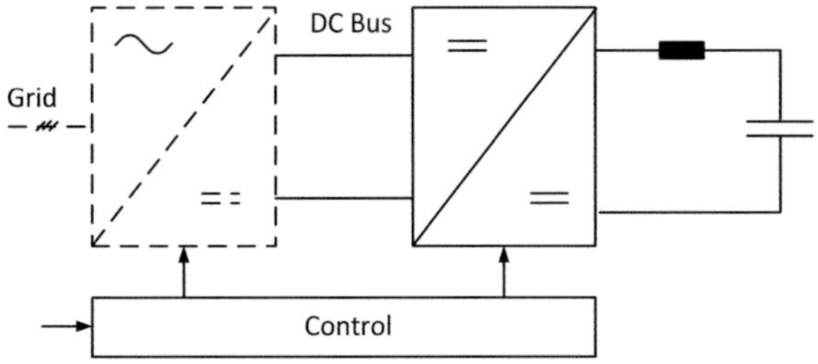

FIGURE 2.5 Capacitive energy storage system.

in Figure 2.5. Classical high-voltage capacitors can be used for energy storage [9], but also more recently developed supercapacitors are characterized by their high capacity.

In the case of a capacitive storage system, the power electronic converter must be able to control the capacitor current. A positive current charges the capacitor, while a negative current causes its discharge. A steady-state idling mode is characterized through a zero current flowing. This is an interesting property regarding energy efficiency.

An example of an application of a high power capacitive storage is described in Reference [9]. This equipment is dedicated to the smoothing of the power demand of the supply of the magnets of a particle accelerator (CERN). The capacity is around 20 MJ, and the maximum power is at 60 MW.

From the beginning of the 21st century, capacitive storage in general draws higher interest due to the appearance on the market of new components called supercapacitors or double-layer capacitors. Chapter 5 gives a deeper description of the components and describes several innovative application examples.

2.2.3 Electrochemical Systems

There is a high diversity of electrochemical technologies as potential solutions in the area of energy storage. Many of them are well established, especially in the domain of mobile applications. Larger systems are today developed for their utilization in connection with public grids [11–14].

Today, an increasing number of modern materials are emerging and appear often with promising characteristics. This progress is seen in the direction of higher energy densities but also with regard to power density. Additionally, more intensive studies on failure and aging mechanisms are aimed at reaching higher numbers of cycles or longer lifetimes of batteries. Many aspects of electrochemical techniques and battery storage technology are presented in Chapter 4.

An electrochemical battery is composed of several elements (cells) connected in series, each consisting of two electrodes. In each cell, an electric current is assumed to flow thanks to ion transport. The positive electrode, called the cathode, and the negative one, called the anode, are generally placed in an electrolyte.

2.2.3.1 Oxidation and Reduction

At the level of electrodes, the chemical reactions are called oxidation and reduction.

An oxidation is a reaction where one atom or one ion loses one or more electrons.

The "actor" of an oxidation is an electron donor, also called a reducer.

A reduction is a reaction where one atom or one ion receives one or several electrons.

The "actor" of the reduction is an electron acceptor, called an oxidant.

A reaction of oxidoreduction (redox reaction) can be described in a generic form:

$$aOx + ne \rightleftarrows b\mathrm{Re}\,d \tag{2.7}$$

where a, n and b appear as coefficients for the balance of the two members of the equation. They depend on the species used. Numerous examples will be presented in Chapter 4.

The *anode* is the electrode where oxidation takes place

The *cathode* is the electrode where reduction takes place

For the system approach point of view, an electrochemical battery can be seen as a voltage source, the charging and discharging variable being the battery current. Figure 2.6 gives a simplified scheme of a battery energy storage system (BESS).

The principal scheme of a BESS is very similar to that of a capacitive storage device. Principally, the positive and negative DC current is provided by a DC-DC converter. The interface to the external world can be defined at the DC bus level or through a front-end converter to the AC grid.

At the lower right side of Figure 2.6, a so-called battery management system (BMS) is represented. This unit has the role of balancing the different elements of the series connection but has also the task of supervising and protecting the overall system.

2.2.4 Flow Batteries

Flow batteries are a new type of electrochemical accumulator composed of two electrodes separated by a proton exchange membrane as can be found in fuel cells or in electrolyzers. Energy storage occurs in such flow batteries within the change of the

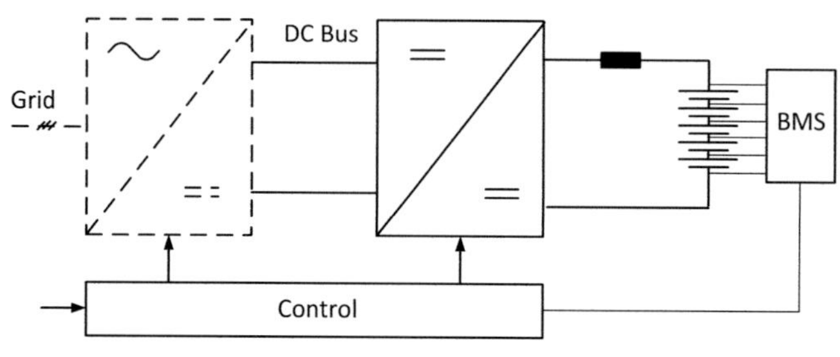

FIGURE 2.6 Battery energy storage system.

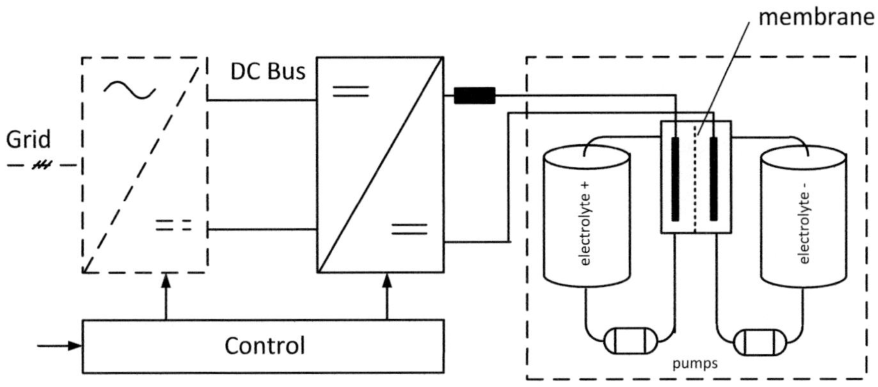

FIGURE 2.7 Redox flow battery storage system.

concentration of ions at the level of two liquids, an anolyte and a catholyte, circulating from two separate reservoirs [15].

Flow batteries allow the skirting of some limitations of classical electrochemical batteries where electrochemical reactions create solid composites accumulated on electrodes where they are generated and where the mass that can be accumulated is perforce limited. The generation of internal stresses on electrodes due to variations of the volumetric density of the active materials is another cause of battery ageing phenomena.

In flow batteries, the chemical compounds that represent the state of charge are in liquid form and are in solution in the two electrolytes.

These electrolytes are pumped from separate reservoirs to the "reactor", which is composed of the electrodes and the membrane.

One main characteristic of flow batteries is that the electrochemical converter (the reactor) is designed for the power level of the accumulator, while the energy capacity is only related to the volume (and mass) of the liquid electrolytes. In Figure 2.7, one example of a flow battery is represented. Because the circulation pumps for the electrolytes need a given amount of power, a more detailed calculation of the efficiency will be given in the dedicated Section 4.2.3 [16, 17].

Electrochemical batteries cover the largest area of today's energy storage applications, and their principles and components will be described in more detail in Chapter 4.

2.2.5 Fuel Cells and Hydrogen Storage

Another electrochemical energy converter is the fuel cell. Combined with a relatively complex chain of subcomponents such as electrolyzer, hydrogen conditioning and storage, the fuel cell represents an interesting energy source for the future. A complete storage system based on hydrogen can be used for storage of large amounts of energy. These systems are discussed in Chapter 9.

REFERENCES

[1] Tanaka, H., An 82 MW variable speed pumped-storage system, *Water Power & Dam Construction*, 40(11), 25–26, November 1991.

[2] Torres, O, *Life Cycle Assessment of a Pumped Storage Power Plant. Master Report*, Norwegian University of Science and Technology, Trondheim, Norway, July 2011.

[3] Lemofouet, S., Investigation and optimization of hybrid electricity storage systems based on compressed air and supercapacitors, Thesis Nr. 3628, EPFL, Lausanne, Switzerland, 2006.

[4] Lemofouet, S., Rufer, A., Hybrid energy storage system based on compressed air and supercapacitors with maximum efficiency point tracking (MEPT), *IEEE Transactions on Industrial Electronics*, 53(4), 1105–1115, 2006.

[5] Reitan, D. K., Bahari-Kashani, M., Superflywheel energy storage and non-synchronous AC/DC/AC electric transmission supplements power system operation, *Proceedings of the IEEE*, 64(10), 1543–1544, October 1976.

[6] Proud, N. J., Kelsall, D. R., Alexander, T. M., A drive system for the Pirouette™ kinetic energy storage system, power electronics and variable speed rives, *IEE Conference Publication*, 429, 342–347, 1996.

[7] Ali, M. H., Wu, B. Dougal, R. A., An overview of SMES Applications in power and energy systems, *IEEE Transactions on sustainable energy*, 1(1), 38–47. 2010.

[8] Nomura, S., Chikaraishi, H., Shimada, R., Design study on series compensated thyristor converters for large scale SMES, *2013 15th European Conference on Power Electronics and Applications (EPE)*, Lille, France, 2013, pp. 1–10, https://doi.org/10.1109/EPE.2013.6634613.

[9] Fahrni, C., Rufer, A., Bordry, F., Burnet, J. P., A novel 60MW pulsed power system based on capacitive energy storage for particle accelerators, *2007 European Conference on Power Electronics and Applications (EPE)*, Aalborg, Denmark, September 2007.

[10] William, R. P. et al., Utility scale energy storage system, US Patent, US 8593012 B2, November 26, 2013.

[11] Miller, N. W., Zrebiec, R. S., Hunt, G., Design and commissioning of a 5 MVA, 2.5 MWh battery energy storage system, *Proceedings IEEE Transmission and Distribution Conference*, Los Angeles, CA, 1996, pp. 339–345.

[12] Miller, N. W. et al., Battery energy storage systems for electric utility, industrial and commercial applications, *Battery Conference on Applications and Advances (11th Annual)*, Long Beach, CA, 1996, pp. 235–240.

[13] Akagi, H., Ota, J., Sato, T., Enhancement of performance, availability and flexibility of a battery energy storage system based on a modular multilevel cascaded converter (MMCC-SSBC), *IEEE Transactions on Power Electronics*, 31(4), 2791–2799, 2016, (IEEE Xplore Early Access).

[14] Chatzinikolaou, E., Rogers, D. J., A comparison of grid-connected battery energy storage system designs, *IEEE Transactions on Power Electronics*, 32(9), 6913–6923, 2017, https://doi.org/10.1109/TPEL.2016.2629020.

[15] Sum, E., Skyllas-Kazacos, M., A study of the V(II)/V(III) redox couple for redox flow cell applications, *Journal of Power Sources*, 15, 179–190, 1985.

[16] Blanc, C., Rufer, A., Multiphysics and energetic modelling of a vanadium redox flow battery, *Proceedings of the International Conference on Sustainable Energy Technologies, ICSET2008*, Singapore, 2008, pp. 696–701.

[17] Blanc, C., Modeling of a vanadium redox flow battery electricity storage system, PhD Thesis No. 4277, EPFL, Lausanne, Switzerland, 2009, http://infoscience.epfl.ch/record129758/files/EPFL_TH4277.pdf. Accessed on September 22, 2017.

3 Comparative Ratings and Properties

3.1 SYSTEM RATINGS

In order to design an appropriate storage device and to choose the right solution in relation to a given application, it is important to have a comparative evaluation of the possible capacity and performance. First, the power range must be defined. Then, with respect to the necessary autonomy, the energy capacity can be defined. Figure 3.1 is a representation of a high number of possible solutions over a large power range of six decades, starting from the 1 kW level up to the GW level. Different storage technologies are represented, from electrochemical batteries to large-scale pumped hydro storage. Compressed air energy storage (CAES), superconductive magnet energy storage (SMES) and flywheels are compared. The vertical axis of the figure illustrates the system autonomy in powers of 10 h. The product of the power multiplied by the time gives the energy capacity. This parameter corresponds to the surface delimited by the horizontal and vertical lines of the represented values in the diagram. A conventional approach considers real systems related to a possible autonomy of up to tens of hours. These technologies are represented in the lower half of Figure 3.1. For longer storage times, such as the so-called seasonal storage, several possibilities exist which are based on complex transformations. In Figure 3.1, the long-term storage is represented up to 10^3 h. This indicative value could correspond to a transformation from solar power to hydrogen defined over 3 months (100 days) and with a collection of up to 10 h/day.

3.2 ENERGY DENSITY

One important parameter related to a given technology is energy density. In Figure 3.2, the energy densities of several techniques are indicated [1]. On the vertical axis, the *weight energy density* is indicated.

$$\text{The symbol used for the weight energy-density is:} \qquad e_m\,[\text{Wh/kg}]$$

This parameter is generally indicated in Wh/kg, or in kWh/ton.

Another parameter for the energy density is the *volume energy density*.

$$\text{The symbol used for the volume energy-density is:} \qquad e_v\,[\text{Wh/dm}^3]$$

This parameter is indicated in kWh/m^3 or in Wh/dm^3.

DOI: 10.1201/9781003736486-3

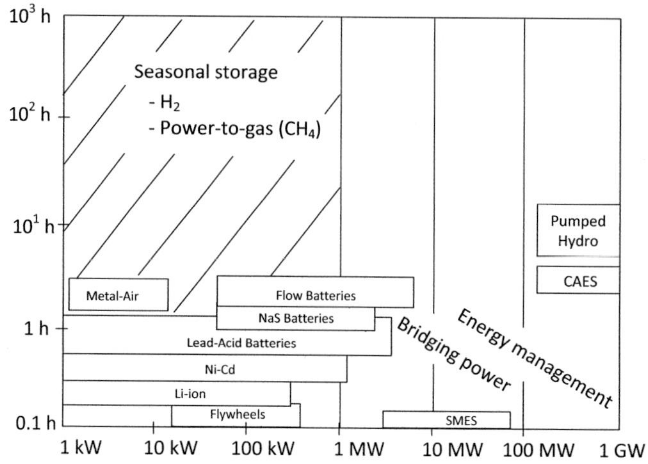

FIGURE 3.1 Overview of storage systems.

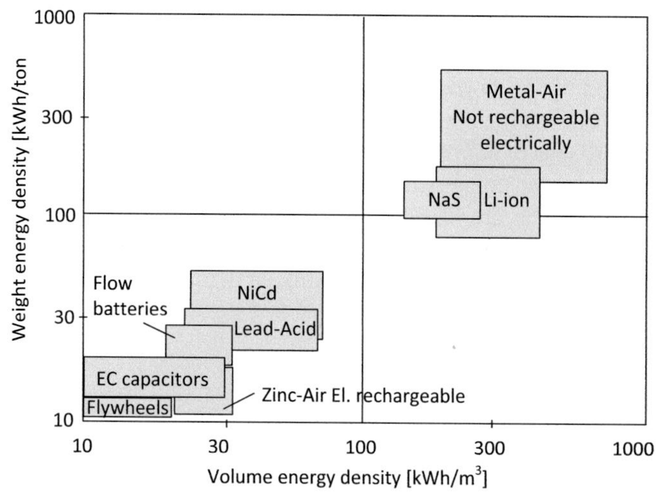

FIGURE 3.2 Weight and volume energy density.

These symbols are chosen according recommendations of the International Electrotechnical Commission (IEC) [2].

The parameters for energy density are important parameters with regard to the choice of a technology for a given application.

Mobile applications are the most concerned by weight energy density. In this context, an extreme case of application is the Solarimpulse project, which uses modern high-performance batteries in order to be able to fly overnight with energy accumulated during the day by photovoltaic (PV) cells placed on the airplane wings [3].

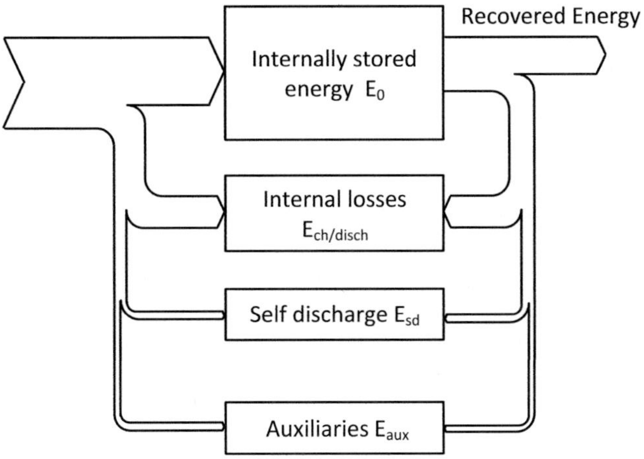

FIGURE 3.3 Global representation of energy amounts and losses.

As will be shown in Section 3.5, the indicated values of energy density generally do not consider any energy efficiency or losses during charging, discharging or self-discharging losses. In reality, the stored energy corresponds to a given value that is not equal to what can be extracted. This last value depends on the power level of the energy exchange.

3.2.1 Usable Energy Density

The usable energy of a storage device is different from the stored energy. This latter amount is also different from the energy needed to realize the storage. In principle, the energy density is calculated on the basis of the amount which is effectively inside the device. This amount can be recovered only under certain limitations. Section 3.5 will show how the usable energy of a storage device depends on the power level of the discharge process.

A global representation of energy amounts and different losses is given in Figure 3.3.

3.3 POWER DENSITY AND SPECIFIC POWER

A second possibility to compare storage devices is to quantify their power capability. This is often given by the power density or by the specific power. The power density is the amount of power (time rate of energy transfer) per unit volume. It is also called the *volume power density.*

The symbol used for the power density is: $p_v\,[\mathrm{W/dm^3}]$

It is expressed in $\mathrm{W/dm^3}$ or in $\mathrm{kW/m^3}$. It is also possible to give the value of the volume power density in $\mathrm{W/m^3}$, as it can be found in some scientific tables [4].

The *power-to-weight ratio* or *specific power* is the power generated by a source divided by the mass.

The symbol used for the *power-to-weight ratio* is: p_m [W/kg]

This parameter is given in W/kg or in kW/ton. For powerful devices such as supercapacitors or thermal generators, the specific power is often given in kW/kg.

As noted in the previous section for energy density, these parameters are only an indication given by the manufacturer to define what is possible or admissible for their use. The indicated values cannot be interpreted in the sense of energy efficiency.

3.4 RAGONE CHART

The Ragone chart is a chart used for performance comparison of various energy storage devices. The represented values on a Ragone chart are the

- Specific energy or weight energy density (in Wh/kg) e_m versus the
- Specific power or power to weight ratio (in W/kg) p_m

The axes of a Ragone chart are logarithmic, which allows comparing the performance of very different devices (for example, extremely high and extremely low power). Figure 3.4 gives an example of a Ragone chart used by one manufacturer for the positioning of a given technology in a general context of energy storage devices (ESDs).

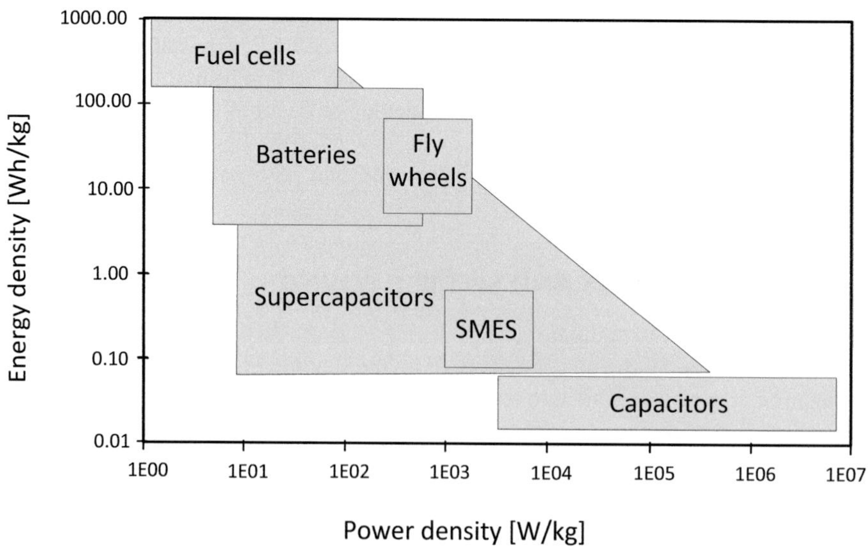

FIGURE 3.4 The Ragone chart.

In this diagram, one can see that the vertical axis is labelled as energy density, more precisely the weight energy density (or specific energy) in Wh/kg [5].

3.5 THEORY OF RAGONE PLOTS

As described in Chapter 1, the energy efficiency of a storage device is related to the different losses. Charging and discharging losses as well as self-discharge losses directly influence round-trip efficiency. As a consequence, the amount of energy which can really be recovered from a fully charged storage device has to be defined depending on the instantaneous power of the energy transfer.

This principle of interdependency between the energy density and the power density has been described under the name of "the theory of Ragone plots" [6].

In this reference, a general circuit is associated with Ragone plots (Figure 3.5). The energy storage device feeds a load with constant power *P*. The ESD contains elements for energy storage. Due to constant power, energy supply occurs only for a finite time $t_{inf}(P)$. The energy available for the load *E* depending on the power *P* defines a Ragone plot.

Consider the general circuit of Figure 3.5. For example, the ESD may consist of a voltage source $V(Q)$, depending on the stored charge Q, an internal resistor R and an internal inductance L. Note that this ESD can describe many kinds of electric power sources. The ESD is connected to a load which draws a constant power $P \geq 0$. Such a load can be realized with an electronically controlled power converter feeding an external user. The current I and voltage U at the load are then related nonlinearly by

$$U = P/I.$$

Provided reasonable initial conditions

$$Q(0) = Q_0 \quad \text{and} \quad \dot{Q}(0) = \dot{Q}_0$$

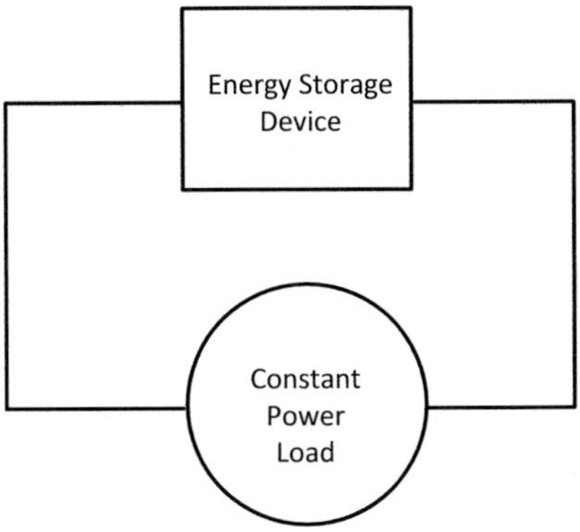

FIGURE 3.5 General circuit associated with Ragone plots. (Adapted from Christen, T., Carlen, M. W., *Journal of Power Sources*, 91, 210, 2000.)

are given, the electrical dynamics is governed by the following ordinary differential equation:

$$L\ddot{Q} + R\dot{Q} + V(Q) = -\frac{P}{\dot{Q}} \tag{3.1}$$

where the dot indicates differentiation with respect to time.

This equation applies not only to electrical ESD but covers many kinds of physical systems (mechanical, hydraulic, etc.). Without making reference to a specific physical interpretation of Equation (3.1), the Ragone curve can be defined as follows. At time t = 0, the device contains the stored energy

$$E_0 = L\dot{Q}_0^2/2 + W(Q_0) \tag{3.2}$$

For $t > 0$, the load draws a constant power P such that $Q(t)$ satisfies Equation (3.1). It is clear that for finite E_0 and P, the ESD is able to supply this power only for a finite time, say, $t_{\text{inf}}(P)$. A criterion is given either by when the storage device is cleared or when the ESD is no longer able to deliver the required amount of power. Since the power is time independent, the available energy is

$$E(P) = P \cdot t_{\text{inf}}(P) \tag{3.3}$$

The curve $E(P)$ versus P corresponds to the Ragone plot.

3.5.1 Ragone Plot of a Battery

In this section, the particular case of an ideal battery is studied. First, and regarding the model leading to Equation 3.1, we assume the condition $L = 0$. Then, the ideal battery with a capacity of Q_0 is characterized by a constant cell voltage $V = U_0$ if $Q_0 \geq Q > 0$ and $V = 0$ if $Q = 0$. In a first step, the leakage resistor R_L is neglected.

Equation 3.1 reads:

$$P = U \cdot I = (U_0 - RI)I$$

where

U is the terminal voltage

$I = \dot{Q}$ is the current.

The solutions of the quadratic equation are

$$I_{\pm} = \frac{U_0}{2R} \pm \sqrt{\frac{U_0^2}{4R^2} - \frac{P}{R}} \tag{3.4}$$

At the limit $P \to 0$, the two branches correspond to a discharge current

$$I_+ \to U_0/R \quad \text{and} \quad I_- \to 0.$$

For the ideal battery, the constant power sink can also be parametrized by a constant load resistance R_{load}.

The two limits belong then to $R_{load} \to 0$ (short circuit) and $R_{load} \to \infty$ (open circuit), respectively.

Clearly, in the context of the Ragone plot, we are interested in the latter limit, such that we have to take the branch with the minus sign, $I \equiv I_-$ in Equation 3.2.

Now the battery is empty at time $t_{inf} = Q_0/I$, where the initial charge Q_0 is related to the initial energy $E_0 = Q_0U_0$. It is now easy to include the presence of an ohmic leakage current into the discussion. The leakage resistance R_L increases the discharge current I by U_0/R_L.

The energy being available for the load becomes:

$$E_b(P) = P \cdot t_\infty = \frac{2RQ_0P}{U_0 - \sqrt{U_0^2 - 4RP + 2U_0R/R_L}} \tag{3.5}$$

Equation 3.5 corresponds to the Ragone curve of the ideal battery. In the presence of leakage, $E_b(0) = 0$. For the extracted energy, there exists a maximum at

$$P = U_0^2/\sqrt{2RR_L}$$

Without leakage $R/R_L \to 0$, the maximum energy is available for vanishing low power $E_b(P \to 0) = E_0$. From Equation 3.5, one concludes that there is a maximum power $P_{max} = U_0^2/4R$ associated with an energy $E_0/2$ (a small correction due to leakage is neglected).

This point is the endpoint of the Ragone curve of the ideal battery, where only half of the energy is available, while the other half is lost at the internal resistance.

Finally, the expression of the Ragone plot is given in dimensionless units using

$$e_b = E_b/Q_0U_0 \quad \text{and} \quad p = 4RP/U_0^2$$

$$e_b(p) = \frac{1}{2}\frac{p}{(1 - \sqrt{1 - p + 2R/R_L})} \tag{3.6}$$

Ragone curves according to Equation 3.5 with and without leakage are shown in Figure 3.6 for the ideal battery. The branch belonging to I_+ is plotted by the dashed curve.

3.5.2 Case of the Capacitor

In the case of an ideal electric capacitor (see Figure 3.7), an ordinary differential equation (ODE) rather than an algebraic equation has to be solved. The electric potential depends linearly on the charge via a capacitance C.

$$V(Q) = Q/C$$

According to the relative complex calculations given in Reference [6], the Ragone curve of a capacitor can be expressed by

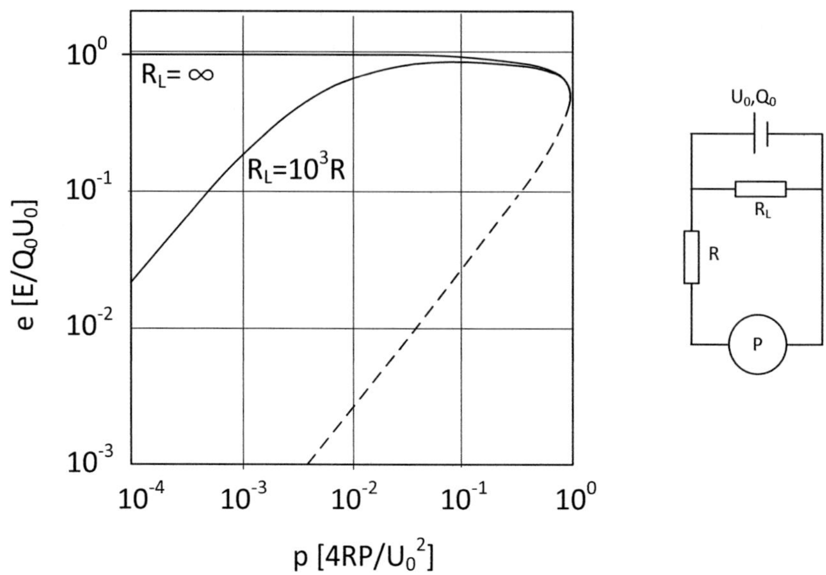

FIGURE 3.6 Ragone curve of the ideal battery. (Adapted from Christen, T., Carlen, M. W., *Journal of Power Sources*, 91, 210, 2000.)

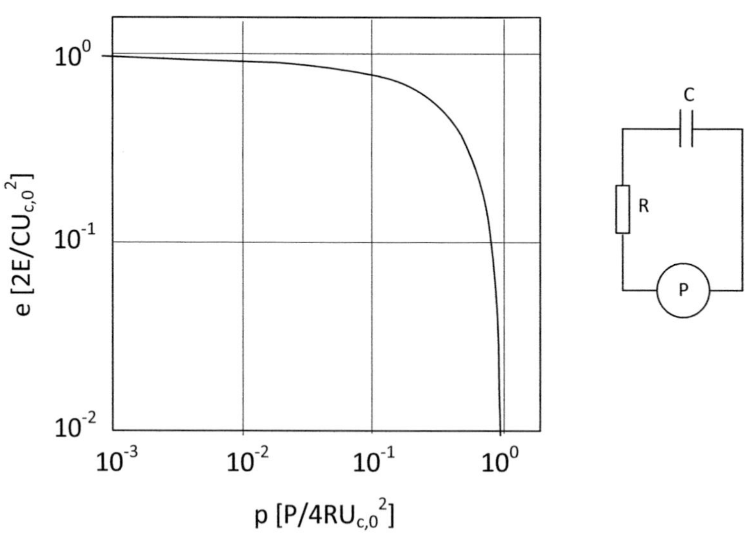

FIGURE 3.7 Normalized Ragone curve for the capacitor. (Adapted from Christen, T., Carlen, M. W., *Journal of Power Sources*, 91, 210, 2000.)

$$E_c(P) = \frac{C}{2}\left(RP\ln\left(\frac{RP}{U_0^2}\right) + U_0^2 - RP \right) \tag{3.7}$$

$$U_0 = \frac{U_{C,0}}{2} + \sqrt{\frac{U_{C,0}^2}{4} - RP} \tag{3.8}$$

where the initial capacitor voltage $U_{C,0}$ is related to the total energy by E_0

$$E_0 = CU_{C,0}^2/2$$

In dimensionless units

$$e_c = 2E_c/CU_{C,0}^2 \tag{3.9}$$

And

$$p = 2RCP/E_0$$

The Ragone curve reads

$$e_c(p) = \frac{1}{4}\left((1+\sqrt{1-p})^2 - p - p\ln\left(\frac{(1+\sqrt{1-p})^2}{p}\right)\right) \tag{3.10}$$

and is represented in Figure 3.7.

3.5.3 Case of Superconductive Magnetic Energy Storage

The Ragone curves of SMES systems are also described in detail in Reference [6]. Figure 3.8 gives the normalized curves for inductive ESDs with Coulomb (C), Stokes (S) and Newton (N) friction. The dashed double-dotted curve corresponds to a SMES with an ohmic bypass $\left(4R/R_b = 0.001\right)$. This resistance R_b is used for modeling of the losses of all freewheeling paths, with a dominant contribution of the freewheeling elements of the power electronic converter (see Figure 2.4).

3.6 MODIFIED RAGONE REPRESENTATION

A similar method for representing the relation between the real energy capacity of a storage device and the charging/discharging power can be found in Reference [7]. In this contribution, the question of the "amplification of power" is addressed.

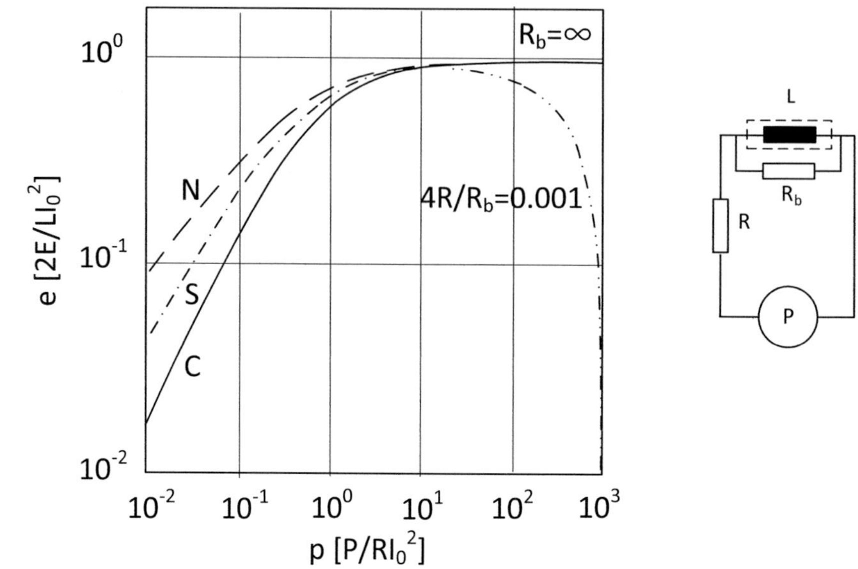

FIGURE 3.8 Normalized Ragone curves for the inductive ESD. (Adapted from [6].)

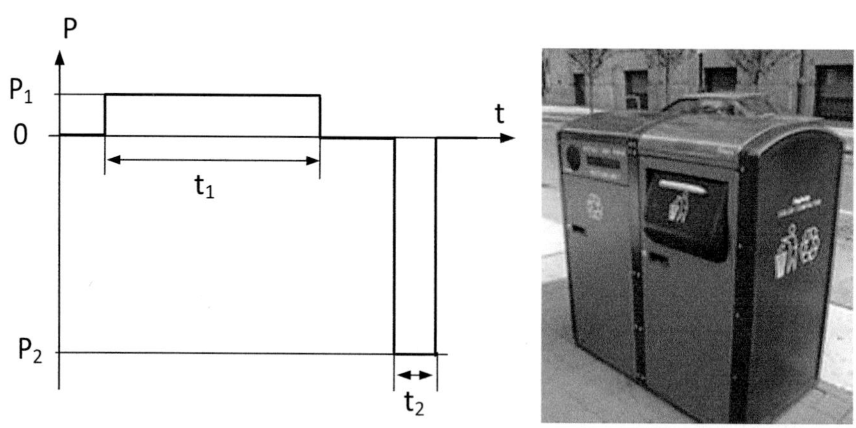

FIGURE 3.9 Principle of power amplification and application for waste compactor.

Power amplification can be illustrated with the example of "waste compactors", such as the known "Big Belly" [8], and is represented in Figure 3.9b. In such special applications, a storage device is used to harvest energy at a low power level (e.g. $P_1 = 10$ W) and to feed an application with a much higher power (e.g. $P_2 = 125$ W, Figure 3.9a). Another example of power amplification through storage is described in Reference [9], where a direct connected supercapacitor is used as an energy buffer for a small PV generator.

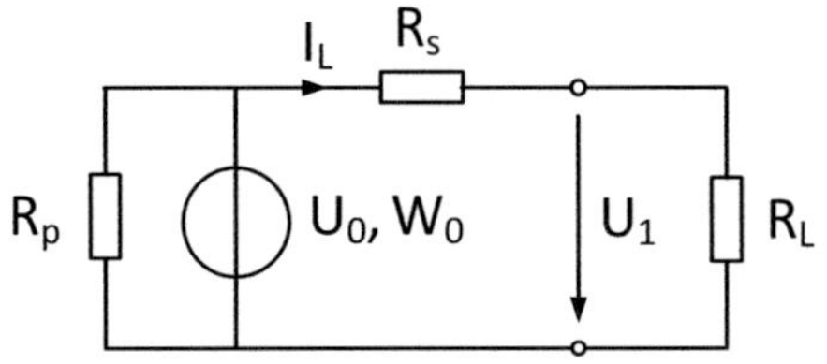

FIGURE 3.10 Equivalent scheme used for the modified Ragone representation.

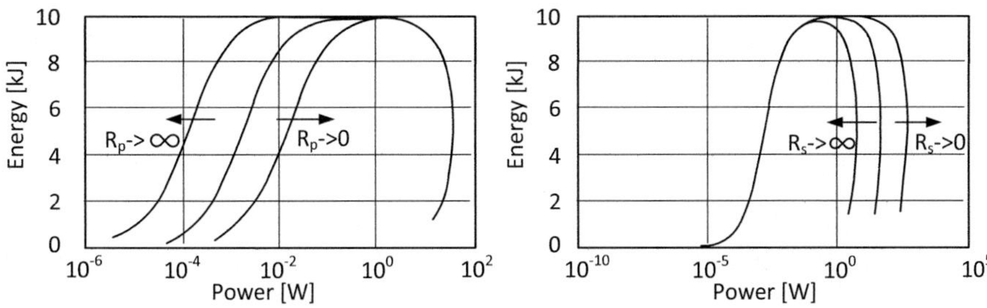

FIGURE 3.11 The modified Ragone representation. (Adapted from Delalay, S., Etude systemique pour l'alimentation hybride–application aux systems intermittents, PhD Thesis No. 5768, EPFL, Lausanne, Switzerland, 2013, https://infoscience.epfl.ch/record/187365/files/EPFL_TH5768.pdf. Accessed on September 22, 2017.)

In Reference [7], the model of the modified Ragone representation (MRR) is used. The ESD is represented as an ideal battery with series resistor R_s and a parallel connected leakage element R_p (Figure 3.10). The energy that can be recovered from the storage device is represented in a function of the transfer power P (logarithmic scale, Figure 3.11). Too small a power in the range of the self-discharge losses results in nearly zero energy recovered (left ends of the curves in Figure 3.11). At the right end of the MRR curve, the effect of a too-high transfer power results in a similar situation of zero recovery.

The MRR given in Figure 3.11 corresponds to the same analysis as already presented in Section 3.5.1 but with a different formulation of the different powers. The boundary between the extracted current and the power provided to the load is given for the specified variables in Figure 3.10:

$$P_L - U_0 I_L + R_s I_L^2 = 0 \tag{3.11}$$

with

$$P_L = R_L I_L^2 \tag{3.12}$$

Additionally, the losses in the self-discharge path and the internal dissipation are considered:

$$P_p = \frac{U_0^{2\ 2}}{R_{p\ L}} \tag{3.13}$$

$$P_s = \frac{1}{2R_s} U_0 (U_0 - \sqrt{U_0^2 - 4P_L R_s}) - P_L \tag{3.14}$$

Finally, for a constant power extracted from the storage device, the actual recovered amount of energy is given by

$$E_{recov} = P_L \frac{W_0}{P_s + P_L + P_p} = \frac{P_L}{U_0} \frac{2R_s W_0}{U_0 \left(1 + \frac{2R_s}{R_p}\right) - \sqrt{\left(U_0^2 - 4R_s P_L\right)}} \tag{3.15}$$

Further calculations for intermittent use of storage devices and also for different steps of the extracted power are given in Reference [7].

3.6.1 Power Amplification in Successive Stages

From Figure 3.9a, the power amplification factor (PAF) can be defined as the ratio of

$$PAF = P_2 / P_1 \tag{3.16}$$

According to its MRR, the storage device used for achieving a given power amplification may be unsuitable in the sense of leading to an insufficient round-trip energy efficiency because of the too-narrow power zone of the MRR or, in other words, where the zone of "good efficiency" is smaller than the desired PAF. In such a case, power amplification must be achieved in two successive steps. In Figure 3.12, a two-stage power amplification system is represented, where the input power is provided by a very small PV generator. The first stage of the power amplification system uses a supercapacitor directly connected to the PV generator. From this supercapacitor storage device, a higher power is used to periodically charge a second accumulation stage where a hydraulic bladder is fed with oil using a small motor and pump. Then,

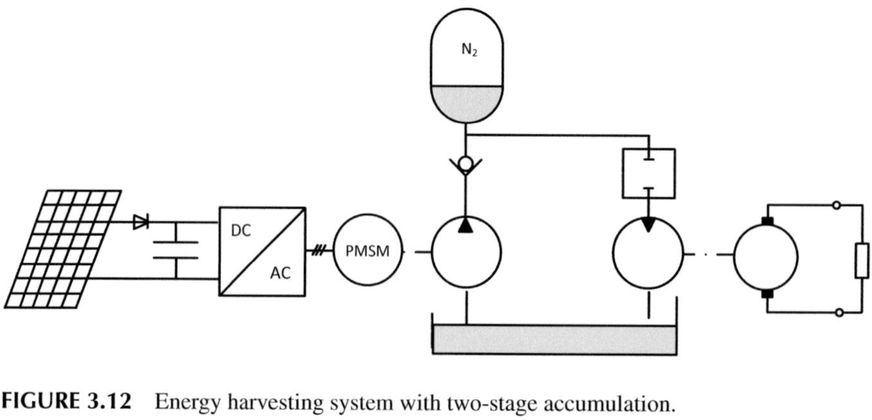

FIGURE 3.12 Energy harvesting system with two-stage accumulation.

for the final application, the hydraulically stored energy can be used with a much higher power level.

In Figure 3.13, the MRRs of the two stages are represented, namely the supercapacitor (black traced line) and the hydraulic accumulator (full line). The dotted curves represent the properties of the PV panels when the solar radiation is varying and the MPPT control circuit sets the operating point (power between 2 and 50 W). The vertical mixed line represents the power needed by the final application (5kW). This power level can be supplied by the hydraulic accumulator, while the supercapacitor shows nearly zero efficiency at that value of the power. The grey zone in Figure 3.13 illustrates the variable power needed for the charging of the hydraulic accumulator due to the variation of its internal gas pressure. The intersections between this grey zone and the MRRs of both the supercapacitor and the hydraulic accumulator define energy transfers between the stages with an acceptable efficiency above 80%. This value comprises the losses of the discharge of the first stage together with the losses of charging the second stage.

It appears in Figure 3.13 that from the point of view of the MMRs of the storage devices only, an acceptable intersection domain between the PV panels and the hydraulic accumulator would be possible (intersection of the PV curves and the hydraulic accumulator's MRR, left from the grey zone). However, the energy efficiency of the needed hydraulic pump also affects the global account. Here, the operating power of this pump must be kept over a minimum value in order to get an acceptable efficiency of that additional intermediary conversion stage.

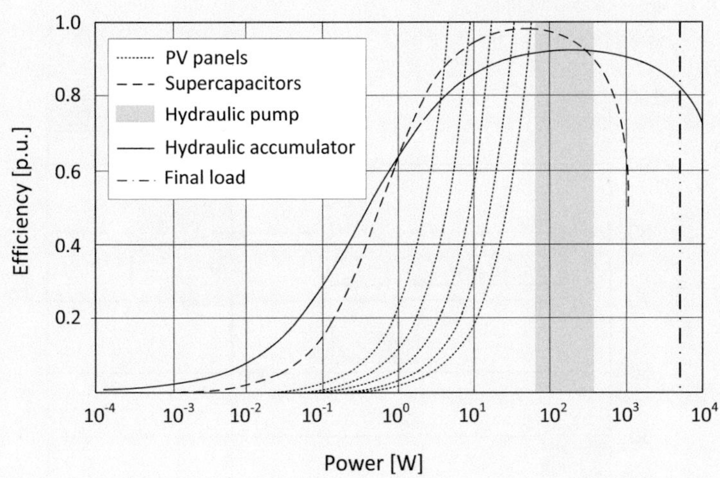

FIGURE 3.13 Modified Ragone representation of a two-stage power amplification system. (Adapted from Delalay, S., Etude systemique pour l'alimentation hybride–application aux systems intermittents, PhD Thesis No. 5768, EPFL, Lausanne, Switzerland, 2013, https://infoscience.epfl.ch/record/187365/files/EPFL_TH5768.pdf. Accessed on September 22, 2017.)

3.7 TYPICAL EFFICIENCIES, LIFETIME AND COSTS

Figure 3.14 illustrates the efficiencies and lifetimes of different storage technologies. These parameters are of high importance, especially in a context where the "energy economy" factors intervene. Classical as well as modern batteries show very good energy efficiencies. But they suffer from limited life cycles or lifetimes in the range of only several hundreds to several thousands of cycles, which is a limiting factor in the domain of renewable energy sources. On the right side of Figure 3.14, technologies with higher numbers of possible cycles are represented. They generally belong to the category of solutions based on reversible physics.

Figure 3.15 indicates the capital costs related to different storage solutions. This representation is related to time and evolution of the costs of the technologies. Especially the costs of new techniques must be periodically reevaluated. The criteria for selection of a given technology must be evaluated together with the technical performance criteria like the lifetime and efficiency, as well as the context of utilization [10–12].

A more realistic economic model for evaluation or comparison of a storage solution should take into account the costs of the storage infrastructure and the lifetime of the system as well as its efficiency (Equation 3.17).

$$Real\_costs = \frac{Capital/Energy}{Lifetime \cdot Efficiency} \tag{3.17}$$

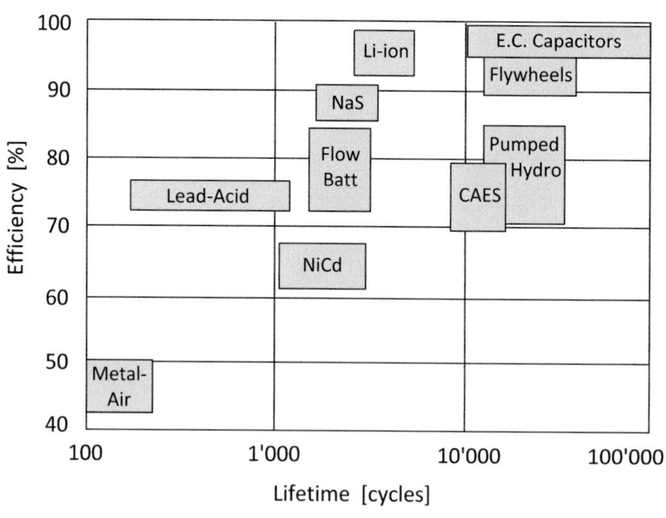

FIGURE 3.14 Efficiency and lifetimes of different storage technologies.

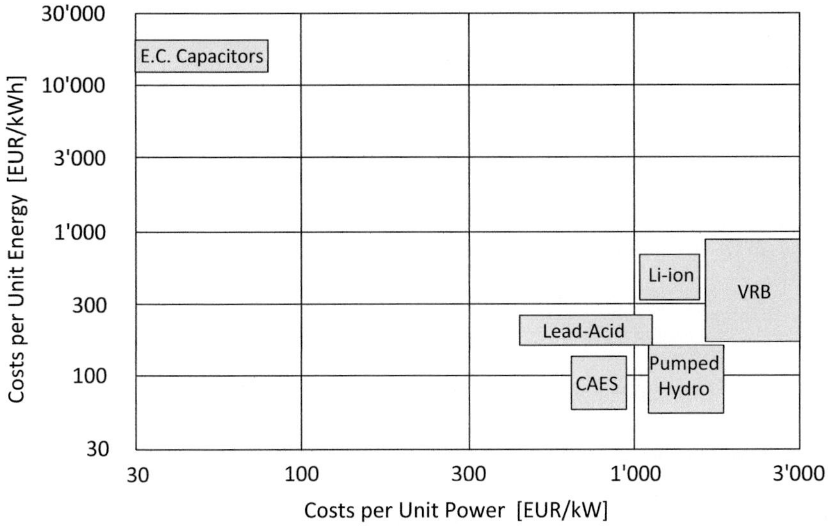

FIGURE 3.15 Capital costs of storage solutions.

Exercise 1: Normal and Fast Charge of Batteries in EVS—The Question of Energy Efficiency

Modern batteries claim high C-rates up to a factor of 8 (short term 15), which could make for possible fast charging in electric vehicle (EV) applications. A system is studied where the energy can be collected from PV panels and prestored during the day. The energy transfer from the local battery to the vehicle battery at the end of the day can then be realized in a longer or shorter time (Figure 3.16). Even if no PV panels are used, the precharge of the local battery is done from grid electricity, and the role of that local battery will be to serve as a buffer, avoiding high power solicitation from the grid during the fast charge (high power) of the EV battery.

However, the batteries present internal resistance that is not negligible, and the energy transfer is affected by losses. The student must investigate the PV collection system and panels, pre-charge and fast transfer mechanism and design suitable accumulators. In particular, the energetic properties must be illustrated (for example, efficiency as a function of transfer time).

Technical Parameters

Local Battery: The local battery is realized with 135 elements of 3.7 V and 0.7 mΩ internal resistance.

No-load voltage of the local battery: 500 V.

The local battery energy capacity is equal to 25 kWh.

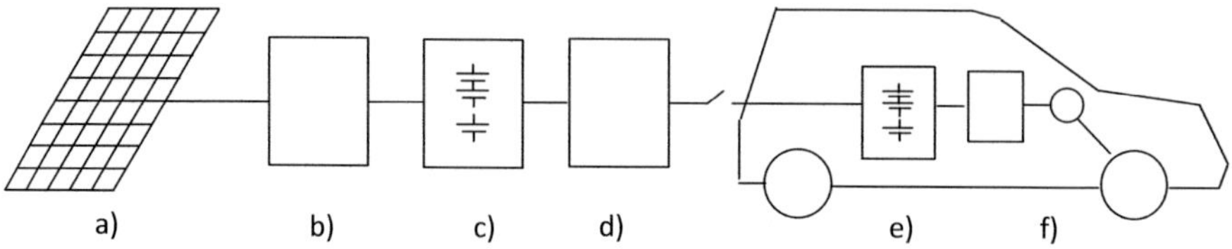

FIGURE 3.16 EV charging from RES (PV): (a) PV panels, (b) MPPT converter, (c) local battery, (d) DC-DC converter, (e) car battery, (f) propulsion system.

From the nominal capacity of the elements (50 Ah), the rated current is defined as 50 A, corresponding to a C-factor equal to 1 (under C = 1 conditions, the charging time is equal to 1 h).

Car Battery: The car battery is realized with 108 elements of 3.7 V and 0.7 mΩ internal resistance (identical elements as the local battery).
No-load voltage of the car battery: 400 V.

Converter Losses: The converter losses are calculated through the conduction loss of the silicon devices with a forward voltage of the devices (transistors and diodes) equal to 1.5 V. More details to be found in Appendix A. 3.1.1.

PV Panels: The PV panel surface is designed according to a charging time of the local battery within 7 hours from SOC 20% to full charge (SOC = 100%).

The required panel surface must be calculated for the following conditions:

$$\varepsilon = 800\ \text{W/m}^2\ \text{(mean value)}$$

$$\eta_{\text{cells}} = 10\%$$

Structure of the System, Converters and Cascaded Conversions

An electric scheme of the system with the different converters should be drawn.

a. Slow charge (7 hours)
   The current in both the PV panels and the local battery should be non-discontinuous. It is smoothed with inductors:

$$L = 9.15\ \text{mH}, R_L = 0.2\Omega$$

b. Fast energy transfer from the local battery to the car battery (C-rate = 8). In order to reduce costs, there is only one converter for both batteries (step-down converter). The current in the car battery is smoothed with an inductor:

$$L = 9.15\,\text{mH},\ R_L = 0.32\Omega$$

Energy Efficiency:

a. Calculate the energy efficiency of a charge from the PV panels (slow charge, 7h). For this case (a), there is a step-up converter cascaded with a step-down converter. Between the two converters, there is a constant DC voltage link based on a buffer capacitor.

   The energy efficiency is calculated on the basis of the different power losses (converters, smoothing inductors, internal losses of the battery).

b. Calculate the energy efficiency of a charge of the car battery (from the local battery, case b), for different charging times (using the C-rates of 2, 4, 6, 8, 10, 12).

   The goal of this exercise is to set in evidence the importance of the different losses, to show what components have the most influence on the efficiency and to show how the losses depend on the charging speed.

SOLUTION OF EXERCISE 1

Model of the Local Battery

The local battery can be modeled through the equivalent scheme of Figure 3.17.

The battery's no-load voltage is

$$U_{b\_loc} = 135 \cdot 3.7V = 499.5V \tag{3.18}$$

The internal resistance is

$$R_{i\_loc} = 135 \cdot 0.7m\Omega = 0.095\Omega \tag{3.19}$$

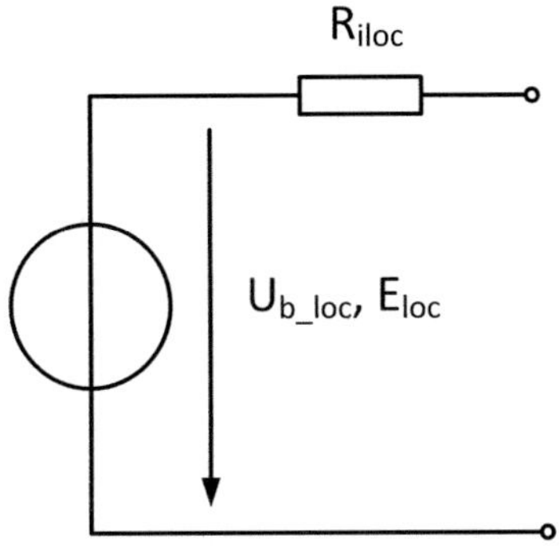

FIGURE 3.17 Simplified model of the local battery.

The battery's energy capacity is

$$E_{loc} = 50Ah \cdot 499.5V = 24.975kWh \tag{3.20}$$

Model of the Car Battery

The car battery can be modeled through the equivalent scheme of Figure 3.18.

The battery's no-load voltage is

$$U_{b\_car} = 108 \cdot 3.7V = 399.6V \tag{3.21}$$

The internal resistance is

$$R_{i\_car} = 108 \cdot 0.7m\Omega = 0.076\Omega \tag{3.22}$$

The battery's energy capacity is

$$E_{car} = 50Ah \cdot 399.6V = 19.980kWh \tag{3.23}$$

PV Panels

The slow charge (7 h) of 80% of the (local) battery capacity defines the charging power:

$$P_{80\%} = \frac{0.8 \cdot 24.975kWh}{7h} = 2.85kW \tag{3.24}$$

The solar (irradiance) power is consequently

$$P_{sol} = \frac{P_{80\%}}{\eta} = \frac{2.85kW}{0.1} = 28.5kW \tag{3.25}$$

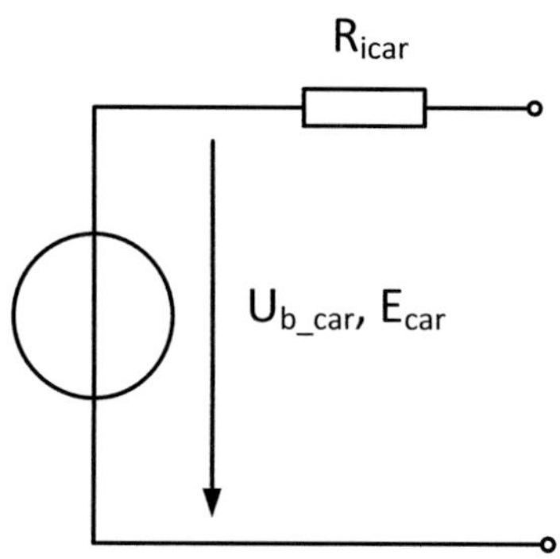

FIGURE 3.18 Simplified model of the car battery.

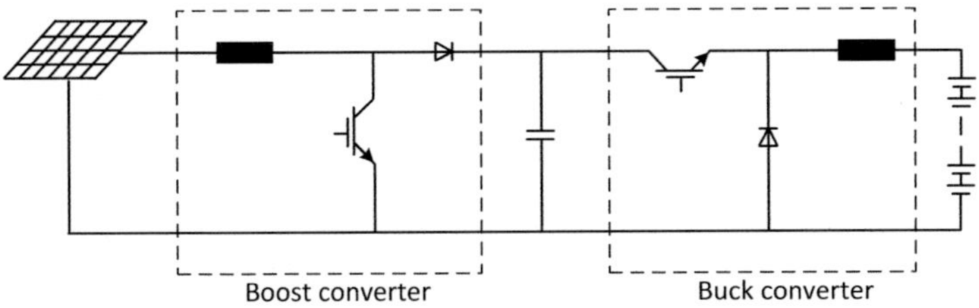

FIGURE 3.19 Charging system of the local battery from PV panels.

For a simplified design of the PV generator, the supposition is made that the solar irradiance is of a constant average value of 800 W/m^2 during 7 hours. The PV panel surface becomes

$$S_{PV} = \frac{28.5kW}{800W/m^2} = 35.7m^2 \tag{3.26}$$

Charging in 7 h

The scheme of the charging system (slow charge) is given in Figure 3.19.

The charging current is calculated as (80% of battery capacity in 7 hours)

$$I_{ch} = \frac{0.8 \cdot 50Ah}{7h} = 5.71A \tag{3.27}$$

In the system represented in Figure 3.19, the power produced by the PV panels is transferred to the intermediate DC circuit with the help of the boost converter. This converter assumes the function of the adaptation of the voltage of the panels to a constant DC voltage. In addition, this converter allows the optimal operation of the PV panels at their point of maximum power (MPPT).

The characteristic curves of the PV panels are represented in Figure 3.20. The voltage of the intermediate DC circuit is also represented (U_{DC} = 550 V). This value allows the boost converter to be operated with any value of the voltage of the PV panels. The represented MPP point corresponds to maximum of power under a solar irradiance of $\varepsilon = 1$ and a temperature $\theta = 0°$.

From the intermediate circuit, the power is transferred to the local battery using the buck converter. This converter assumes the transfer of power under constant current control.

The typical waveforms of the boost and buck converters can be found in Appendix A.3.1 and A.3.2.

The use of the cascade of a boost and buck converter implies slightly reduced energy efficiency due to the double conversion. But the main advantage is that the current at both the input and the output side is nondiscontinuous.

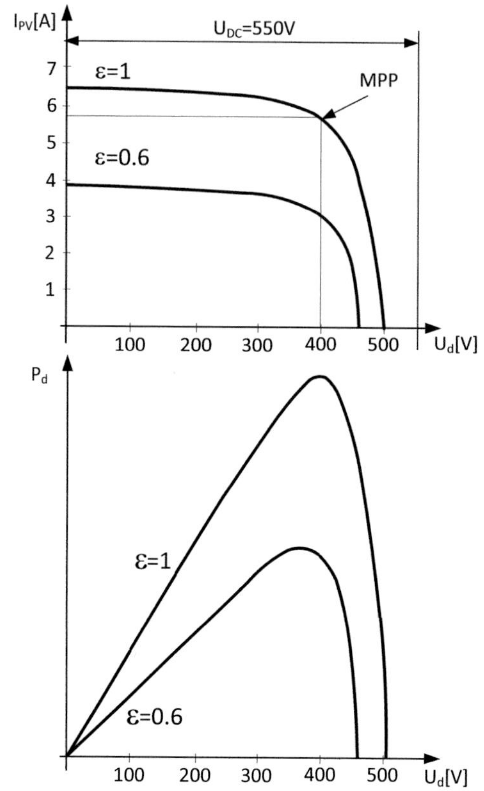

FIGURE 3.20 Characteristics of the PV panels.

The charging system with buck and boost converters is affected by power losses in the following elements (simplified estimation):

- Ohmic losses in the inductors
- Conduction losses in the power semiconductors of the boost and buck converters
- Ohmic losses in the battery

For estimation of the losses, it is considered that the PV panels are operated at a voltage level corresponding approximately to the voltage level of the local battery. As a consequence, one can suppose the input current of the boost converter is identical to the output current of the buck converter.

The losses in the inductor of the boost converter are calculated according to:

$$Loss_{Lboost} = R_L \cdot I^2 = 0.2\Omega \cdot (5.71A)^2 = 6.5W \tag{3.28}$$

The losses in the buck inductor are calculated as follows:

$$Loss_{Lbuck} = R_L \cdot I^2 = 0.2\Omega \cdot (5.71A)^2 = 6.5W \tag{3.29}$$

For the conduction losses in the converters, Equations A.6 and A.7 are used.

$$P_{cond} = P_{onT} + P_{onD} = U_{CEon} I_T + U_{Don} \cdot I_D = 1.5V \cdot 5.71A = 8.5W \tag{3.30}$$

(The on-state voltages of the diode and transistor are identical, and the on-state durations of these elements are complementary. Their sum is equal to 1 (see A.8).)

The ohmic losses in the battery are given by

$$Loss_{battery} = R_{iloc} \cdot I^2 = 0.095\Omega \cdot (5.71A)^2 = 3.1W \tag{3.31}$$

The total transfer losses are then

$$\begin{aligned} Loss_{transf} &= Loss_{Lboost} + Loss_{Lbuck} + P_{cond\_boost} + P_{cond\_bucl} + Loss_{batt} \\ &= 6.5W + 6.5W + 8.5W + 8.5W + 3.1W = 33.1W \end{aligned} \tag{3.32}$$

The efficiency of the charging process is further

$$\eta = \frac{P_{ch}}{P_{ch} + Loss_{transf}} = \frac{2850W}{2850W + 33.1W} = 0.989 \tag{3.33}$$

Fast Charge from the Buffer

It is assumed that the car battery as well as the buffer (local) battery can be overloaded with a factor of 8C.

The car battery is then

$$I_{car\_batt} = 8 \cdot 50A = 400A \tag{3.34}$$

For the fast transfer from the local battery to the car battery, the scheme given in Figure 3.21 is chosen.

Figure 3.21 shows the use of a single buck converter. This choice is made on the basis of economic considerations.

To estimate the energy efficiency of the fast charge, the following elements are considered:

- Ohmic losses in the local battery
- Ohmic losses in the car battery
- Conduction losses in the power semiconductors

The ohmic losses in the inductor are neglected.

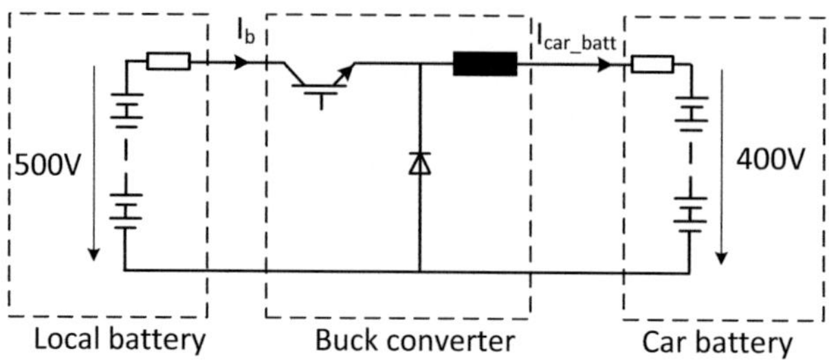

FIGURE 3.21 Scheme for the fast charge from the buffer battery.

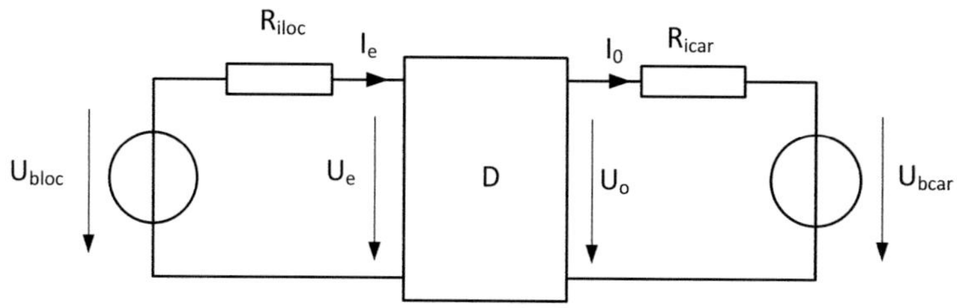

FIGURE 3.22 Equivalent scheme of the fast charge.

For the calculation of losses in the local battery, its current must be calculated. This current depends on the duty cycle of the buck converter corresponding to the ratio of the input to the output voltages.

$$I_e = (U_o / U_e) \cdot I_o = D \cdot I_o \ \text{(see Equation A.5)} \tag{3.35}$$

The scheme of Figure 3.22 can be used as follows:

$$(U_{bloc} - R_{iloc} \cdot I_e) \cdot D = U_{bcar} + R_{icar} \cdot I_o \tag{3.36}$$

$$U_{bloc} \cdot D - D^2 R_{iloc} \cdot I_o = U_{bcar} + R_{icar} \cdot I_o$$

$$-D^2 R_{iloc} \cdot I_o + U_{bloc} \cdot D - (U_{bcar} + R_{icar} \cdot I_o) = 0 \tag{3.37}$$

$$D = \frac{-U_{bloc} \pm \sqrt{(U^2{}_{bloc} - 4R_{iloc} I_o (U_{bcar} + R_{icar} \cdot I_o)}}{-2R_{iloc} I_o} \tag{3.38}$$

With the numerical values:

$$D = \frac{-500V \pm \sqrt{(500V)^2 - 4 \cdot 0.095\Omega \cdot 400A(400V + 0.076\Omega \cdot 400A)}}{-2 \cdot 0.095\Omega \cdot 400A} = 0.926 \tag{3.39}$$

The second solution of the quadratic equation corresponds to a value of D = 15, which is incompatible with the normal operation of a buck converter. It must be ignored.

The current in the local battery is

$$I_e = D \cdot I_o = 0.926 \cdot 400A = 370A \tag{3.40}$$

And the losses in the local battery:

$$Loss_{l\_batt} = R_{iloc} \cdot I^2{}_e = 0.095\Omega \cdot (370A)^2 = 13032W \tag{3.41}$$

The losses in the car battery:

$$Loss_{car\_batt} = R_{icar} \cdot I^2{}_0 = 0.076\Omega \cdot (400A)^2 = 12160W \tag{3.42}$$

The conduction losses in the semiconductors:

$$Loss_{converter} = U_{CEon} \cdot I_T + U_{Don} \cdot I_D = 1.5V \cdot 400A = 600W \tag{3.43}$$

The total losses related to the fast charge are consequently:

$$\begin{aligned} Loss_{total} &= Loss_{lbatt} + Loss_{car\_batt} + Loss_{converter} = \\ &= 13032W + 12160W + 600W = 25792W \end{aligned} \tag{3.44}$$

Transferred Power

The power related to the fast charge is calculated. This value corresponds to the power that is really accumulated in the car battery.

$$P_{fast\_ch} = U_{bcar} \cdot I_o = 400A \cdot 400V = 160000W = 160kW \tag{3.45}$$

Finally, the efficiency of the fast charge is given by

$$\eta_{fast\_ch} = \frac{P_{fast\_ch}}{P_{fast\_ch} + Loss_{total}} = \frac{160kW}{160kW + 25.8kW} = 0.86 \tag{3.46}$$

This value of efficiency corresponds to a charging current of 400 A. With this value of current, the car battery is charged within a time equal to

$$t_{charge} = \frac{C\_capacity}{I_{charge}} = \frac{50Ah}{400A} \cdot 60\min/h = 7.5\min \tag{3.47}$$

Figure 3.23 shows the value of the efficiency as a function of the battery charging time.

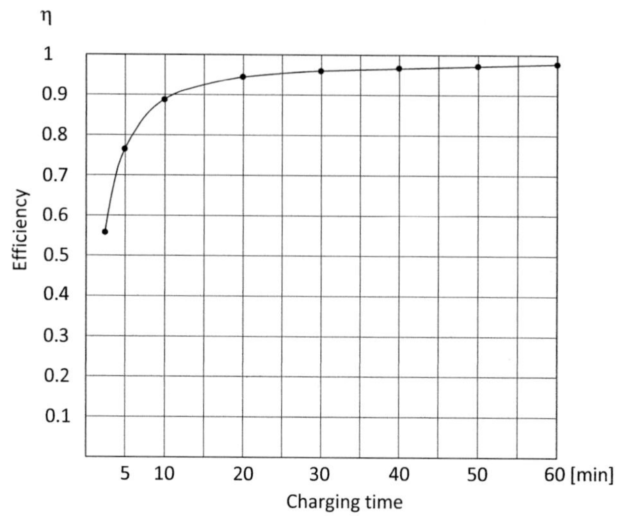

FIGURE 3.23 Efficiency as a function of the charging time.

Exercise 2: Kinetic Energy Recovery System

A kinetic energy recovery system (KERS) is a power assistance system based on the recovery of a moving vehicle's kinetic energy under braking (Figure 3.24). The recovered energy is stored in a reservoir for later reuse under acceleration. Such systems have been developed for race cars based on different storage technologies. A first example using supercapacitors is described in Reference [13] and was developed under the label Formula S2000.

In this study, a supercapacitor-based storage system was chosen. It makes it possible to recover successive small amounts of braking energy before allowing the driver to benefit from an additional acceleration power through a function called "push-to-pass".

The profile of the power recovered into the storage device is represented in Figure 3.25, together with the typical power impulse reused for acceleration. The voltage of the supercapacitors U_{SC} is also represented.

The power profile shows braking recovery impulses of 10 kW followed by an acceleration power of –30 kW. The duration of the additional acceleration is specified to 6 s.

The objective of the exercise is to evaluate the energy efficiency of the described design using supercapacitive storage for its discharge and to compare the result with another design using a Li-ion battery.

1. Calculate the maximum and minimum energy efficiency of the described solution using supercapacitors.

   The maximum efficiency corresponds to the case where the supercapacitors are fully charged and the minimum to the case where the supercapacitor voltage is the lowest and where simultaneously the discharge current is the highest when the discharge occurs under constant power.

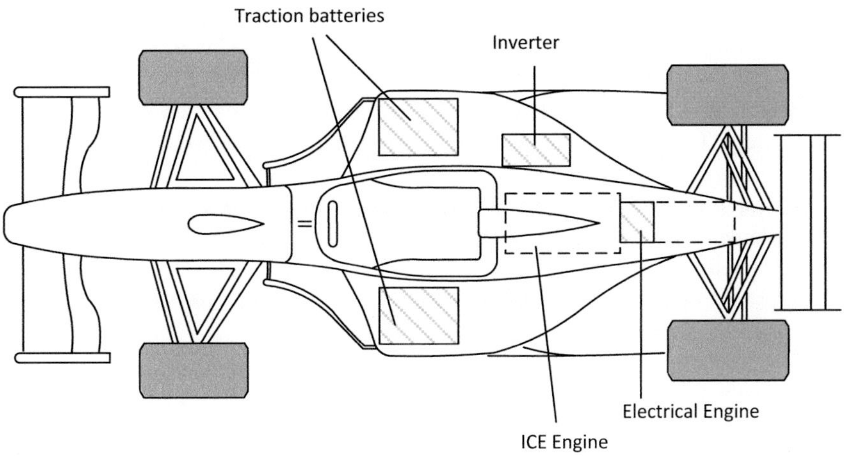

FIGURE 3.24 Power-assisted race car (Formula S2000).

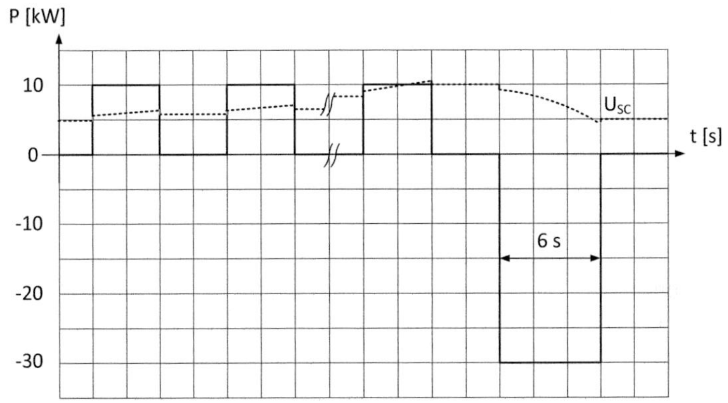

FIGURE 3.25 Power profile.

TECHNICAL DATA:

- Supercapacitors:
  Maxwell, $C = 310$ F, $R_i = 0.4$mOhm, $U = 2.7\,V$
- Acceleration power (boost power): 30 kW
- Energy capacity of the storage device: $\mathrm{E} = 180000\ \mathrm{J}$ for a discharge down to a minimum voltage of 50% (selected for an additional power of 30 kW during 6 s)

1. With the goal to illustrate the influence of the power level on the efficiency of the discharge, repeat the calculation of the minimum and maximum efficiency for a discharge power of 15kW, 6 kW and 3 kW.

   The values of the efficiency can now be represented graphically in a MRR (logarithmic scale for the power level). For the different power levels, the

different values of the efficiency can be represented by vertical segments illustrating the domain between the minimum and maximum values.
For the logarithmic scale: log10 (2) ~= 0.3, log10 (5) ~= 0.7

2. Calculate the energy efficiency of a storage device realized with a Li-ion battery. The technical data of the Li-ion battery can be found in Table 3.1.

The design of the storage device must especially take into account that the Li-ion battery elements have a limited discharge current. This will lead to an oversize of the energy capacity.

Represent the properties of this storage device (Li-ion) in a MRR (P = 30, 15, 6, 3 kW).

Reference [10] indicates the difficulty of getting real values on the internal resistance of commercial batteries and includes a method for evaluating this parameter.

SOLUTION TO EXERCISE 2

1. Design with Supercapacitors

The energy stored in a supercapacitor charged at its maximum voltage is calculated according to

$$E = \tfrac{1}{2} C \cdot U_{\max}^2 = \tfrac{1}{2} \cdot 310F \cdot (2.7V)^2 = 1129J \tag{3.48}$$

The remaining amount of energy after discharge down to 50% of the voltage is

$$E = \tfrac{1}{2} C \cdot U_{\min}^2 = \tfrac{1}{2} \cdot 310F \cdot (1.35V)^2 = 882J \tag{3.49}$$

The extracted amount of energy becomes

$$E_{disch} = \tfrac{1}{2} C \cdot U_{\max}^2 - \tfrac{1}{2} C \cdot U_{\min}^2 = 1129J - 882J = 847J \tag{3.50}$$

The number of supercapacitors to be used for a total capacity of 180 kJ is then

$$N_{sc} = \frac{E_{tot}}{E_{disch\_sc}} = \frac{180000J}{847J} = 212.5 \Rightarrow 213 \tag{3.51}$$

The maximum voltage of the storage device (all elements connected in series) becomes

TABLE 3.1
Parameters of the Li-Ion Battery

V	C	I_{dchmax}	Ri	m
3.3 V (2. . .3.6V)	2.3 Ah	120 A (10 s)	10 mΩ	70 g

$$U_{max} = N_{sc} \cdot U_{max\,sc} = 213 \cdot 2.7V = 575V \tag{3.52}$$

For a power delivery of 30 kW, the current in the supercapacitors is

$$I_{min\,30} = \frac{30kW}{575V} = 52.2A \tag{3.53}$$

At the end of the discharge with constant power, the current takes the value of

$$I_{max\,30} = \frac{30kW}{575V \cdot 0.5} = 104.4A \tag{3.54}$$

For the calculation of the efficiency, the following relation is used

$$\eta = \frac{U_{sc} \cdot I_{sc} - R \cdot I_{sc}^2}{U_{sc} \cdot I_{sc}} = \frac{U_{sc} - R \cdot I_{sc}}{U_{sc}} \tag{3.55}$$

The voltage drop on the internal resistance of the supercapacitors must be calculated.

For the whole storage device (213 elements in series)

$$U_{Rsc\_tot\_min} = R_{sc} \cdot I_{sc} \cdot N_{sc} = 0.4m\Omega \cdot 52.2A \cdot 213 = 4.45V \tag{3.56}$$

for the lowest current

and

$$U_{Rsc\_tot\_max} = R_{sc} \cdot I_{sc} \cdot N_{sc} = 0.4m\Omega \cdot 104.4A \cdot 213 = 8.9V \tag{3.57}$$

for the highest one

The corresponding efficiencies are then

$$\eta_{max\_30} = \frac{U_{sc} - R \cdot I_{sc}}{U_{sc}} = \frac{575V - 4.45V}{575V} = 0.992 \tag{3.58}$$

$$\eta_{min\_30} = \frac{U_{sc} - R \cdot I_{sc}}{U_{sc}} = \frac{575V \cdot 0.5 - 8.9V}{575V \cdot 0.5} = 0.969 \tag{3.59}$$

For a discharge at reduced power, only the minimal value of the efficiency is calculated. It has to be evaluated at the lower limit of the supercapacitor's voltage after discharge.

The extracted energy after a 15 kW impulse during 6 sec is

$$E_{extr\_15} = 15000W \cdot 6s = 90000J \tag{3.60}$$

leading to the state of energy of one element of

$$E_{sc\_fin\_15} = 1129J - \frac{90000J}{213} = 706J \tag{3.61}$$

the corresponding voltage of the elements being

$$U_{sc\_min\_15} = \sqrt{\frac{2E_{sc\_fin\_15}}{C}} = \sqrt{\frac{2 \cdot 706J}{310F}} = 2.13V \tag{3.62}$$

At this point, the discharge current is

$$I_{max\_15} = \frac{15000W}{213 \cdot 2.13V} = 33A \tag{3.63}$$

leading to an efficiency of

$$\eta_{min\_15} = \frac{U_{sc} - R \cdot I_{sc}}{U_{sc}} = \frac{2.13V - 0.4 \cdot 10^{-3}\Omega \cdot 33A}{2.13V} = 0.993 \tag{3.64}$$

The efficiency is recalculated for the lowest values of the power (6 and 3 kW) and is represented in Figure 3.26.

Figure 3.26 shows that the design of the storage device on the basis of supercapacitors presents a MRR where the efficiency is of high value (above 96%) within a full decade of the power level of the discharge, between 10 and 100% of the nominal power.

2. Design with a Li-Ion Battery

Parameters of the battery element:

U = 3.3 V
C = 2.3 Ah

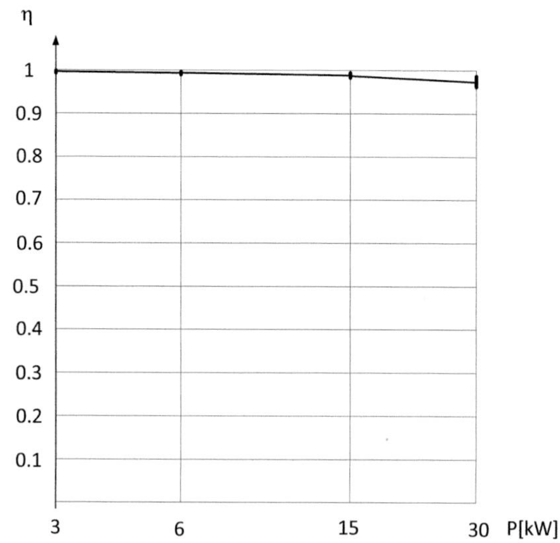

FIGURE 3.26 Efficiency (MRR) of the discharge of the supercapacitor-based storage device

The energy content of such an element is

$$E_{batt} = U \cdot C = 3.3V \cdot 2.3Ah = 3.3 \cdot 2.3 \cdot 3600Ws = 27324J \tag{3.65}$$

For the 30 kW/6 s discharge capacity, the number of needed elements would be

$$N_{batt\_theoric} = \frac{180000J}{27324J} = 6.58 => 7 \tag{3.66}$$

with a voltage of the whole battery of

$$U_{stack\_theoric} = 7 \cdot 3.3V = 23.1V \tag{3.67}$$

the corresponding current of the 30 kW discharge is

$$I_{stack\_theoric} = \frac{30kW}{23.1V} = 1298A \tag{3.68}$$

This value of current is around 10 times the admissible battery current!

The design of the battery pack must be oversized. In order to reduce the current to an acceptable value, the number of elements is multiplied by 10, leading to

$$U_{stack} = 70 \cdot 3.3V = 231V \tag{3.69}$$

$$I_{stack} = \frac{30kW}{231V} = 129.8A \tag{3.70}$$

The efficiency of the discharge at 30 kW becomes

$$\eta_{batt\_30} = \frac{U_{stack} - R \cdot I_{stack}}{U_{stack}} = \frac{231V - 70 \cdot 10 \cdot 10^{-3}\Omega \cdot 129A}{231V} = 0.6 \tag{3.71}$$

The calculations for the current and efficiency at reduced power give

P [kW]	I [A]	η [p.u.]
15	65	0.8
6	26	0.92
3	13	0.96

The values of the efficiency are represented in Figure 3.27.

The representations of the efficiencies in Figures 3.26 and 3.27 are a good illustration of the performances of the storage devices and highlight the differences between the two technologies. In opposition to the properties of the supercapacitor-based design, where the efficiency is over 96% in the whole range of operation (power), the design with the Li-ion battery presents problematic energy efficiency values for the upper range of the operation power (0.6 at rated power).

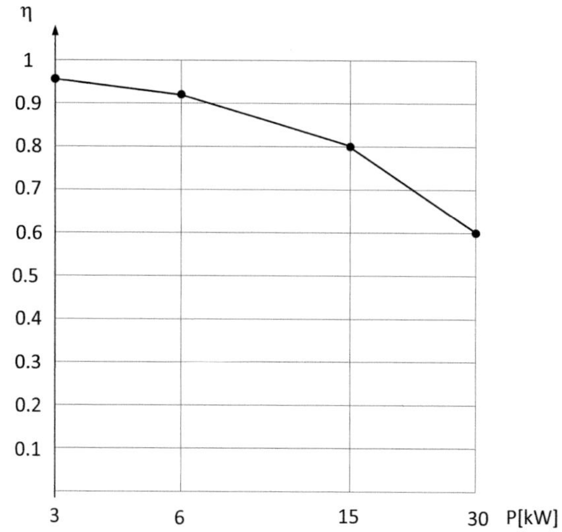

FIGURE 3.27 Efficiency (MRR) of the discharge of the Li-ion battery based storage device.

REFERENCES

[1] Brunet, Y., *Energy Storage*, ISTE Ltd/John Wiley & Sons, Inc., London, UK/Hoboken, NJ, 2011.

[2] International Electrotechnical Commission (IEC), *Letter Symbols Including Conventions and Signs for Electrical Technology, IEC 60027-2-19*, January 8, 2019.

[3] Rebeaud, M., *Solar Impulse, Encyclopedia Universalis* [Online], www.universalis.fr/encyclopedie/solar-impulse/. Accessed on July 21, 2017.

[4] Vaclac, S., *Power Density: A Key to Understanding Energy Sources and Uses*, The MIT Press, Cambridge, MA, 2015.

[5] Mouries, G., Condensateurs utilisés en électronique de puissance, Techniques de l'ingénieur, Art. Ref. D3010 V1, August 10, 2001.

[6] Christen, T., Carlen, M. W., Theory of ragone plots, *Journal of Power Sources*, 91, 210–216, 2000.

[7] Delalay, S., Etude systémique pour l'alimentation hybride—application aux systèmes intermittents, PhD Thesis Nr. 5768, EPFL, Lausanne Switzerland, 2013, http://dx.doi.org/10.5075/epfl-thesis-5768.

[8] Poss, J., Satwicz, J., Richmond, B., Taylor, M., Solar powered compaction apparatus, US Patent, US 7481159 B2, Seahorse Power Company, Needham, MA, January 27, 2009.

[9] Barrade, P., Delalay, S., Rufer, A., Direct connection of supercapacitors to photovoltaic panels with on-off maximum power point tracking, *IEEE Transactions on Sustainable Energy*, 3(2), 283–294, 2012.

[10] Culcu, H. et al., Internal resistance of cells of lithium battery modules with FreedomCAR model, *EVS24 and World Electric Vehicle Journal*, 3, 1–9, 2009.

[11] DiOrio, N., Dobos, A., Janzou, S., *Economic Analysis Case Studies of Battery Energy Storage with SAM. Technical Report, NREL TP-6A20-64987*, National Renewable Energy Laboratory (NREL), Golden, CO, November 2015.

[12] Klaus, D., *Redox-Flow_Batterie: Weniger als 1000 Euro pro kWh Kapazitat*, Forschung, November 17, 2016, https://www.energyload.eu.

[13] Rufer, A., Barrade, P., Correvon, M., Weber, J.-F., Multiphysic modelling of a hybrid propulsion system for a racecar application, *EET-2008 European Ele-Drive Conference. International Advanced Mobility Forum*, Geneva, Switzerland, March 11–13, 2008.

4 Electrochemical Energy Storage

4.1 INTRODUCTION

Electrochemical storage systems, often called electrochemical batteries, are closed systems able to perform a reversible conversion from chemical to electrical energy. Such transformations can be done with good efficiency, depending on the operating conditions. Usual values of the energy efficiency are situated around 80 to 90%.

Similar to other storage systems, electrochemical batteries are characterized by their energy density, which is between 30 and 200 Wh/kg. As a comparison, hydrocarbon-based sources present a much higher value of the energy density, on the order of 10 kWh/kg.

As opposed to the primary cells like the ordinary flashlight battery, storage batteries are also called secondary batteries and are capable of being recharged. Their electrode reactions can be processed in either direction.

4.1.1 System Structure

Electrochemical accumulators are charged by converting the electrical energy into chemical and are discharged by an inverse transformation. Each element of a battery is made of a so-called negative active mass able to provide electrons to the external circuit during discharge and a positive active mass able to accept the electrons from the return path of the external circuit.

Figure 4.1 is a schematic representation of an electrochemical cell. The indicated direction of electrons and ion flows corresponds to the battery discharge process.

The transfer of electrons from the active masses to and from the external circuit is possible due to the presence of current collectors assuming electronic conduction. The positive and negative active masses are of different chemical composition and have between them an electric separator that prevents the passage of electrons from the positive to the negative side. The internal flow of the electric current is possible due to the displacement of electric charges in the form of ion transfer from one active mass to the other. The separator must be an ionic conductor.

Current collectors are generally metallic, and the active masses are porous solids, made from powders, and are soaked by a liquid ion conductor.

4.1.2 Elementary Principle

The active masses of an electrochemical cell each contain a couple of Redox, namely a set of chemical species in reduced form (Red) able to provide electrons and to be transformed into another set of chemical species (Ox), able to accept electrons.

DOI: 10.1201/9781003736486-4

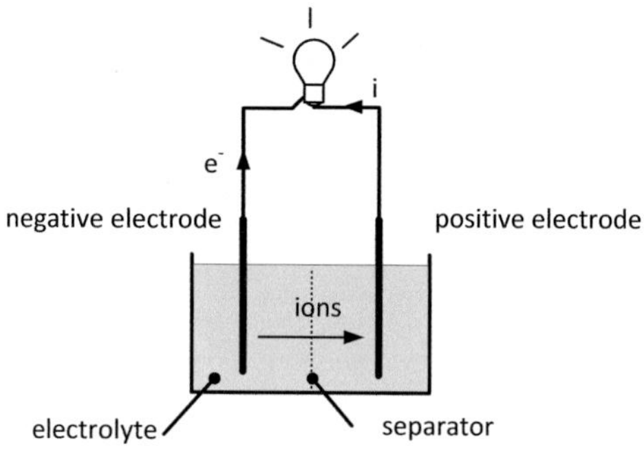

FIGURE 4.1 Electrochemical cell connected to a load.

A Redox couple is characterized through its potential determined by the Nernst law [1].

$$E = E^0 + (RT/nF)\cdot \ln[(Ox)/(Red)] \tag{4.1}$$

In this relation, (Ox) and (Red) designate the concentrations of the oxidized and reduced forms of the Redox couple: $R = 8.32$ Joule/Kelvin/mole, T is the absolute temperature, 298K at 25°C, n is the number of electrons implicated in the reaction, and F is the Faraday constant: 96500 C/mole.

An electrochemical generator is transforming chemical energy (free enthalpy) into electrical energy provided to the external system in the form of an electrical current.

The negative terminal of an elementary cell is connected to the negative active mass that contains a reducer material (electron source) Red_1 capable of undergoing an electrochemical oxidation

$$Red_1 \longrightarrow Ox_1 + ne^- \tag{4.2}$$

At zero current, the E_1 potential of the negative terminal is controlled by the Nernst equation related to the Red_1/Ox_1 couple. Symmetrically, the positive terminal is connected to the positive active mass containing an oxidant material (electron acceptor) Ox_2 capable of undergoing an electrochemical reduction

$$Ox_2 + ne^- \longrightarrow Red_2 \tag{4.3}$$

At zero current, the potential E_2 of the positive terminal is controlled by the Nernst equation related to the Red_2/Ox_2 couple.

The global chemical reaction is then

$$Red_1 + Ox_2 \longrightarrow Ox_1 + Red_2 \tag{4.4}$$

The voltage of the elementary cell is equal to the difference of the positive and negative potentials.

The transformation of one mole of reactant in these conditions releases an electrical energy of *n* Faraday, namely a maximum of $n \times 96500\ \mathrm{C} \times (\mathrm{E}_2 - \mathrm{E}_1)$ Joules. This maximum corresponds to the variation of the standard free enthalpy of the reaction (4.4) [2].

In a general manner, the electrode where the reduction takes place is called the cathode. The electrode where the oxidation takes place is called the anode. In the case of a cell working as a generator (discharge), the cathode will be the positive terminal and the anode the negative terminal. On the contrary (charge), the cathode will be the negative terminal and the anode the positive terminal.

An elementary example of an electrochemical battery is given by the lead-acid battery. In the discharged state, both the positive and the negative plates become lead sulfate ($PbSO_4$), and the electrolyte loses much of its dissolved sulfuric acid and becomes primarily water. The discharge process is driven by the conduction of electrons from the negative plate back into the cell at the positive plate through the external circuit.

The negative active mass contains a Redox couple as

$$Pb + HSO_4^- \rightarrow PbSO_4 + H^+ + 2e^- \qquad (4.5)$$

The negative active mass potential is equal to $E° = -0.356$ V.

At the positive active mass, another Redox couple corresponds to

$$PbO_2 + HSO_4^- + 3H^+ + 2e^- \rightarrow PbSO_4 + 2H_2O \qquad (4.6)$$

The potential at the positive side is $E° = 1.685$ V.

The global reaction is

$$Pb + PbO_2 + 2HSO_4^- + 2H^+ \longrightarrow 2PbSO_4 + 2H_2O \qquad (4.7)$$

The cell voltage corresponds to the difference of potentials and is equal to $E° = 2.05$ V.

The sum of the molecular masses of the reactants is 642 g/mol (calculated from the right-hand-side terms of Equation 4.7), so theoretically a cell can produce two Faraday of charge (192.971 coulombs from 642 g of reactants, or 83 Ah/kg). For a 2 V cell, this comes to a theoretical value of 167 Wh/kg of reactants. In reality, a lead-acid cell gives only 30–40 watt-hour per kilogram, due to the mass of the water and other constituent parts.

In the fully charged state, the negative plate consists of lead and the positive plate lead dioxide. The electrolyte is of concentrated sulfuric acid.

During discharge, H^+ ions produced at the negative plate move into the electrolyte solution and then are consumed into the positive plates. HSO_{4^-} is consumed at both plates. The reverse occurs during charge.

During charge and discharge, the potentials of the active masses vary depending on the state of chemical conversion. This leads to a variation of the cell voltage in the open circuit state. In the real case of a lead-sulfuric acid battery, the cell voltage varies from 2.15 V fully charged to 1.6 V while discharged (open circuit).

4.2 DIFFERENT TYPES OF ACCUMULATORS

Several possibilities exist for accumulators like accumulators in an aqueous, acid or alkaline medium; hot accumulators; or accumulators in an organic medium.

The aqueous medium presents the advantage of a higher ionic conductivity, but the possible choices for the active masses are limited. Non-aqueous media permit higher cell voltages and a higher number of possible active masses.

4.2.1 Accumulators with Aqueous Electrolytes

The acid or alkaline aqueous electrolyte is a medium with a relatively good ionic conduction, but the cell voltage is limited by the possible decomposition of water into oxygen and/or hydrogen produced through the oxidant (or reductant) character of the active masses.

4.2.1.1 Lead-Acid Accumulator

Invented by Gaston Planté in 1858, the lead-acid accumulator is today one of the most used accumulators and must be considered the cheapest electrochemical storage solution. The principle of operation and the used chemical species have already been described in Section 4.1.2.

As also already mentioned, the acid concentration of the electrolyte brings the maximum of conductivity at the fully charged state of the cells. This brings a high availability of power with up to 10 times the rated current (a 30 Ah battery can deliver 300 A, delivering 3.6 kW when the battery voltage is of 12 V).

At its discharged state, the lead-acid battery has a relatively weak acid concentration that limits the available power.

With each discharge, lead sulfate is accumulated on the electrodes and is normally dissolved by the next charge. With a too long discharge, at a too-low state of charge, or operation at too high a temperature, crystals of lead sulfate appear and are no longer dissolved by the charge. The capacity of the battery and its power level are strongly reduced.

In practice, five main categories of lead-acid batteries are available.

1. *Starter batteries (with grating)*: Developed and produced for the automotive industry for the start-up of internal combustion engines. They are designed to be kept permanently charged and to deliver high currents rapidly. They are widely available and at very low costs (0.2 EUR/Wh).
2. *Drive batteries (flat plates)*: They are similar to starter batteries but present larger capacity. They are used in small electric vehicles like autonomous carts or fork-lifts. They are designed to be charged and discharged on a daily basis and can be operated at low SoC. Their cost is around 0.5 EUR/Wh.

3. *Batteries with gel electrolyte (flat plates):* These batteries do not need any maintenance and can be operated in any position. They are typically used in small professional equipment (radio communication, shining buoys, etc.). Their capacity is in the order of 100 Ah. Their cost is around 0.6 EUR/Wh.
4. *Absorbent glass mat (AGM) batteries*: In AGM batteries, the sulfuric acid is absorbed by a very fine fiberglass mat, making the battery spill-proof.
   AGM batteries have a very low internal resistance and are able to deliver high currents on demand and offer a relatively long service, even when deeply cycled.
5. *Stationary batteries (tubular plates)*: These batteries are used in security supply technology. They use a frame structure consisting of a series of vertical spines connected to a common bus bar. The tubular design keeps the active material together and presses it onto the collecting grid. The active lead oxide mass is held in microporous nonconductive tubes (gauntlets), which are placed over the individual spines. Assembling the spines, gauntlets, lead oxide and cap together makes a positive plate (Figure 4.2).
   In all lead acid cells, the negative plates are of the flat pasted type.

4.2.1.2 Alkaline Accumulators

The non-acid electrolyte batteries were proposed initially by Waldemar Junger in 1899 and by Thomas Edison in 1901. These alkaline accumulators were primary batteries and based on a zinc/manganese pair of active materials.

Chemical reactions in alkaline batteries produce electricity when the manganese dioxide cathode is reduced and the zinc anode becomes oxidized.

At the MnO_2 cathode, water (H_2O) is consumed and hydroxyl ion (OH^-) is produced by the following reaction:

$$2MnO_2 + H_2O + 2e^- \longrightarrow Mn_2O_3 + 2OH^- \tag{4.8}$$

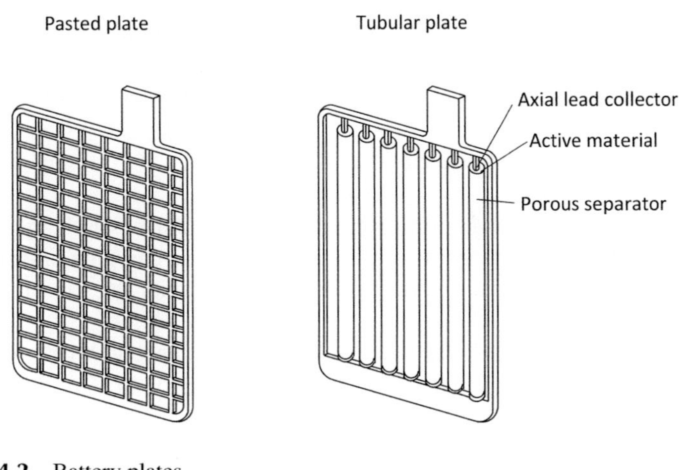

FIGURE 4.2 Battery plates.

At the same time, the anode is consuming hydroxyl ions and producing water:

$$Zn + 2OH^- \longrightarrow ZnO + H_2O + 2 \cdot e^- \quad (4.9)$$

The overall reaction being

$$Zn + 2MnO_2 \longrightarrow ZnO + Mn_2O_3 \quad (4.10)$$

The electrons generated during the reaction are used to power devices in the electric external path during discharge. The rate of the reaction is dependent on the quality of the raw materials and availability of water and hydroxyl ions during the reaction.

A battery is designed to keep the cathode and anode separated to prevent the reaction from occurring. The stored electrons only flow when the circuit is closed.

The evolution of alkaline batteries has gone through the use of other metals like nickel (Ni), a totally insoluble metal in alkaline medium. Its oxyhydroxyde form NiOOH layer is used on its positive electrode.

The electrochemical reaction at the positive active mass side is

$$NiOOH + H_2O + e^- \longrightarrow Ni(OH)_2 + OH^- \quad (4.11)$$

The mode of operation of the positive electrode can be considered a proton intercalation into the layered crystalline structure of the NiOOH during discharge and disinsertion during charge.

The negative active mass has been primarily based on zinc metal soaked in a potassium hydroxide solution like in the original zinc/manganese battery.

4.2.1.3 The Nickel-Iron Battery

One of the highly manufactured alkaline batteries is the nickel-iron battery developed in Europe and in the USA.

At the positive active mass, the reaction is identical to the reaction given through Equation 4.11.

The negative active mass consists of iron, and its electrochemical reaction is

$$Fe + 2OH \longrightarrow \mathrm{Fe}(OH)_2 + 2e^- \quad (4.9)$$

The open-circuit voltage is 1.4 V, dropping down to 1.2 V during discharge. The electrolyte mixture of potassium hydroxide is not consumed in charging or discharging, so unlike a lead-acid battery, the electrolyte-specific gravity does not indicate the state of charge.

Nickel-iron batteries were used intensively during the Second World War. In the USA, Ni-Fe batteries were profitably made from about 1903 to 1972 by the Edison Storage Battery Company.

The limited value of the energy density (19–25 Wh/kg) of the Ni-Fe battery was the reason of its wide use for railroad signaling, forklifts, and standby power applications.

However, its robustness and low aging effects are attracting new manufacturers and users (PV plants) to this technology nowadays.

4.2.1.4 Nickel–Cadmium Accumulator

Due to corrosion effects of zinc in the basic solution, Zn/NiOOH elements were soon replaced by Ni-Cd elements that rapidly became high-performance rechargeable alkaline batteries.

The electrochemical reactions of Ni-Cd batteries are (discharge)

$$Cd + 2OH^- \longrightarrow Cd(OH)_2 + 2e^- \tag{4.12}$$

at the negative electrode and

$$NiOOH + H_2O + e^- \longrightarrow Ni(OH)_2 + OH^- \tag{4.13}$$

at the positive electrode, the global reaction being

$$Cd + 2NiO(OH) + 2H_2O \longrightarrow 2Ni(OH)_2 + Cd(OH)_2$$

Ni-Cd batteries have specific energy of 40 to 60 Wh/kg and can easily overcome deep discharge and recharge.

They have a nominal voltage of 1.2 V per cell and can be stored in their discharged state. One of the most interesting parameters is the high power capability, up to 1 kW/kg.

Ni-Cd batteries can be recycled but present the disadvantage of the toxicity of cadmium, what has made this technology obsolete.

Another typical behavior of Ni-Cd cells is their memory effect. This effect corresponds to situations where the battery gradually loses its maximum energy capacity if it is repeatedly recharged after being only partially discharged.

4.2.1.5 Ni-MH Accumulator

In Ni-MH accumulators, the problematic cadmium in the cell has been replaced by a metallic alloy able to store hydrogen. It works similarly to the Ni-Cd accumulator and has the following characteristics

- 1.2 V cell voltage
- 40% higher energy density than that of Ni-Cd cells
- Very low memory effect

The electrochemical reactions in a Ni-MH cell are (discharge)

$$MH + OH^- \longrightarrow M + H_2O + e^- \tag{4.14}$$

at the negative electrode and

$$NiOOH + H_2O + e^- \longrightarrow Ni(OH)_2 + OH^- \tag{4.15}$$

at the positive electrode.

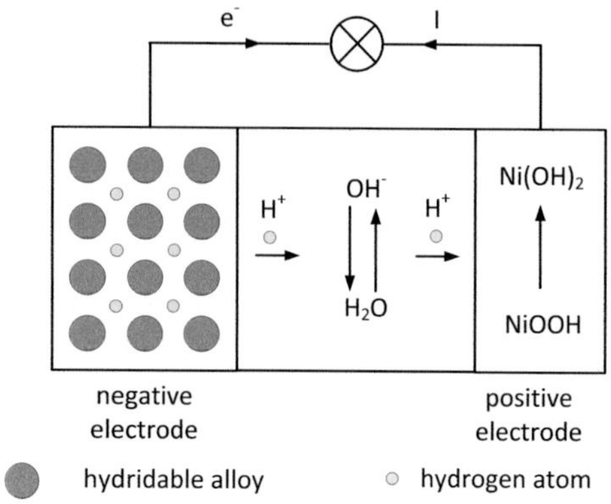

FIGURE 4.3 Discharge of a Ni-MH cell. (Adapted from Odru, P., *Le stockage de l'énergie*, 2nd edition, Dunod, Paris, France, 2013.)

A widely used hydratable metallic alloy is the nickel-lanthanum alloy LaNi5. Its reaction with hydrogen is given through

$$LaNi_5 + x \cdot H_2 \longrightarrow LaNi_5H_{2x} \tag{4.16}$$

where $1/2 < x < 5/2$.

The structure of a Ni-MH element is represented schematically in Figure 4.3.

The application domain of Ni-MH batteries extends from portable tools and toys (replacement for Ni-Cd batteries) to the automotive industry (hybrid cars).

4.2.2 Accumulators with Non-Aqueous Electrolytes

From the beginning of the 80s, the idea of utilizing lithium (Li), one of the highest electropositive elements (–3.02 V), has led to the development of chemical compounds able to insert and disinsert Li at the negative electrode. Because of the high reactivity of Li versus H_2O, these developments have induced the change from aqueous to non-aqueous electrolytes. Such electrolytes consist of lithium salts dissolved in an organic solvent [3].

One should distinguish lithium metal technology where the negative electrode is composed of metallic lithium from lithium-ion technology where lithium remains in an ionic state due to the use of insertion/disinsertion compounds both at the negative and positive electrodes.

4.2.2.1 Lithium-Metal Accumulators

In lithium-metal batteries, the negative electrode (the anode) is made of metallic lithium. During discharge, the lithium ions migrate from this electrode through the electrolyte to the cathode, where they mix with a host material (for example, $LiMn_2O_4$).

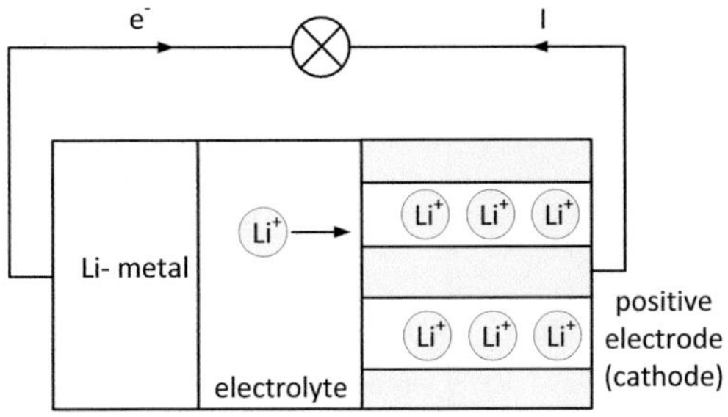

FIGURE 4.4 Lithium-metal cell. (Adapted from [4].)

The electrons produced at the anode supply the external circuit and return to the cathode. During charging, the Li+ ions migrate in the opposite direction while electrons are being supplied by the external (Figure 4.4) [4].

4.2.2.2 Lithium-Metal Polymer Cells

Lithium-metal polymer technology uses a lithium metal electrode at the negative side. The positive electrode is based on vanadium oxide. The electrolyte is a (ion)-conductive polymer membrane based on poly-oxide ethylene (POE). A lithium polymer battery costs around 0.1 EUR/Wh, and its energy density is given as 110 Wh/kg (Batscap). The temperature range for good performance is between 60 and 150°C without security problems.

4.2.2.3 Lithium-Ion Accumulator

A lithium-ion cell is different from a lithium metal cell at the side of the negative electrode, where the pure lithium is replaced by a compound able to insert and disinsert lithium ions (Figure 4.5). The lithium-ion concept uses two materials that allow the reversible exchange of lithium ions. The negative electrode (anode) is a thin layer of graphite in which atoms of lithium have been inserted (LiC_6). For the positive electrode (cathode), a lithium oxide of a transition metal such as $LiCoO_2$ can be used. The liquid electrolyte is usually a hexafluorophosphate of lithium ($LiPF_6$) mixed with a solution (carbonate mixture) [3].

Lithium-ion cells present very high energy density, up to 200 Wh/kg. Their typical potential is 3.7 V.

4.2.2.4 Lithium-Iron Phosphate Cells

Lithium-iron phosphate (LFP) is widespread. The positive electrode active material $\left(LiFePO_4\right)$ is associated with a negative graphite electrode. Today, these cells are used more often than the classic Li-ion cells and have replaced the classical Li-ion cells based on lithium cobalt dioxide $\left(LiCoO_2\right)$. The energy density and the potential of these cells are slightly lower than those of classical Li-ion cells (130 Wh/kg, 3.3 V).

These batteries need a battery management system (BMS).

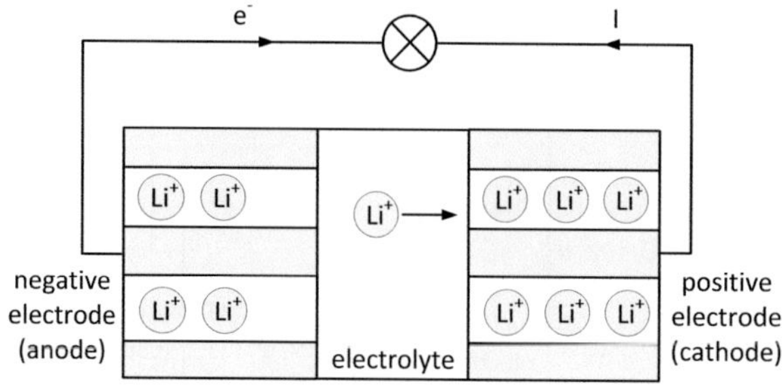

FIGURE 4.5 Lithium-ion cell. (Adapted from [4].)

4.2.2.5 Lithium-Ion Polymer (Li-Po) Cells

The mode of operation of these cells is similar to that of other lithium-ion cells. The electrolyte is made of gelled polymer. Their energy density is higher than other Li-ion cells, but their supervision and balancing circuits are more complex. The maximum cell voltage (over 4 V) and also the minimum cell voltage (around 2.7 V) must be monitored carefully.

4.2.2.6 Lithium-Titanate Cells (LTO)

Lithium-titanate (LTO) batteries are also called SCiBs (supercharge ion batteries). This technology is characterized by a high charging and discharging rate. It is based on the use of a nanocrystalline technology. The specific energy is given between 30–110 Wh/kg, and the cell voltage is between 1.9 and 2.4 V. The specific power can go up to 5000 W/kg. These batteries can operate between –40°C and +55°C.

The cycle durability is given at 6,000 cycles, and some manufacturers even indicate more than 10,000 cycles.

Main manufacturers: Toshiba, Altairnano, Leclanché.

4.2.2.7 Choosing between Competitive Battery Technologies

The constant evolution in battery technology and performances addresses the question of the adequate choice to be made in the design of an application. A very general principle produces competition between proven techniques and innovative ones, where even if the performance advantages are evident, the life cycle and ageing phenomena cannot be evaluated due to the lack of hindsight and experience. A common example is given by classical lead-acid batteries, which have been used for decades and can be considered tried-and-true technology. They are in competition with smaller lithium-ion batteries which have in addition longer life cycles [5].

Several comparison criteria can be used, such as:

- The specific energy
- The initial costs
- The life cycle
- The cost per life cycle

- The state of charge window for low ageing
- The available power for low ageing
- The recommended charging time

All batteries are sensitive to excessive discharging and extreme temperature. Lead-acid batteries lose potential cycles if they are discharged below 50% of their state of charge (SOC) or if discharged faster than C/8. On the other hand, lithium-ion batteries can be discharged to about 80% SOC and at a rate of C/2 without any long-term damage. Table 8.1 shows common characteristics of different types of batteries: lead-acid, NiCad, and NiMh and two different types of lithium-ion, graphite NMC and LTO technologies.

	"Conventional"			Li-Ion	
	Lead-Acid	**NiCad**	**NiMh**	**Graphite/NMC**	**LTO**
Specific energy	40 Wh/kg	40–60 Wh/kg	30–80 Wh/kg	120–200 Wh/kg	70–80 Wh/kg
Initial costs	150 EUR/kWh	500 EUR/kWh	600 EUR/kWh	120–250 EUR/kWh	400–800 EUR/kWh
Life cycle	200–300	1000–1500	300–800	500–4000	15000–25000
Cost per life cycle	0.15–0.7 EUR/kWh	0.3–0.5 EUR/kWh	0.75–2.0 EUR/kWh	0.35–0.04 EUR/kWh	0.024–0.04 EUR/kWh
Disch cond. for low ageing	C/3	C/2	C/5	1C	2C
Charging rate	C/4–C/8		C/10	C/2	2C
DOD window	50%		80–100%	80%	80%

The indicated values in Table 8.1 are indicative values obtained from different sources. They show available values of specific manufacturers.

A more global figure of the evolution of the price of lithium-ion batteries is given in Figure 4.6.

4.2.2.8 Solid-State Batteries

A solid-state battery uses a solid electrolyte for the conduction of ions from the anode to the cathode. The solid electrolyte replaces the conventional liquid or polymer gel electrolytes as used in classical Li-ion batteries. The solid-state batteries present higher energy density than typical lithium-ion or lithium polymer batteries.

Solid-state batteries can use metallic lithium for the anode and oxides or sulfides for the cathode materials. The solid electrolyte is realized with materials as ceramics or solid polymers.

The structure of a solid-state electrolyte battery is shown in Figure 4.7.

Solid-state batteries are developed as thin film elements with a typical energy density of 300–900 Wh/kg or as bulk elements with 250–500 Wh/kg.

Due to their principal advantage of higher energy density, they are promised an important expansion in the domain of electric vehicles to reach extended ranges. The

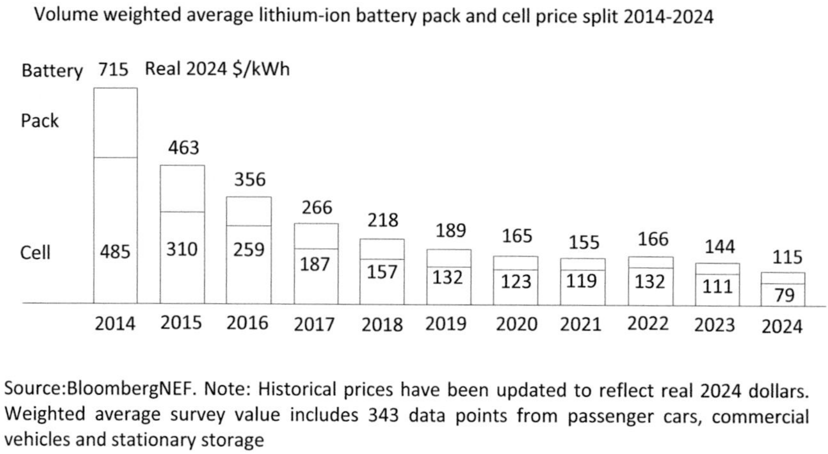

FIGURE 4.6 Evolution of lithium-ion battery prices over the last 10 years. (Adapted from Bloomberg New Energy Finance.)

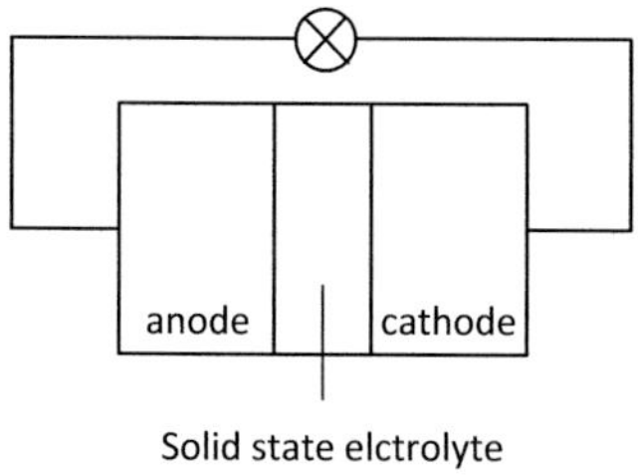

FIGURE 4.7 Structure of the solid-state electrolyte battery.

currently produced solid-state batteries have another important advantage, which is their very high cycle durability (10,000–100,000 cycles).

4.2.2.9 Long-Term Prospects for the Use of Electrochemical Batteries

The need to replace the consumption of fossil fuels by renewable sources for the reduction of CO_2 emissions and to limit the consequences of climate change has been confirmed by the scientific community and a large majority of the world population. The consequences of the integration of fluctuating and unpredictable electricity production such as photo-voltaic generators or windmills implies that tremendous amounts of energy storage facilities must be added to grid systems in a central or decentralized configuration. Additionally, the conversion of conventional ICE vehicles to electric propulsion will contribute to an explosion of the demand in material resources and minerals. Generally, equitable social development over the world will increase the global energy demand and make the problem even more difficult to solve.

4.2.2.10 The Scenario of Second Life and Recycling

A scenario with two possible ways of solving the problem is widely recognized where first the batteries of electric vehicles could be reused in applications with reduced performance requirements in terms of specific power, for example. The concept of second life batteries can be realized with the replacement of damaged elements in a large battery pack or by totally dismantling them and rebuilding new modules specifically designed for the new application. The re-use of battery elements in a second life application has the potential to largely reduce the demand for new materials and has in addition a significant economic advantage.

Second, the concept considers recycling batteries for new usage of the basic materials. Several recent developments propose an alternative to the older concept of melting battery parts where the needed energy demand is strongly reduced. In different new concepts, similar process steps can be found, such as dismounting, discharging and disassembling battery modules and reducing them to small pieces of approximately 1 cm in a shredder. To avoid thermal reactions, the process takes place under water (e.g., Batrec AG) or nitrogen (e.g., Librec AG). The shavings are then ground in an impact or hammer mill. Ferrous parts are magnetically removed, plastic films and platinum pieces from the battery electronics are blown into an air separator and the copper and aluminum granules are separated from each other according to their density. The black, powdery mass contains graphite and the active materials nickel, manganese, cobalt and lithium. The separated and cleaned materials are packaged by type and transported to the metal and battery industry for use.

4.2.3 Large-Size Accumulators

4.2.3.1 Sodium-Sulfur (NaS) Batteries

Sodium-sulfur (NaS) battery technology is known for its very high energy density, excellent cycle life, low-cost materials and high efficiency. However, these cells need high temperatures to operate due to the use of molten sodium. NaS batteries were developed from the middle of the 60s (Ford Motor Company, United States; Brown Boveri, Germany; and NGK in Japan). The packaging of the battery consists of a high-performance thermal insulation enclosure (operating temperature around 350°C). Other difficulties of realization are related to the quality of the insulator (brittle ceramics). Safety concerns stem from the need to protect the highly reactive sodium from moisture.

The specific energy is between 150 and 760 Wh/kg (high-temperature version).

The volume energy density is around 150 Wh/L.

The specific power is given at 200 W/kg.

The nominal cell voltage is equal to 2 V.

These batteries have a very good cycle durability (4000–5000 cycles).

Very large facilities have been realized (50 MW/300 MWh NGK [6]).

The electrolyte of a NaS cell consists of a beta alumina tube and contains liquid sodium. The external envelope is of sulfur (Figure 4.8).

During discharge, sodium ions converted from sodium in the negative electrode pass through the electrolyte and combine with sulfur in the positive electrode to form

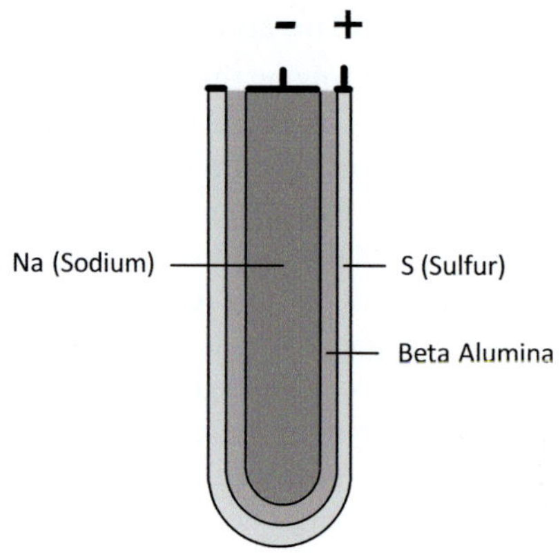

FIGURE 4.8 Sodium-sulfur cell.

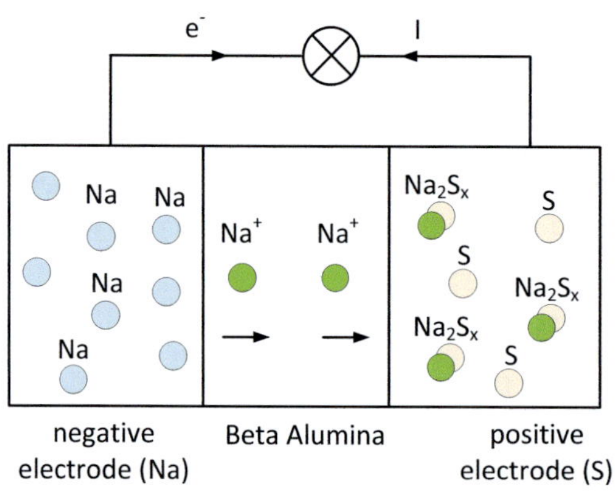

FIGURE 4.9 Discharge in a NaS cell.

sodium polysulfide. The electrons flow through the external path. The structure and reactions in a NaS cell are represented in Figure 4.9.

The electrochemical reactions in a NaS cell are (discharge)

$$2Na \longrightarrow 2Na^{+} + 2e^{-} \qquad \text{at the negative electrode} \tag{4.17}$$

$$4S + 2Na^{+} + 2e^{-} \longrightarrow Na_2S_4 \qquad \text{at the positive electrode} \tag{4.18}$$

The global reaction is:

$$2Na + 4S \longrightarrow Na_2S_4 \tag{4.19}$$

Othe chemistries are possible

FIGURE 4.10 A 34 MW NaS storage facility in Japan. (Courtesy of Japan Wind Development Co, Ltd, Tokyo, Japan and NGK Insulators Ltd, Nagoya, Japan.)

$$2Na + 3S \longrightarrow Na_2S_3 \tag{4.20}$$

or

$$2Na + (SSCH_2CH_2)_n \longrightarrow Na_2SSCH_2CH_2 \tag{4.21}$$

for lower melting temperatures (90–100°C).

An example of a large energy storage plant based on NaS batteries is shown in Figure 4.10.

4.2.3.2 Vanadium Redox Flow Battery

As already introduced in Section 2.2.4, the vanadium redox flow (VRB) is a very special type of accumulator where the positive and negative active masses correspond to two electrodes in contact with two separated electrolytes, namely an anolyte and a catholyte separated by a proton exchange membrane, as can be found in fuel cells or electrolyzers. The energy storage occurs in such flow batteries in the change of the concentrations of ions at the level of two liquids, the anolyte and the catholyte circulating from two separated reservoirs [7, 8]. A schematic representation is given in Figure 4.11.

4.2.3.3. Chemical Reactions of the VRB Battery

The vanadium ions V^{4+} and V^{5+} are in fact vanadium oxides, respectively VO^{2+} and VO_2^+. So, in order to maintain the charge balance and stoichiometry, water

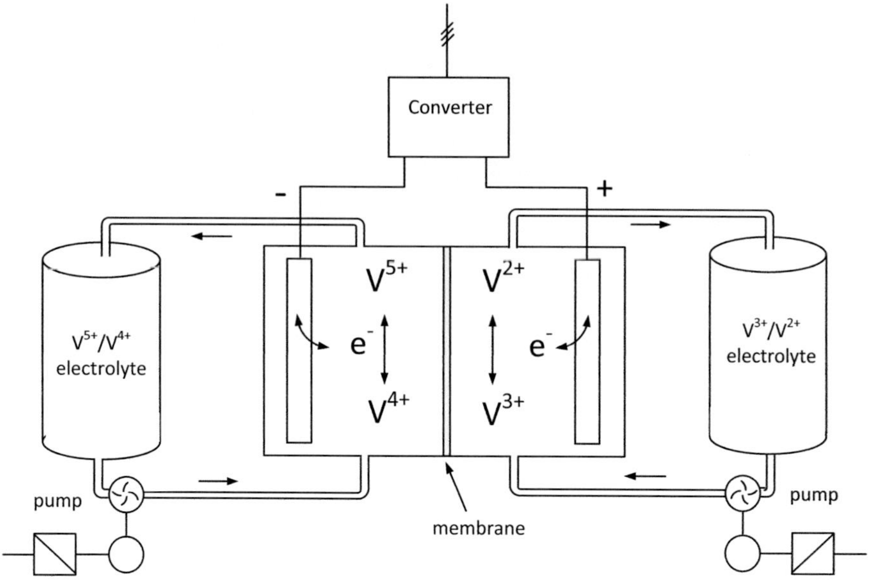

FIGURE 4.11 Vanadium redox flow battery.

molecules H_2O and protons H^+ must be introduced in the cathodic reaction. The chemical reactions can then be formulated as

$$\text{Cathode:} \quad VO_2^+ + 2H^+ + e^- \rightleftarrows VO^{2+} + H_2O \tag{4.22}$$

$$\text{Anode:} \quad V^{2+} \rightleftarrows V^{3+} + e^- \tag{4.23}$$

$$\text{The total reaction being} \quad V^{2+} + VO_2^+ + 2H^+ \rightleftarrows VO^{2+} + V^{3+} + H_2O \tag{4.24}$$

In Figure 4.11, a simplified notation for the cathodic reaction is used. In fact, Equation 4.22 can be written differently, taking into account the full ionic notation. The complete transformations are noted further in Figure 4.12.

Flow batteries have the advantage of a possible design with separated power level (mainly related to the membrane/electrode surface) and energy capacity (related to the volume of electrolytes). They present long cycle life (there is no solid-to-solid phase transition) and fast response time. There is no need for an equalization circuit during charging. Compared to other batteries like Li-ion batteries, redox flow batteries have higher power densities but lower energy density. They are mainly dedicated to stationary applications. The cell voltage is typically 1.4 V.

VRBs exist in different sizes, for example from 10 kW/40 KWh to 200 kW/1600 kWh (Figure 4.13).

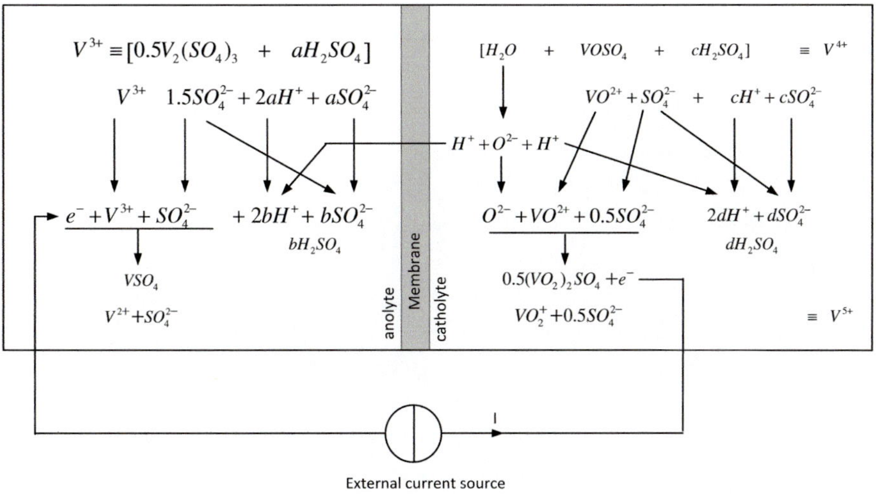

FIGURE 4.12 Full ionic transformation in a VRB battery. (Adapted from Blanc C., Modelling of a vanadium redox flow battery electricity storage system, Thesis EPFL No. 4277, EPFL, Lausanne, Switzerland, 2009, https://infoscience.epfl.ch/record/129758/files/EPFL_TH4277.pdf. Accessed on September 22, 2017.)

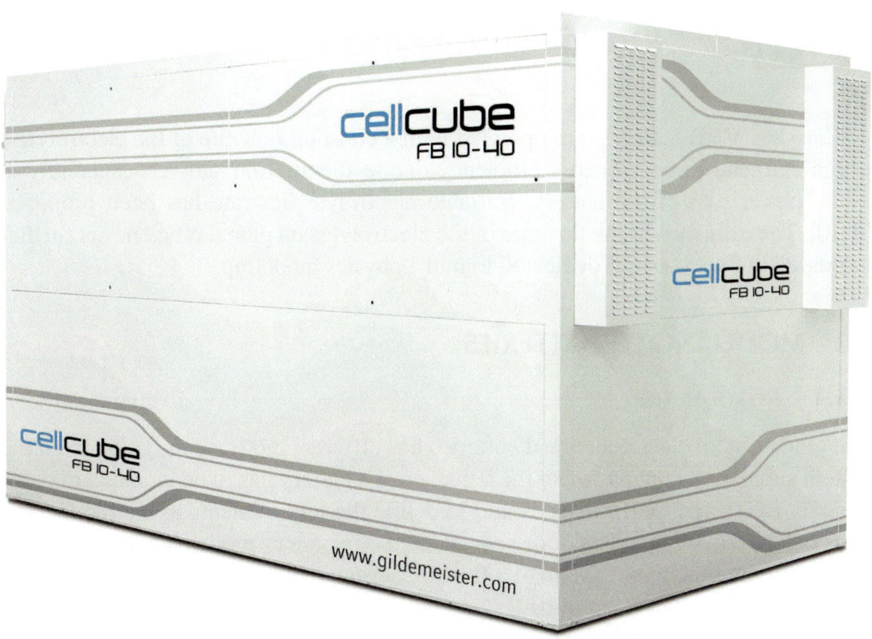

FIGURE 4.13 Vanadium redox flow battery. (Courtesy Cellcube/Gildemeister energy storage GmbH, Würzburg, Germany.)

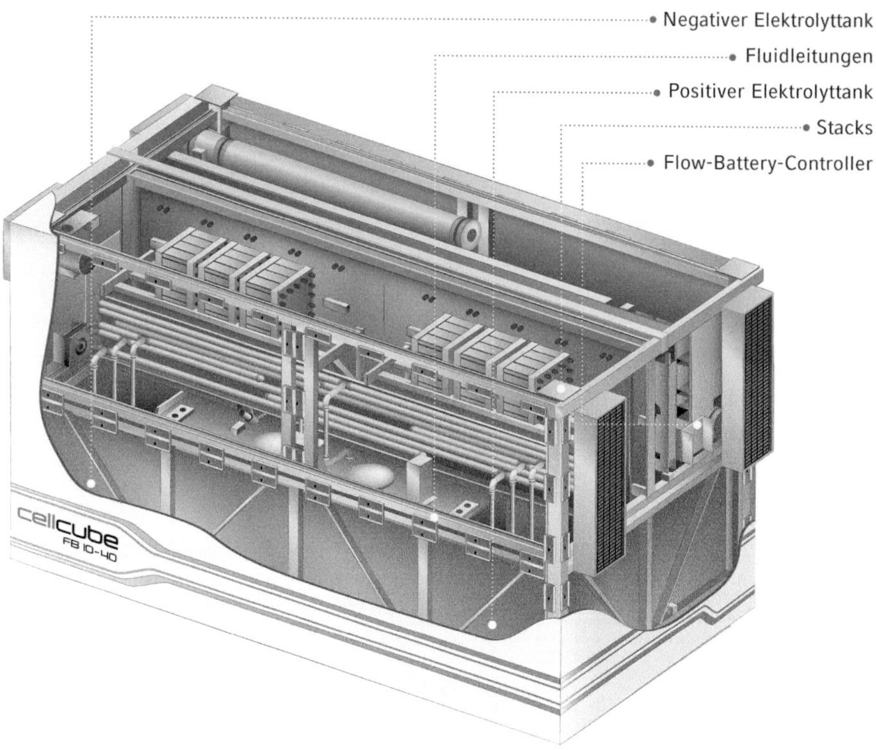

FIGURE 4.13 (Continued)

Classical VRB batteries are operated with a constant flowrate of the electrolytes, which can lead to poor energy efficiency due to the auxiliary losses, especially at a low power rate of the battery. Variable electrolyte flowrate has been proposed [9, 10]. The influence of the flowrate of the electrolytes on global efficiency is further discussed in Section 4.3.5 dedicated to multi-physics modeling.

4.3 MODELING OF BATTERIES

4.3.1 Introduction

The modelling of electrochemical batteries has different motivations, going from the system simulation context where the behavior of a battery has to be modeled in order to study the system performances and to define the requirements to interconnected power electronics converters. One requirement of a battery model is to complete the BMS with the function of evaluating the SoC, also called the "gas gauge" function.

Another important issue of battery modeling is to complete the electrochemical/electric part of the battery model with a simulation of the thermal conditions of operation.

4.3.2 Generic Models of Batteries

Circuit-based models of batteries describe their electrochemical phenomena, using mainly electrical active and passive components [11–13].

4.3.2.1 R_{int} Model

The R_{int} model is one of the simplest models of a battery, proposed in the mid-1990s by the National Renewable Energy Laboratory (NREL).

This model is composed of an ideal voltage source representing an ideal and constant open circuit voltage of the battery U_{OC}, connected in series with a resistor. The corresponding circuit is given in Figure 4.14.

This model can be used in applications where the energy source is considered infinite and where the transient phenomena are neglected.

4.3.2.2 Thévenin Battery Model

The Thévenin battery model is a completed form of the R_{int} model where a parallel RC dipole is added in order to represent the transient effects related to the battery discharge. The equivalent scheme of the Thévenin model is given in Figure 4.15.

The circuit equation from Figure 4.15 gives for the output voltage of the battery U_b

$$U_b = U_{oc} - R_0 \cdot I - R_p \cdot I_p \tag{4.25}$$

R_p is called the polarization resistance and I_p the polarization current.

The circuit analysis gives for the polarization current

$$\frac{dI_p}{dt} = \frac{I - I_p}{\tau} \tag{4.26}$$

with $\tau = R_p C_p$

A discrete solution for (4.26) is [13]

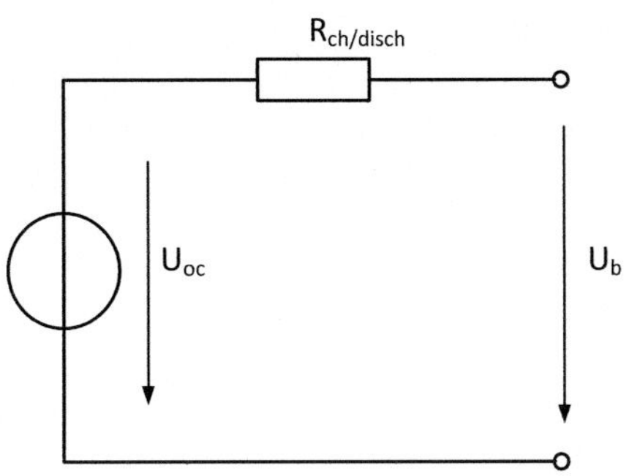

FIGURE 4.14 R_{int} model, equivalent circuit.

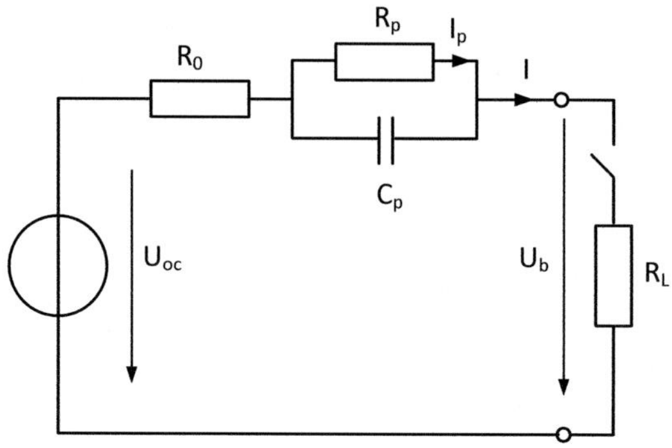

FIGURE 4.15 Thévenin model, equivalent circuit.

$$I_{p,i} = \left(1 - \frac{1-\exp(-\Delta t/\tau)}{-\Delta t/\tau}\right) \cdot I_i + \left(1 - \frac{1-\exp(-\Delta t/\tau)}{-\Delta t/\tau}\right) \cdot I_{i-1} + \exp(-\Delta t/\tau) \cdot P_{p,i-1} \tag{4.27}$$

For the output voltage of the battery, it comes from (4.27) and (4.25)

$$U_b = U_{oc} - R_0 \cdot I - R_p \cdot I_{p,i} \tag{4.28}$$

The Thévenin model can be represented through a structural diagram, as illustrated in Figure 4.16.

U_{oc}, R_o and R_p are the main parameters of the battery model. U_b is the battery output voltage, and I is a feedback signal representing the load current.

4.3.2.3 RC Network Battery Model

The RC network battery model is an extended form of the Thévenin model where additional RC networks are connected in series, the idea being to model additional internal voltage drops of the battery occurring with different time constants. Another way of considering the RC network model is to represent the transient behavior of the battery by a truncated exponential series [12]. The number of RC dipoles must be chosen in dependency with the required accuracy of the model or the error between the simulated and measured curves of a given charge/discharge profile.

The equivalent scheme of the RC network battery model is given in Figure 4.17.

The equivalent circuit of the RC network can be analyzed using the state space differential equation:

$$\begin{bmatrix} \dot{U}_{C1} \\ \dot{U}_{C2} \\ \cdot \\ \dot{U}_{Cn} \end{bmatrix} = \begin{bmatrix} \frac{1}{R_1C_1} & 0 & .. & 0 \\ 0 & \frac{1}{R_2C_2} & .. & 0 \\ 0 & .. & .. & 0 \\ 0 & 0 & .. & \frac{1}{R_nC_n} \end{bmatrix} \cdot \begin{bmatrix} U_{C1} \\ U_{C2} \\ \cdot \\ U_{Cn} \end{bmatrix} + \begin{bmatrix} \frac{1}{C_1} \\ \frac{1}{C_2} \\ .. \\ \frac{1}{C_n} \end{bmatrix} \cdot I \tag{4.29}$$

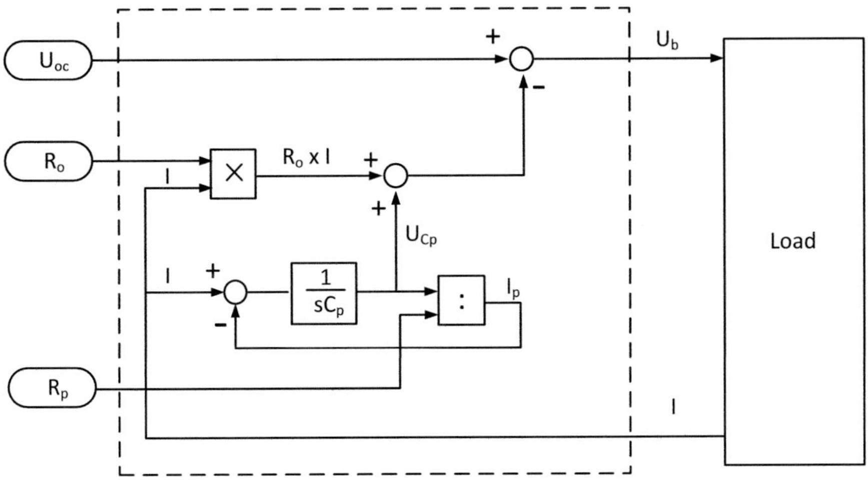

FIGURE 4.16 Structural diagram of the Thévenin model.

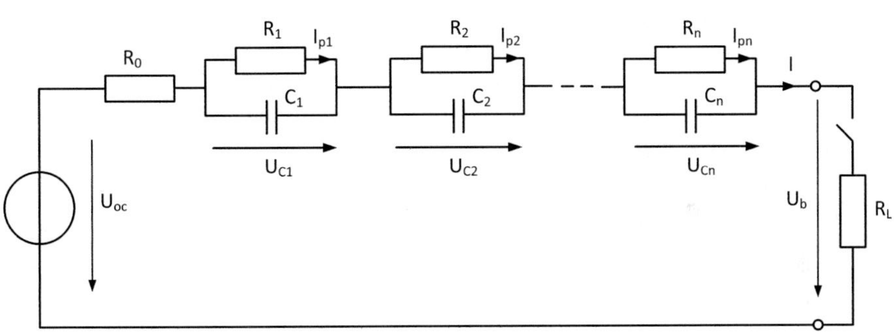

FIGURE 4.17 RC network model.

where $U_{C1}, U_{C2} \ldots U_{Cn}$ are the voltages at the RC dipoles; $C_1, C_2, .. C_n$ the capacitors of the RC branches; and I is the load current.

The output voltage U_b becomes

$$U_b = U_{oc} - R_0 \cdot I - U_{C1} - U_{C2} .... - U_{Cn} \tag{4.30}$$

A similar structural diagram as made for the Thévenin model (Figure 4.15) can be established for the RC network model.

4.3.2.4 Model of the Extracted Charges

Many parameters and variables of a battery are not constant and vary depending on temperature, SoC and state of health (SoH). While in many cases in practice, the state of health is not considered for the models, the main changes of state of a battery are determined by the state of charge and the temperature. The battery SoC is determined through the charge/discharge process and can be modeled through the evaluation of the extracted charges

$$Q_e = \int_0^t i(t)dt \tag{4.31}$$

From the extracted charges Q_e and referring to the nominal capacity C_N of the battery, the SOC can be calculated:

$$SoC = \frac{C_N - \int_0^t i(t)dt}{C_N} \tag{4.32}$$

Then, parameters and variables like the open-circuit voltage U_{oc} can be represented as functions of the SoC. The discrete values are generally determined experimentally for a SoC between 1 and 0 in steps, typically of 0.1 intervals.

Table 4.1 gives an example of recorded values in an identification run for the open circuit voltage U_{oc}.

The values indicated in Table 4.1 illustrate an example of a LFP battery [12]. Further parameters needed for the Thévenin model are estimated and given in Table 4.2.

From Tables 4.1 and 4.2, one can see the difficulty of interpreting the variations of the estimated parameters, especially for the R_p and C_p values. This is favors the choice of a simple model structure.

The structural diagram of a Thévenin model completed by a simulation of the extracted charges and a U_{oc} characteristic as a function of the SoC is given in Figure 4.18.

TABLE 4.1
Open Circuit Voltage Identification

SoC [%]	100	90	80	70	60	50	40	30	20	10
U_{oc}	3.42	3.33	3.33	3.31	3.29	3.29	3.29	3.27	3.24	3.20

TABLE 4.2
Parameter Identification

SoC [%]	100	90	80	70	60	50	40	30	20	10
R_0 [Ω]	0.169	0.13	0.15	0.13	0.14	0.15	0.14	0.14	0.13	0.15
R_p [Ω]	0.029	0.053	0.029	0.053	0.024	0.014	0.038	0.063	0.053	0.048
C_p [F]	618	337	618	337	741	1236	463	285	337	370

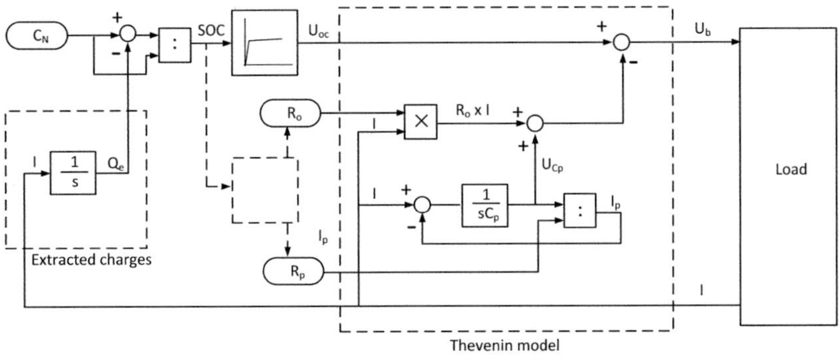

FIGURE 4.18 Thévenin model completed by a model of extracted charges.

4.3.3 Thermal Behavior of a Battery

The parameters of a battery strongly depend on the temperature. The internal parameters such as resistance and open circuit voltage but also the battery capacity are temperature dependent but are also influenced by the charging/discharging current.

The battery capacity is a parameter with important signification for the specific application. As an example, the range of an electric vehicle is given in ideal conditions (270 km for Nissan Leaf 40KWh, WLTP), but in reality this value can decrease down to around 140 km in winter. As a consequence, thermal models should be made and illustrate the possible performance of a battery in real conditions regarding not only the ambient temperature but also the internal thermal conditions due to the consumption of current.

Another motivation for the development of a thermal model is related to security and the admissible operation conditions. The temperature of the battery elements can be measured directly and monitored, but the cell internal temperature must be identified on the basis of a thermal model.

Typical curves for a LFP battery and their parameter dependency on temperature and are represented in Figures 4.19 and 4.20. Additionally, the influence of the charging/discharging current is indicated.

4.3.4 Thermal Modeling

The thermal modeling of a battery can be done with the help of a thermal equivalent circuit. In such a circuit, a thermal resistor R_{thCA} (case to ambient) will be used for steady state conditions. The battery temperature or the temperature of the cell case or heat sink depends on the ambient temperature and on the heat flux from the battery to the ambient air. The global heat flux, including heat conduction, heat convection and heat radiation, is represented in the thermal equivalent circuit and corresponds to the transferred thermal power. This power produces in steady state conditions a battery temperature according to the relation

$$\theta_b = \theta_A + R_{thCA} \cdot P_{thb} \tag{4.33}$$

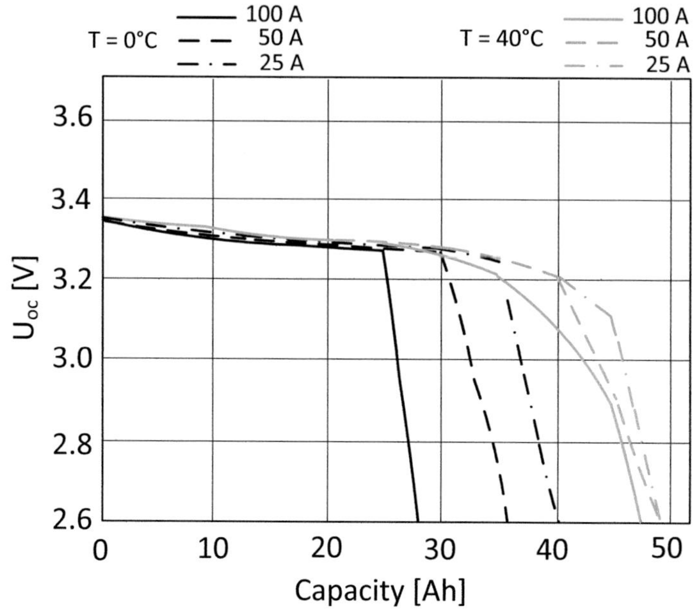

FIGURE 4.19 Battery open circuit voltage and capacity (extracted charges). (Adapted from [14].)

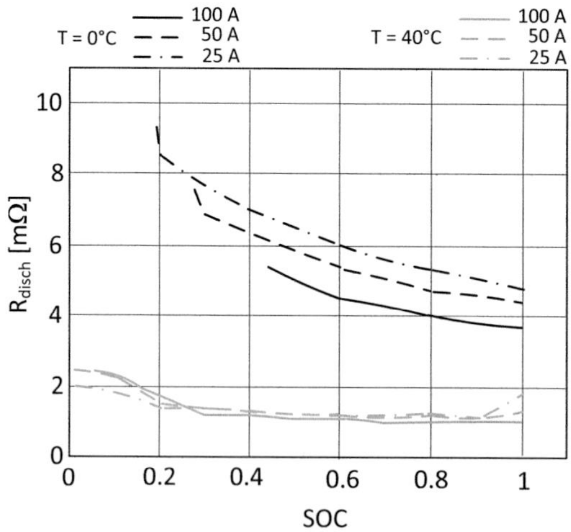

FIGURE 4.20 Battery discharge resistor as a function of the SoC and temperature [14].

The thermal equivalent circuit is represented in Figure 4.21. In this circuit, the input variable corresponds to the dissipated power inside of the battery P_{thb}. Within the thermal equivalent circuit, a power is represented through a current, and a temperature is represented as a voltage. In addition to the thermal resistance R_{thCA}, the thermal capacity of the battery is modeled through the thermal capacitance C_{thb} [14].

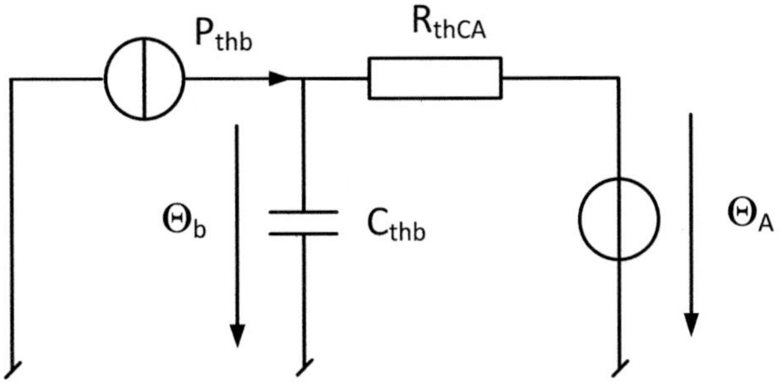

FIGURE 4.21 Thermal equivalent circuit of a battery.

The thermal capacitance of a battery C_{thb} describes the evolution of the temperature of the battery itself depending on the accumulated heat and transferred heat to the ambient:

$$C_{thb}\frac{d\theta_b}{dt} = P_{thb} - \frac{\theta_b - \theta_A}{R_{thCA}} \tag{4.34}$$

C_{thb} is proportional to the mass of the battery and depends on its specific heat capacity c_{v}

$$C_{thb} = m \cdot c_v \tag{4.35}$$

The solution of the differential equation (4.34) is of the following form

$$\theta_b(t) = A + B \cdot \mathrm{e}^{-(t/\tau)} \tag{4.36}$$

$$\text{where } \tau = R_{thCA} \cdot C_{thb} \tag{4.37}$$

A and B are constants

At $t = 0$, the battery temperature is identical to the ambient temperature θ_{A}

$$\theta_b(0) = \theta_A = A + B \tag{4.38}$$

$$\text{For } t \to \infty \; \theta_b(\infty) = A = P_{thb} \cdot R_{th} + \theta_A \tag{4.39}$$

Finally, the expression for the battery temperature becomes

$$\theta_b(t) = \theta_b(\infty) - [\theta_b(\infty) - \theta_b(0)] \cdot e^{-t/\tau} \tag{4.40}$$

Independently of the initial value of the battery temperature $\theta_b(0)$, the final value $\theta_b(\infty)$ only depends on the dissipated power P_{thb}, the thermal resistance R_{th} and the ambient temperature θ_A.

$$\theta_b(\infty) = P_{thb} \cdot R_{th} + \theta_A \tag{4.41}$$

4.3.4.1 Weighted Average Value of *Cv*

For the evaluation of a weighted average value of the C_v of a battery, an example of calculation is given in [14]. This calculation gives an equivalent specific heat capacity for one cell c_{ve}, using Equation (4.42)

$$c_{ve} = \sum x_i \cdot c_{vi} \tag{4.42}$$

where

x_i is the mass fraction of the cell material components
C_{vi} is their corresponding specific heat capacities.

The mass fractions of the materials and the corresponding c_{vi}s are given in Table 4.3. The composition of such a cell is given in Figure 4.22 [14].

For a battery with n elements in a global framework where all elements are cooled by a common convection air flow, a dedicated model can be built, such as the example of Figure 4.23. In this model, each cell is modeled with its own power dissipation

TABLE 4.3
Mass Fraction and Specific Heat Capacity of a LFP Cell

Composition of the LFP Cell	Mass Fraction x_i [%]	C_{vi} [J/kgK]
Aluminum	0.18	900
Copper	0.11	385
Separator (PP, PE, PVDF, PET)	0.07	1551
LFP ($LiFePO_4$)	0.24	841
Graphite	0.19	700
Electrolyte	0.2	1375
Stainless steel	0.01	477
Entire specific heat capacity	1	927

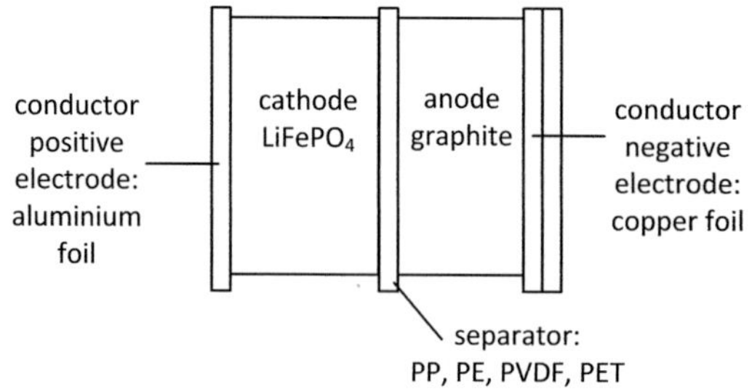

FIGURE 4.22 Composition of the LPF cell. (Adapted from Waheed, A. et al., Modelling and validation of a three-dimensional thermoelectric model of a 50 Ah lithium-iron-phosphate battery cell at three different ambient temperatures in the new European driving cycle (NEDC), in *Automotive Meets Electronics*, VDE Verlag GMBH, Offenbach/Dortmund, Germany, March 1–2, 2016.)

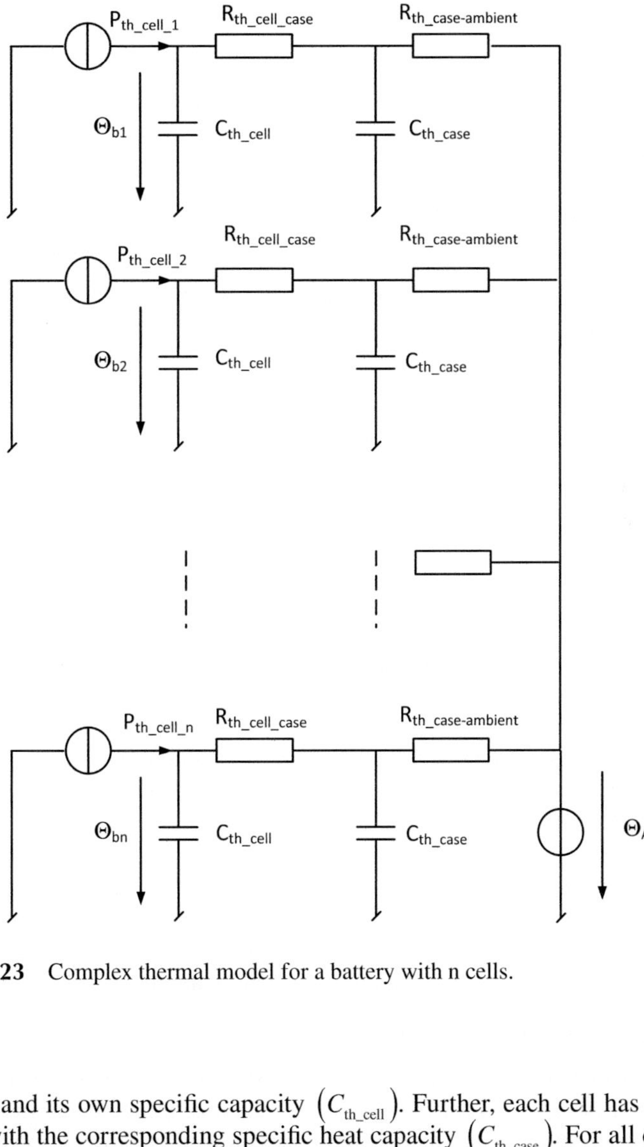

FIGURE 4.23 Complex thermal model for a battery with n cells.

$(P_{\text{th\_cell\_n}})$ and its own specific capacity $(C_{\text{th\_cell}})$. Further, each cell has a separated package with the corresponding specific heat capacity $(C_{\text{th\_case}})$. For all cells, a separate thermal resistance for the transfer from the cell to its case $(R_{\text{th\_cell\_case}})$ is represented, as well as one separate resistance for the transfer from the case to the ambient.

In many batteries and especially in large systems (typically in the automotive sector), the cooling system uses forced convection (cooling fans). The thermal resistance from a local heat sink to the external ambient temperature strongly depends on the convection velocity. Figure 4.24a illustrates the properties of a thermal resistance with forced air cooling [15].

Figure 4.24b indicates the behavior of air cooling with natural convection. In this representation, the thermal resistance is not constant but decreases with the transferred power. This comes from the fact that the air is heated up and produces a slow (vertical) movement along the cooling fins or the external surface.

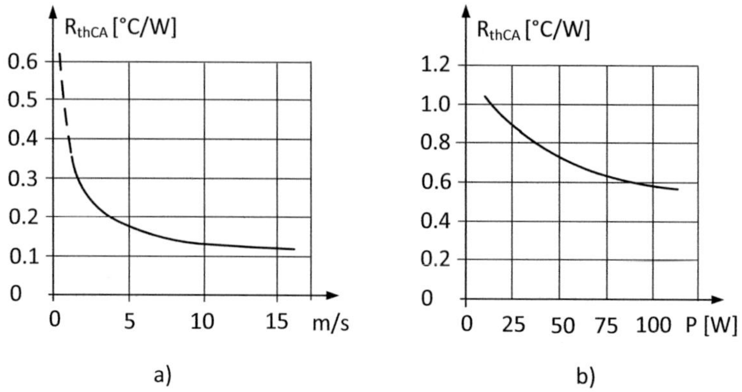

FIGURE 4.24 Dependency of the thermal resistance (a) on air velocity (forced convection), (b) on power (natural convection). (Adapted from Buhler, H., *Electronique de puissance, Traité d'électricité*, vol. 15, Presses Polytechniques Universitaires Romandes (PPUR), Lausanne, Switzerland.)

4.3.5 Multiphysics Modeling

Complex energy storage systems like the vanadium redox flow battery include several auxiliary devices like pumps for the circulation of the electrolytes or involve multiple subcomponents of different physical natures. Such systems should be investigated in detail, first in order to be operated in good conditions and to avoid critical situations where aging phenomena can occur, second to obtain operating conditions where acceptable energy efficiency can be reached.

Typically, in large battery systems with forced air cooling or liquid cooling, the energy consumed by the auxiliaries can strongly influence the global energy efficiency of the storage system, especially when the power level of the stored or recovered energy is low. Similar consequences will be discussed in this section for a VRF battery considering the consumption of its electrolyte pumps.

4.3.5.1 Global Multi-Physics Model

Figure 4.25 shows an overall multi-physics model of a vanadium redox flow battery where beside the VRB battery control system, the considered main subsystems are the electrochemical stack model, the reservoir and electrolyte model and the mechanical model where the needed mechanical power is evaluated [9, 10].

For this overall model, the input (command) variables are the reference current and the reference power.

The main output variables are the stack voltage U_{stack} and the VRB battery power.

As internal variables, the vanadium concentration is considered, together with the related state of charge SoC. As other internal variables, the stack current $\left(I_{stack}\right)$, the stack power $\left(P_{stack}\right)$ the electrolyte flowrate (Q) and the mechanical power $\left(P_{mech}\right)$ are considered.

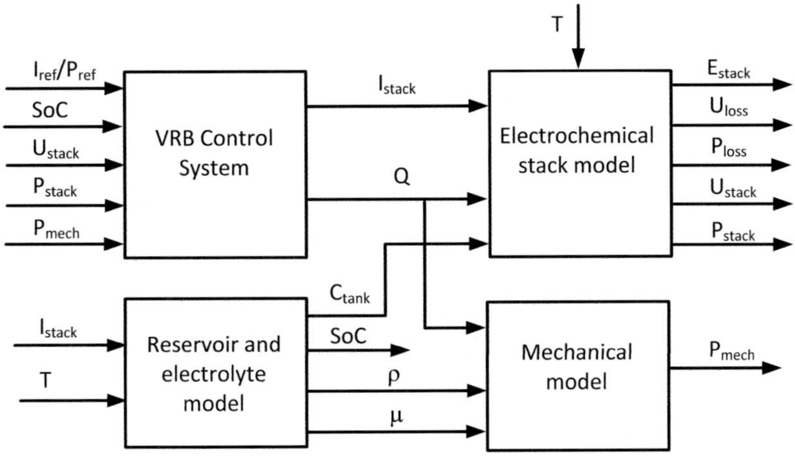

FIGURE 4.25 The overall multi-physics model of a VRB battery. (Adapted from Blanc, C., Modelling of a vanadium redox flow battery electricity storage system, Thesis EPFL No. 4277, EPFL, Lausanne, Switzerland, 2009, https://infoscience.epfl.ch/record/129758/files/EPFL_TH4277.pdf. Accessed on September 22, 2017.)

4.3.5.2 Electrochemical Stack Model

Inside the electrochemical stack model, a detailed model is made as a decomposition of the stack voltage into a sum of internal potentials and overpotentials (Figure 4.26). The stack voltage takes into account the number of cells that multiplies the sum of the different potentials (Equation 4.43)

$$U_{stack} = N_{cell}\left(E - \sum \eta_i\right) [\text{V}] \tag{4.43}$$

The stack model takes into account the operating variables, the electrical current, the vanadium concentrations, the proton concentration, the temperature and the electrolyte flowrate.

The internal voltage drops (called the overpotentials) are responsible for internal losses. They are:

- The activation potential η_{act} associated with the energy needed to initiate a charge transfer
- The concentration overpotential η_{conc} caused by concentration differences between the bulk solution and the electrode surface
- The ohmic losses η_{ohm} occur in the electrodes, between the bipolar plates and the collector plates
- The ionic losses η_{ion} occur in the electrolytes and the membranes

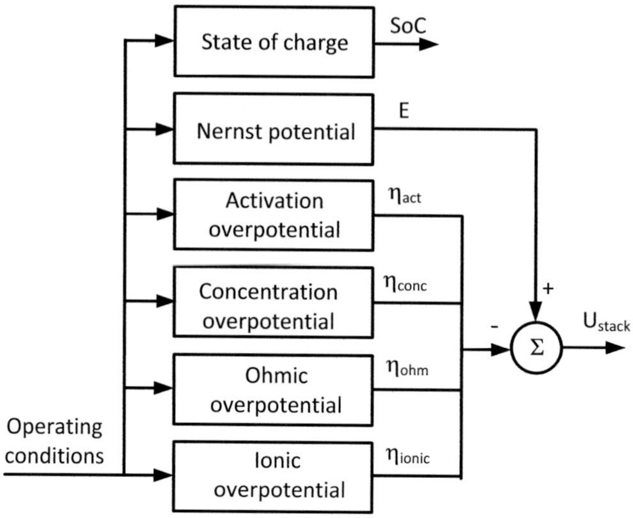

FIGURE 4.26 Structural diagram of the stack model. (Adapted from Blanc, C., Modelling of a vanadium redox flow battery electricity storage system, Thesis EPFL No. 4277, EPFL, Lausanne, Switzerland, 2009, https://infoscience.epfl.ch/record/129758/files/ EPFL_TH4277.pdf. Accessed on September 22, 2017.)

The sum of overpotentials is written as

$$U_{loss}(t) = \eta_{act}(t) + \eta_{conc}(t) + \eta_{ohm}(t) + \eta_{ion}(t) \ [\text{V}] \tag{4.44}$$

Figure 4.27 illustrates the variation of the stack voltage during a charge and discharge process for different stack currents.

In Figure 4.28, a completed and simplified stack model is represented. The input variables are the electrolyte flowrate Q and the stack current. The output variables are the stack voltage U_{stack} and the SoC.

The internal variables of this stack model are the vanadium and proton concentrations $\left(c_{\text{tank}}, c_{\text{cell}}, c_{\text{H+}}\right)$, the equilibrium voltage (E) and the internal losses (U_{loss}).

4.3.5.3 Example of a Small Stand-Alone System

The electrochemical model is able to predict the behavior of the stack, for example, in a small stand-alone system. Such a simple system includes a photovoltaic generator, the VRB battery and the load (Figure 4.29).

In Figures 4.30 and 4.31, a daily cycle of the stand-alone system is represented. In Figure 4.30, the power delivered by the PV system is represented (black curve), together with the power profile of the load (dotted line). The stack power (grey curve) is also represented. The curves are represented with a sampling time of 1 hour for simplification.

Figure 4.31 illustrates the variation of the stack voltage and the stack current during the same cycle.

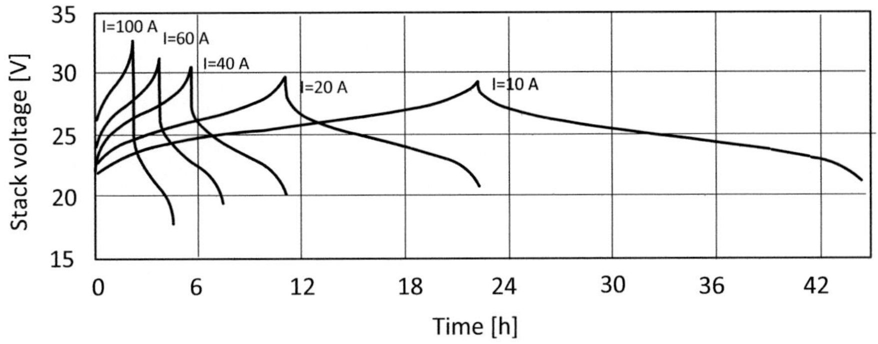

FIGURE 4.27 Stack voltage during charge and discharge. (Adapted from Blanc, C., Modelling of a vanadium redox flow battery electricity storage system, Thesis EPFL No. 4277, EPFL, Lausanne, Switzerland, 2009, https://infoscience.epfl.ch/record/129758/files/EPFL_TH4277.pdf. Accessed on September 22, 2017.)

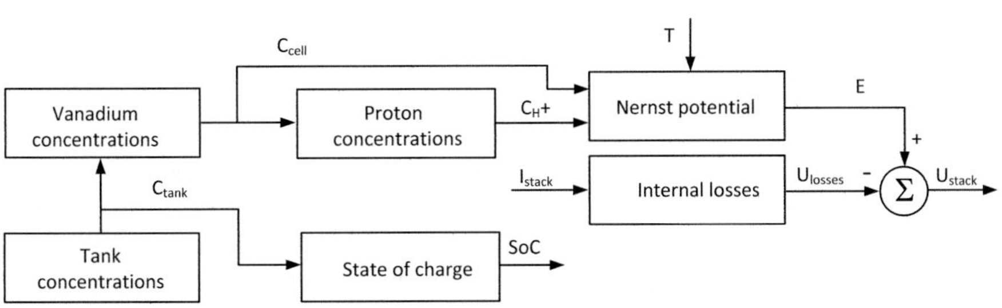

FIGURE 4.28 The completed and simplified stack model. (Adapted from Blanc, C., Modelling of a vanadium redox flow battery electricity storage system, Thesis EPFL No. 4277, EPFL, Lausanne, Switzerland, 2009, https://infoscience.epfl.ch/record/129758/files/EPFL_TH4277.pdf. Accessed on September 22, 2017.)

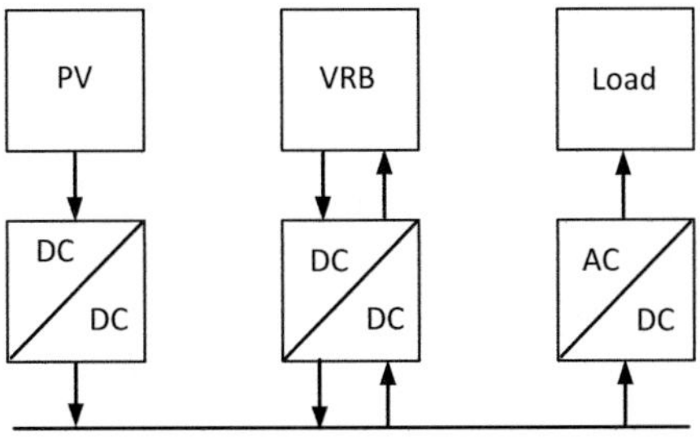

FIGURE 4.29 Small stand-alone system. (Adapted from Blanc, C., Modelling of a vanadium redox flow battery electricity storage system, Thesis EPFL No. 4277, EPFL, Lausanne, Switzerland, 2009, https://infoscience.epfl.ch/record/129758/files/EPFL_TH4277.pdf. Accessed on September 22, 2017.)

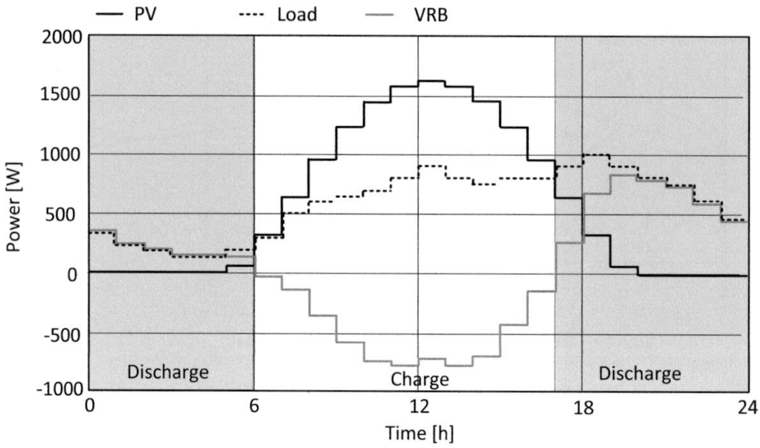

FIGURE 4.30 Daily cycle of the stand-alone system (power). (Adapted from Blanc, C., Modelling of a vanadium redox flow battery electricity storage system, Thesis EPFL No. 4277, EPFL, Lausanne, Switzerland, 2009, https://infoscience.epfl.ch/record/129758/files/EPFL_TH4277.pdf. Accessed on September 22, 2017.)

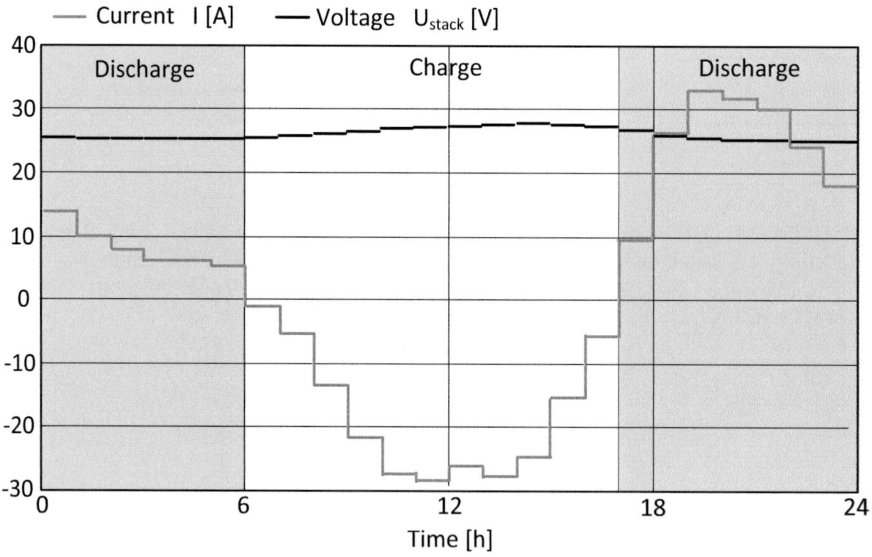

FIGURE 4.31 Voltage and current in the stack.

Finally, Figure 4.32 represents the variations of the vanadium concentrations in the electrolyte tanks.

4.3.5.4 Mechanical Model

The mechanical model as a part of the overall model serves to determine the power required for the electrolytes to flow. Indeed, the mechanical losses constitute the

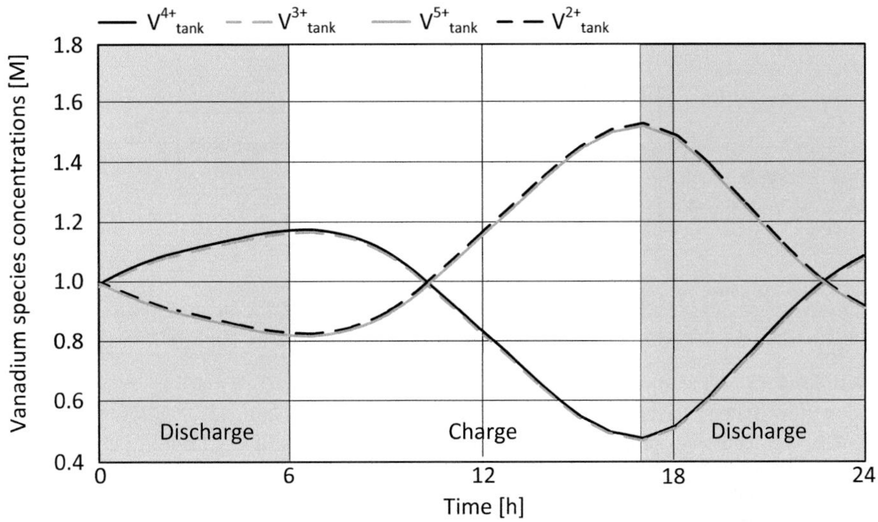

FIGURE 4.32 Variations of the vanadium concentrations in the tanks. (Adapted from [10]).

main part of the auxiliary losses and are therefore necessary to define the battery energetic performance.

The mechanical losses P_{mech} depend on

- The electrolyte flowrate Q
- The electrolyte properties such as the density ρ and the viscosity μ
- The dimensions and the characteristics of the hydraulic circuit

The model of the mechanical part is represented in Figure 4.33.

The input variables of the mechanical model are the electrolyte flowrate (Q), the density of the electrolytes (ρ) and their viscosity (μ).

The output variable is the mechanical power P_{mech}.

The internal variables and parameters of the mechanical model are the Reynolds number (R_e), the head losses (h_f) and (h_m), the flow resistance $(\tilde{R})$ and the pressure drops (Δp_{pipes} and Δp_{stack}). The mechanical power is the sum of the mechanical losses in the pipes (P_{mech_pipes}) and in the stack (P_{mech_stack}). For the calculation of the consumed electric power for the auxiliaries, the efficiencies of the pumps and driving machines must be considered.

4.3.5.5 Constant and Variable Flowrate of the Electrolytes

The simplest operation of a VRB battery can be defined with constant flowrate of the electrolytes. For such a strategy, the flowrate must provide enough electroactive species to the stack under the worst operating conditions: low SoC during the discharge and high SoC during the charge. The consequence is that when the electrolytes have a high concentration, or if the stack current is low, unnecessary mechanical power is spent for the pumps.

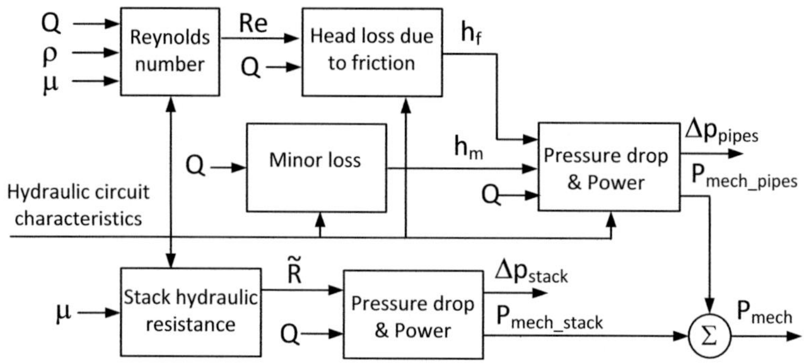

FIGURE 4.33 The mechanical model. (Adapted from Blanc, C., Modelling of a vanadium redox flow battery electricity storage system, Thesis EPFL No. 4277, EPFL, Lausanne, Switzerland, 2009, https://infoscience.epfl.ch/record/129758/files/EPFL_TH4277.pdf. Accessed on September 22, 2017.)

The constant and large mechanical losses lead to poor global efficiency of the VRB battery.

A second strategy for the operation of the battery is to constantly adapt the flowrate to the actual operating conditions to minimize the mechanical losses. The adaptation of the electrolyte flowrate is realized depending on

- the stack current
- the concentration of vanadium (i.e. SoC)

More sophisticated control strategies for the variation of the electrolyte flowrate are given in References [9, 10].

4.3.5.6 From a Positive to a Negative Energy Efficiency

In a more general consideration, the round-trip efficiency of a storage system can be defined as the ratio of the recovered energy during discharge divided by the amount of energy spent during charge.

$$\eta_{roundtrip} = \frac{P_{VRBdisch} \cdot t_{disch}}{P_{VRBch} \cdot t_{ch}} \tag{4.45}$$

When the discharge time is identical to the charging time $\left(t_{disch} = t_{ch}\right)$, Equation 4.45 can be simplified (normal round-trip efficiency; see Section 1.3.4).

$$\eta_{roundtrip} = \frac{P_{VRBdisch}}{P_{VRBch}} \tag{4.46}$$

The power of the auxiliaries is then considered. For the VRB battery, the mechanical power for the electrolyte pumps has to be subtracted from the stack power (the

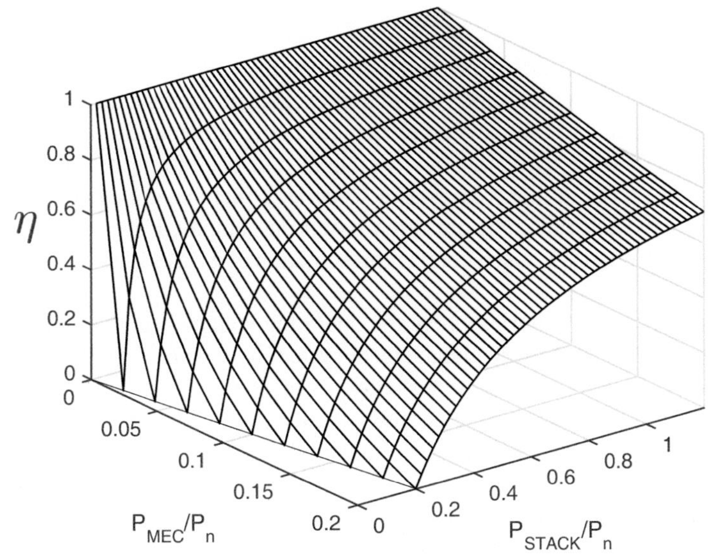

FIGURE 4.34 Efficiency of a VRB battery accounting the auxiliary power (pumps).

battery itself is powering its auxiliaries), and it must be added to the stack power during charge (the external source is powering the auxiliaries). The round-trip efficiency becomes

$$\eta_{roundtrip} = \frac{\left(P_{stackdisch} - P_{mech}\right)}{\left(P_{stackch} + P_{mech}\right)} \tag{4.47}$$

From Equation 4.47, it becomes evident that the round-trip energy efficiency can become negative (for example, $P_{\text{mech}} > P_{\text{stackdis}}$).

The variation of the round-trip efficiency depending on the charging/discharging power related to the nominal power of the battery P_{n} and the related mechanical power $P_{\text{mech}}/P_{\text{n}}$ is represented in Figure 4.34. The operating range of the battery power is between zero and 1.2 [p.u.], and the range of the needed related auxiliary power is between zero and 0.2 [p.u.]. In Figure 4.34, the negative values of the efficiency are not represented, because they would correspond to a nonsense system.

4.4 BATTERY MANAGEMENT SYSTEMS

4.4.1 Motivation

Modern batteries are optimized for their high energy density and power density. Often the technology is sensitive to high solicitation and there is a need for an intelligent battery management system (BMS). Its tasks are mainly to increase the battery life time or to reduce the battery aging effects due, for example, to operating of the cells outside of their recommended voltage range.

4.4.2 Main Functions of a Battery Management System

The main functions of a BMS are cell voltage monitoring, cell voltage balance, temperature monitoring and monitoring of the exchanged current. In some cases, the battery BMS includes power switching devices in the interconnection to the load and to the battery charger. Switching devices can also be used for protection purposes.

The BMS serves in many cases as a so-called "fuel gauge", evaluating the SoC of the battery. Such functions are often based on complex battery models, as was described in Section 4.3. A schematic representation of a BMS is given in Figure 4.35 [16, 17].

For large capacity batteries where the operating voltage is in the range of several hundreds of volts, the battery management system is organized according to a modular concept. Smaller units as represented in Figure 4.33 are connected in series. For this purpose, the BMS circuits are generally equipped with serial communication interfaces. A simplified example of a complete BMS structure is represented in Figure 4.36.

Many industrial components for BMS systems exist or are realized as private systems for protecting intellectual property or sensitive knowledge [16, 17].

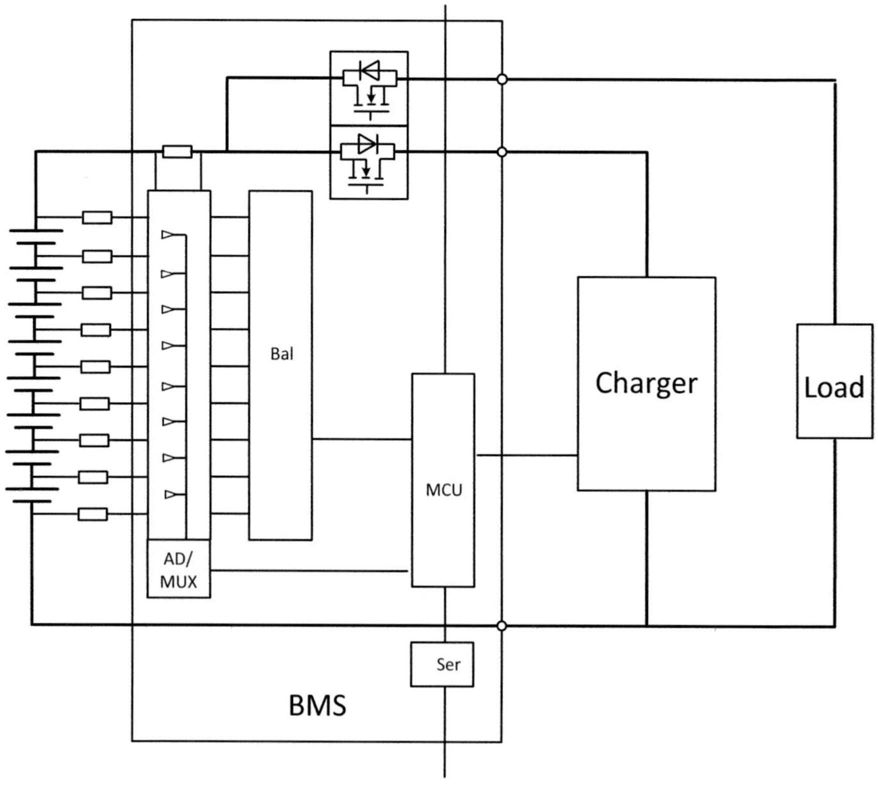

FIGURE 4.35 Block diagram of a BMS.

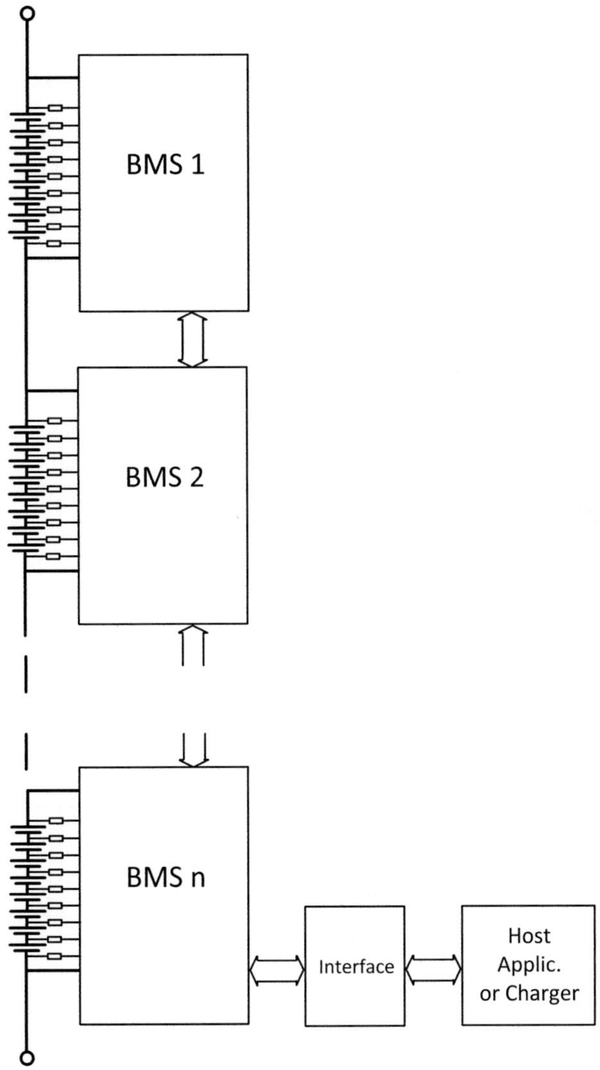

FIGURE 4.36 Simplified diagram of a larger BMS system.

4.4.3 Cell Balancing Circuits

Cells can be connected in series or parallel to form a battery pack. A parallel connection increases the battery pack's current capability, while the series connection increases the overall voltage. Due to manufacturing differences, non-identical ageing phenomena or other differences in the rates of cells charge or discharge, battery packs can show a spread distribution of the individual cell voltages across the pack.

A simplistic means of determining the SoC of a battery pack is to monitor each cell's voltage to a reference voltage level. The first cell voltage reaching the reference level would determine the charging limit. If the battery pack has a weaker-than-average cell, in which voltage reaches the reference limit first, the rest of the cells would stay in a not fully charged state. In the discharge process, the weaker cell discharges faster, reaching the lower discharge limit first.

The consequence in such a case of monitoring is a loss or inefficient use of the real battery stack capacity. To get around such disadvantages, cell balancing circuits have to be added to the BMS.

4.4.3.1 Active Dissipative Cell Balancing Circuits

One way to reach equalized cell voltage levels for the full SoC of the battery stack is to slow the charge of the weakest cell during the charge cycle. This can be achieved by activating a bypass circuit composed of a field effect transistor (FET) and current limiting resistors placed across all cells of the stack.

By locally limiting the current that effectively charges the cells, the maximum of the battery charge capacity can be achieved while all cells reach the fully charged limit simultaneously. The scheme of a battery stack with a BMS completed by an active dissipative balancing circuit is given in Figure 4.37.

The dissipative balancing circuits cause additional losses while charging. But the dispersion of the charge and discharge rate of the cells is normally maintained in a

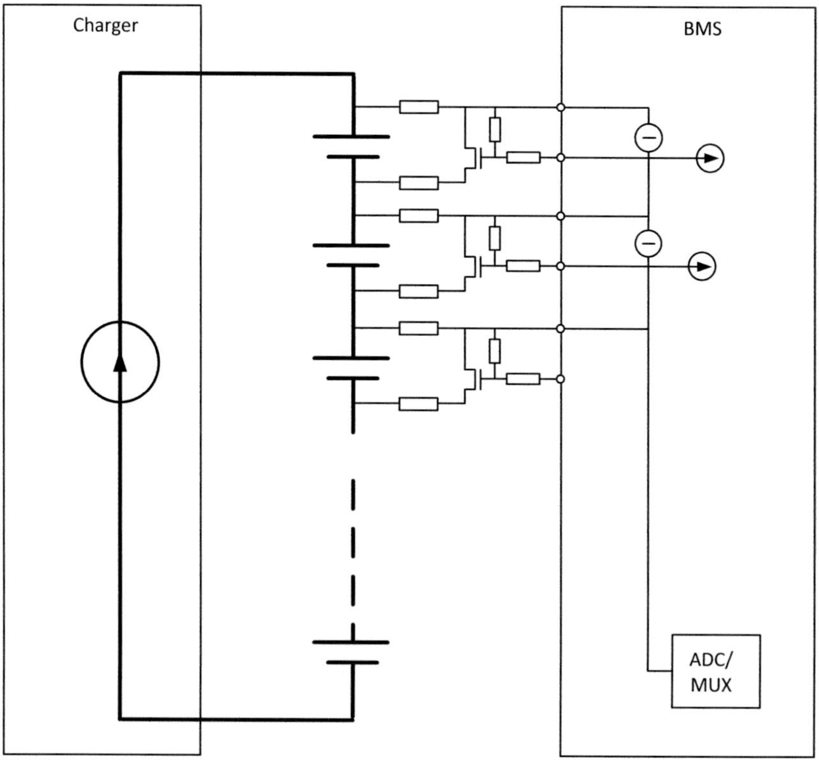

FIGURE 4.37 Dissipative cell balancing circuit.

limited range, and the balancing current of a weak cell is therefore only a fraction of the normal charging current. Additionally the balancing dissipation will concern only one or two cells during the charging process.

4.4.3.2 Active Non-Dissipative Cell Balancing Circuits

Especially during the discharge process, the cells of a battery pack can be balanced by implementing a charge displacement scheme. A charge displacement scheme is achieved by taking charge via inductive coupling or capacitive storage from the strong cells and injecting the stored charge into the weak cell [16].

Figure 4.38 gives an overview of different non-dissipative balancing schemes based on capacitive storage. For this principle, the charge transfer from the battery

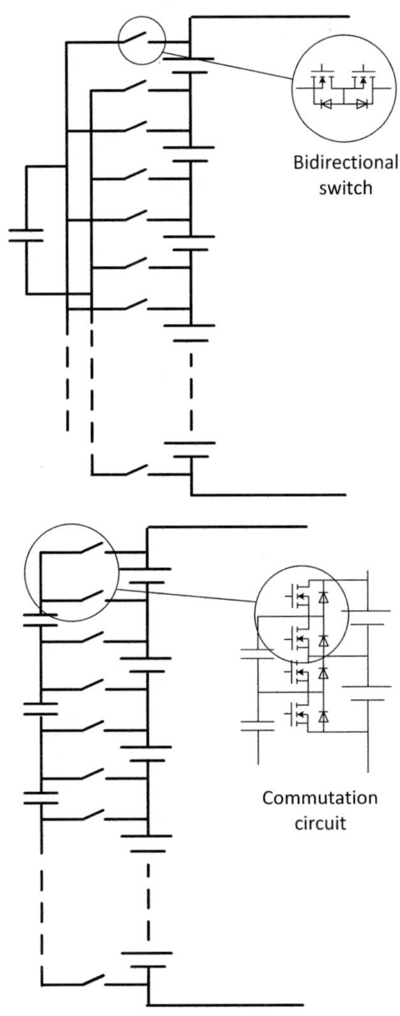

FIGURE 4.38 Capacitor based active balancing circuits. (a) Flying capacitor and (b) Switched capacitors.

to the capacitor or from the capacitor to the next battery cell is defined through Equation 4.48 [18]:

$$i_C(t) = \frac{\Delta U}{R_{tr}} \cdot e^{-t/\tau} \tag{4.48}$$

where R_{tr} is the equivalent resistance of the transfer circuit, mainly defined by the capacitor and battery internal resistance, added to the transistor on-state equivalent resistor. ΔU is the voltage difference of the cell and capacitor, C the capacitance and τ the time constant $\tau = R_{tr}C$.

In Figure 4.38a, the so-called "flying capacitor" balancing circuit is represented. The balancing charge is moved from one battery cell to another by connecting the flying capacitor to them with the help of bidirectional switches. In Figure 4.38b the topology of the so-called switched capacitors is represented. The figure also represents the details of the commutation circuit for a real implementation.

Figure 4.39 shows the principle of an active balancing circuit using inductive storage. The topology is known as "current diverter" or buck-boost converter. The operating principle is described in Section 5.6.5 [19].

4.5 AGING OF ELECTROCHEMICAL BATTERIES

4.5.1 Introduction

Electrochemical batteries used in many application domains as intermittent renewable energy sources or electrical propulsion systems for cars must have low maintenance requirements, great durability and high reliability. The question of environmental impact is also indirectly addressed through the durability aspects, together with the availability of raw material resources.

Due to the specific operating conditions, batteries must have adapted design and properties, but their long-term behavior is often not included in the development concepts.

As a consequence, the degradation mechanisms of batteries have been analyzed in different contexts of use and within dedicated research facilities [20]. Larger international research programs have established complete description and modeling methods, validated battery models, lifetime modeling and aging assessment of thermal parameters [21].

The next sections will describe the global performance parameters and aging phenomena of two kinds of batteries, namely lead-acid batteries and lithium-ion batteries. Lead-acid batteries are today the most-used storage elements in relation to RES like photovoltaics, and lithium-ion batteries are expected to become the standard technology in electric transportation, after their universally recognized superiority in portable applications due to their high energy density.

4.5.2 Global Performance Parameters

The global performance parameters of a battery can be defined through

- The battery capacity
- The battery power capability

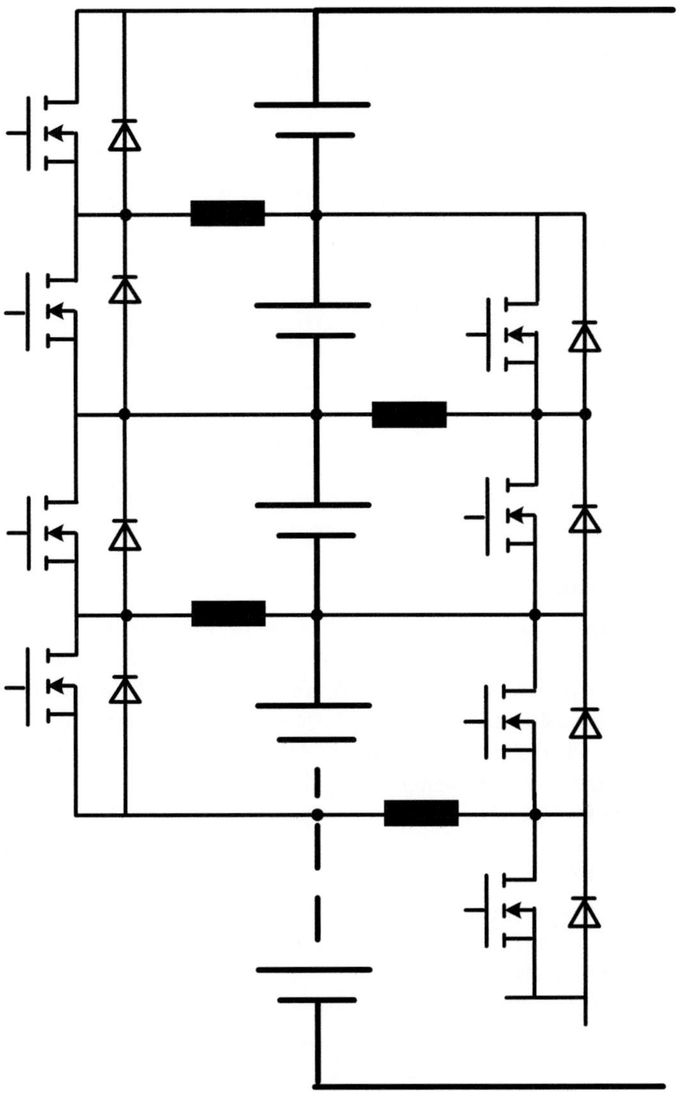

FIGURE 4.39 Active balancing with a current diverter (buck-boost converter).

The battery capacity evolution describes the total available charge or discharge capacity while the battery is subject to various operation conditions (stress conditions) and thus degradation.

The power capability evolution describes the ability of a cell to deliver power over the course of the lifetime of the cell. This parameter is related to the internal resistance R_{int} of the cell.

Cell degradation occurs both during operation (cycling ageing) and during stationary time periods (calendar ageing), when no current is flowing through the cell. These are two independent factors in the degradation of the cell and are modeled independently.

The stress conditions are the external parameters that influence cell performance degradation. They can be classified as follows:

- Cycling ageing stress factors
- Temperature
- Depth-of-discharge (DOD)
- Middle state-of-charge (mid-SoC)
- Charge current rate
- Discharge current
- Calendar ageing stress factors
- Storage temperature
- Storage state-of-charge

4.5.3 Degradation Mechanisms of Lead-Acid Batteries

The degradation of lead-acid batteries can be different depending on their application context, but several basic phenomena are recognized and can be listed as follows [20].

4.5.3.1 Stratification

Stratification is the formation of regions or layers of different density in the battery electrolyte due to non-uniform concentration of the sulfuric acid in the vertical direction of the battery plates. As a consequence, the current distribution over the plate becomes inhomogeneous, the capacity is reduced and the voltage and current characteristics change. The stratification of the electrolyte is not a degradation process itself, but it accelerates other ageing mechanisms such as sulfation and corrosion in the top and bottom of the plates, being most common in the positive plate.

4.5.3.2 Sulfation

During the discharging process, the active mass becomes lead sulphate. If the recharge does not occur in sequence, lead sulphate crystals form and accumulate at the plates. During the following recharge, small lead sulphate crystals dissolve, but large crystals are not totally removed and grow, reducing the active surface of the electrodes, and finally reduce the battery capacity over time since there is less active mass available for the charge and discharge reactions.

4.5.3.3 Corrosion

In normal battery operation, the lead grid of the positive plate corrodes, reducing the cross-section of the lead grids. The internal resistance is therefore increased. Then, a layer of oxide and lead sulphate is formed between the bars and the active material, which also increases the contact resistance. The corrosion rate depends on the density of the electrolyte, the float voltage, the temperature, the alloy used for the grid and the quality of the manufacturing process. Corrosion occurs more intensively when the element voltage is below 2.0 V or above 2.4 V.

4.5.3.4 Active Mass Degradation

During operation of a lead-acid battery, the electrodes suffer from strong mechanical stress. Throughout the discharging process, more than 50% of the active material is transformed into lead sulphate. Because lead sulphate has a volume per mol 1.89 times greater than lead dioxide, the large variations in volume cause the loss of active material because it becomes loose and the drag of the acid can separate it from the electrode. This effect is stressed by increase in DoD during the daily cycles of photovoltaic systems.

4.5.3.5 High Temperature

In general, any chemical or electrochemical process is accelerated by an increase of the temperature. The same happens with batteries, accelerating the degradation process, mainly corrosion, sulfation, excessive generation of gasses and self-discharge. It is estimated that for every increase of 10°C in temperature, the battery life is reduced by half, and it is suggested to keep the operating temperature between 10°C and 25°C.

4.5.4 Lithium-Ion Batteries' Aging Effects

The ageing effects of a lithium-ion battery represent a very complex system involving many components. Anode, cathode, separator and electrolyte interact while charged and discharged and during storage. In this complex process, it is not possible to separate different aging mechanisms, but it is possible to determine the most dominant ageing effects for the components of the cell [21].

4.5.4.1 Anode Side

The anode is the best-known component in lithium-ion batteries and consists in most cases of carbon materials like graphite or hard carbon. For these materials, the dominant aging effects are basically known and can be separated into three parts:

1. Surface film formation on the particles
2. Degradation of the active material
3. Lithium metal deposition on the anode surface

The materials used for the anode of a lithium-ion battery have a typical voltage level of 0.05–0.5 V against metallic lithium when charged and discharged. At this potential, the commonly used electrolytes (carbonates) are not stable and decompose on the anode surface within the first charge. The formed layer is called a solid electrolyte interface (SEI) and is necessary for the function of the battery. If this layer is formed well, it prevents further decomposition of the electrolyte on the anode surface. Within this process, mobile lithium between the electrodes is involved and also integrated into the SEI. Loss of mobile lithium leads to a capacity fade. Unfortunately the SEI doesn't stop after the first cycle but continues over the whole lifetime, which makes it the most dominant ageing effect in a lithium-ion battery.

Beside film formation, graphitic anode materials also undergo a volume change of 10% in the intercalation and deintercalation process of lithium. Volume change in active materials is equivalent to mechanical stress inside the particles and the SEI and results in cracks. The consequences are loss of active lithium by new SEI formation and loss of active material by contact loss. Additionally, a thicker surface layer on the anode also results in an impedance rise, which is equivalent to power fade of the battery.

The third main ageing effect at the anode side is a lithium metal deposition, which is also known as "plating". Within the process of lithium plating, the intercalation reaction at the anode is too slow and the potential too low so that the lithium ions tend to deposit on the particles in metallic form. The low intercalation rate and the low potential at the anode can be caused by low temperatures, high charging currents and inhomogeneous current and potential distribution. Deposited lithium reacts directly with the electrolyte and also forms a layer at the metallic surface. This consumes active lithium and electrolyte, which results in capacity and power loss.

4.5.4.2 Cathode Side

The aging effects at the cathode, especially new materials, are less known but can also be summarized within three groups of effects:

1. Ageing of the active material
2. Surface film formation
3. Interaction of aging products with the negative electrode

The active materials of the cathode of modern lithium-ion batteries are layered metal oxides, with a variation of the used metals. The changes of these materials and the influence on the capacity loss are very small compared to the capacity loss over the lifetime. Similar to the anode, mechanical stress can cause loss of active material, and change of the composition can change the lattice structure of the crystals, which both lead to a capacity loss and an increase of impedance. Similar to the anode, the stability of the electrolytes influences the surface film formation on the cathode side. However, this effect plays a minor role for the total aging of the cell.

For some materials like manganese spinel or manganese rich nickel-manganese-cobalt (NMC) layered oxides, the dissolution of the transition metal manganese plays a major role in the interaction of the anode. The manganese ions migrate to the anode and destroy the SEI layer, which leads to new SEI formation, resulting in accelerated capacity loss and impedance increase.

4.5.4.3 Identification of the Influencing Factors

All the described ageing effects occur at the same time but have different accelerators and rates. State of the art for the identification of the influencing factors is based on accelerated ageing tests. These tests are performed with variation of operating conditions like temperature, current rate, DoD and SoC. The acceleration in comparison to the application is realized by deleting pauses and rest times. For studying the aging mechanisms at storage conditions, the cells are stored under defined and controlled environmental conditions. Capacity fade and impedance increase are monitored.

4.5.4.4 Example of Calendric and Cycling Aging Behavior

The calendric ageing behavior of a first-generation graphite lithium-ion cell has been recorded for 161 batteries at 47 different operating conditions and is represented in Figure 4.40. Even if these results are related to one specific commercial cell, they are representative for many different batteries of the same technology [21].

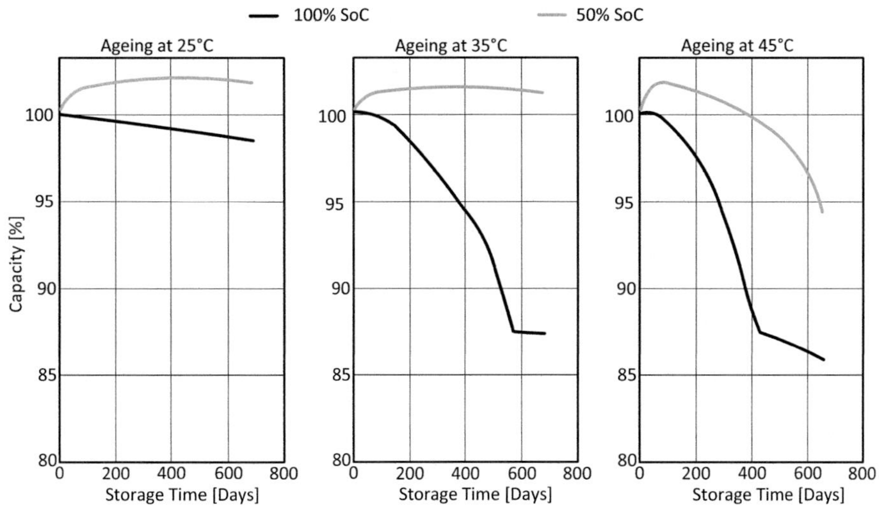

FIGURE 4.40 Calendric ageing behavior of a first generation lithium-ion cell. (Adapted from Timmermans, J.-M. et al., Batteries 2020-Lithium-ion battery first and second life ageing, validated battery models, lifetime modelling and ageing assessment of thermal parameters, *EPE16-ECCE Europe, European Conference on Power Electronics and Applications*, Karlsruhe, Germany, September 6–8, 2016.)

The capacity loss is faster for cells at higher temperatures. This indicates that cells should not be stored at high temperatures when not used.

In Figure 4.40, one can see that the lower values for the SoC (higher DoD) lead to much faster capacity degradation. A result from these ageing curves is that, for example, an electric vehicle should not be parked for a long time at a high SoC and only be charged just before the use to achieve maximum battery life.

Besides the pure storage of a cell, the cycling of the battery adds additional capacity loss. Figure 4.41 shows the remaining capacity curve of cycled cells where the cell temperature is kept at 25°C during cycling.

The DoD has a big influence on the lifetime. Cells that are charged and discharged completely in every cycle are ageing much faster than cells that are only discharged partially. As a conclusion, one can say that discharging the cell more deeply before recharging is not beneficial for the lifetime.

4.5.5 Cycling Ageing Model

The description of the cycling ageing of a lithium ion cell is established on the basis of experimental tests and recordings. The following cycling stress factors are considered in this model:

1. Temperature (°C)
2. DOD (%)
3. Mid-SOC (%)

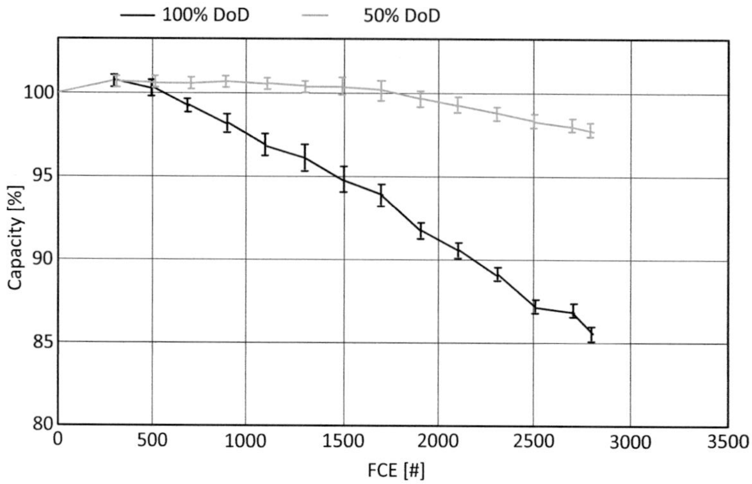

FIGURE 4.41 Remaining capacity curves of cyclic aged cells at 25°C for different values of the DOD. FEC denotes full equivalent cycles and provides a same Ah-throughput value for different depth-of-discharge conditions. (Adapted from Timmermans, J.-M. et al., Batteries 2020-Lithium-ion battery first and second life ageing, validated battery models, lifetime modelling and ageing assessment of thermal parameters, *EPE16-ECCE Europe, European Conference on Power Electronics and Applications*, Karlsruhe, Germany, September 6–8, 2016.)

The general layout of the cycling ageing model can be found in Figure 4.42. As can be seen, the model consists of two stages, which are used in a sequential manner.

The first stage calculates the influence of the temperature and DoD with the assumed conditions:

1. Mid-SoC: 50%
2. Charge rate: C/3
3. Discharge rate: 1C

The second stage calculates the deviation in capacity loss due to a mid-SOC other than 50%.

The final result combines (adds) the results from first and second stages, resulting in the capacity loss due to DOD, mid-SOC, temperature and cycle number.

The model itself is constructed in two successive actions:

1. Extend the measured input data using a 2D polynomial.
2. Use the extended input vectors as input for a 3D polynomial surface fit.

The 2D polynomial extrapolation of measured data is visualized in Figure 4.43.

The final 3D surface allows interpolation between measured DoD levels, as well as extrapolation towards the end of life (EoL) of the cell. As these surfaces are

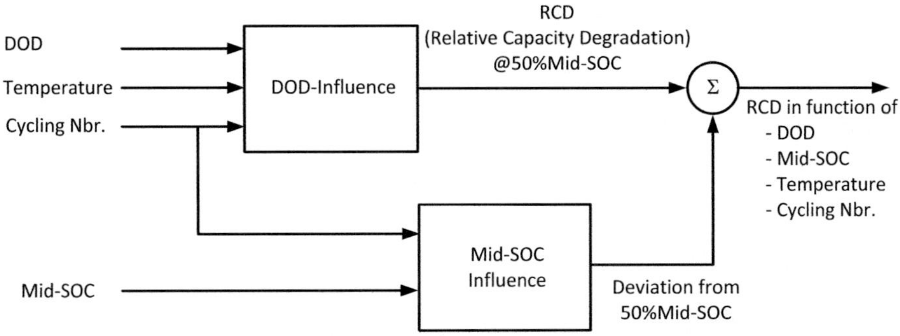

FIGURE 4.42 General layout of the Batteries2020 Cycle Ageing model. (Adapted from Timmermans, J.-M. et al., Batteries 2020-Lithium-ion battery first and second life ageing, validated battery models, lifetime modelling and ageing assessment of thermal parameters, *EPE16-ECCE Europe, European Conference on Power Electronics and Applications*, Karlsruhe, Germany, September 6–8, 2016.)

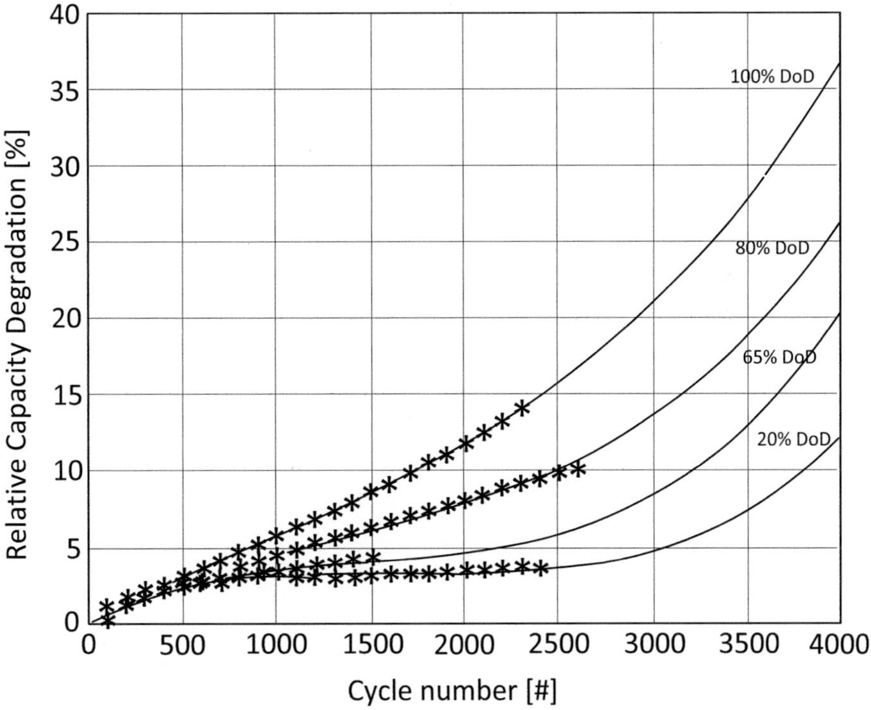

FIGURE 4.43 2D extrapolation of measured data. (Adapted from Timmermans, J.-M. et al., Batteries 2020-Lithium-ion battery first and second life ageing, validated battery models, lifetime modelling and ageing assessment of thermal parameters, *EPE16-ECCE Europe, European Conference on Power Electronics and Applications*, Karlsruhe, Germany, September 6–8, 2016.)

constructed for three input temperatures, (25, 35 and 45°C), these are also used for interpolation between temperatures [21].

4.5.6 Calendar Aging Model

For establishing a calendar aging model, the following stress factors are considered:

- Storage temperature (°C)
- Storage SoC (%)

As the influence of these parameters is often quite hard to predict due to the fact that the time needed to perform tests that show sufficient calendar ageing is in most cases prohibitively long, the predictions are supported by two electrochemical equations:

1. The Arrhenius equation, describing the evolution of the reaction rate depending on the temperature

$$A(T) = a \cdot e^{(b/T)} \tag{4.48}$$

2. The Tafel equation, which relates the rate of an electrochemical reaction to the over-potential

$$\Delta V = A \cdot \ln(i/i_0) \tag{4.49}$$

The following empirical equation was found to fit the input measurement points:

$$f(t, SoC, T) = a \cdot tb \cdot SoCc + d \cdot t \cdot SoC \tag{4.50}$$

with

$$a(T) = a_1 \cdot T^{a_2} + a_3 \tag{4.51}$$

$$b(T) = b_1 \cdot T^{b_2} + b_3 \tag{4.52}$$

$$c(T) = c_1 \cdot T^{c_2} \tag{4.53}$$

$$d(T) = d_1 \cdot T^2 + d_2 \cdot T + d_3 \tag{4.54}$$

The parameters a_i to d_j are given in Table 5.4.1

TABLE 4.4.1
Numerical Values for Equations 4.51 to 4.54

$a_1 = -7.007$	$b_1 = 3.459e{-}14$	$C_1 = 0.003542$	$d_1 = -2.457e{-}06$
$a_2 = -0.6682$	$b_2 = 7.815$	$C_2 = 1.395$	$d_2 = -0.0001313$
$a_3 = 0.5259$	$b_3 = 0.3124$		
$d_3 = 0.001928$			

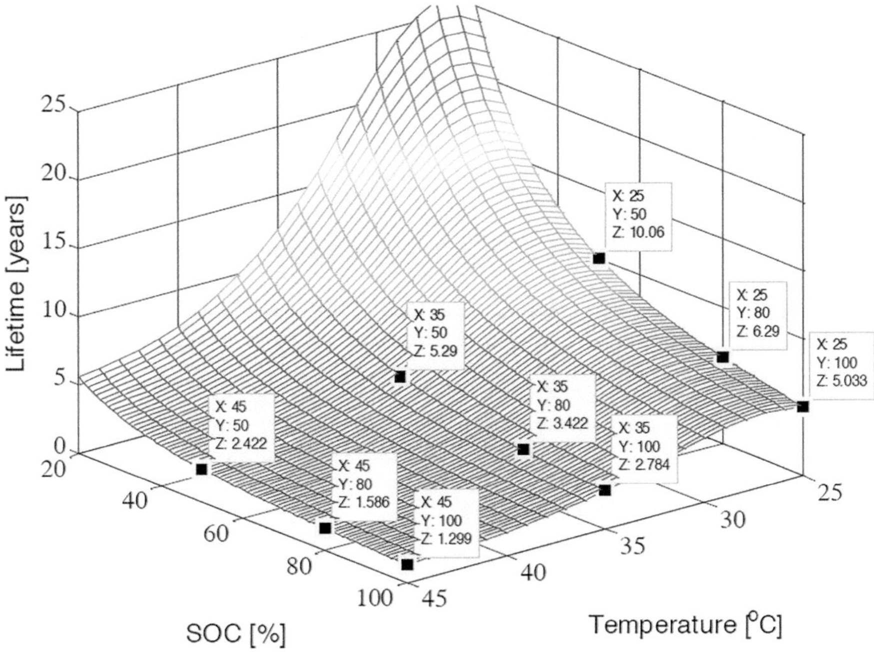

FIGURE 4.44 Calendar lifetime at different conditions (SOC and temperature) until 20% capacity degradation. (Courtesy [21]/EPE.)

The function is represented in Figure 4.44 where the lifetime is expressed in years and is estimated for an EoL criterion of 20% capacity loss, in function of storage SoC from 20% to 100% and temperature from 25°C to 45°C.

4.5.7 Complete Aging Model

The schematic overview of the complete ageing model can be seen in Figure 4.45.

Exercise 3: Battery for an Electric Car: Design and Thermal Behavior

An electric vehicle travels in south Italy (ambient temperature 40°C). Its electrochemical battery has a capacity of 20 kWh and is discharged during a 1-hour use from 100% SoC down to 20% SoC. After this travel, the car is recharged with a fast charging station at a current corresponding to 6C.

The objective of the exercise is to illustrate the design and thermal behavior of the car battery under the following conditions

- During the 1-hour run, the battery is cooled by natural convection through the walls of the battery housing.
- During the fast recharge, the battery is cooled with forced convection through cooling channels, as represented in Figure 4.46.

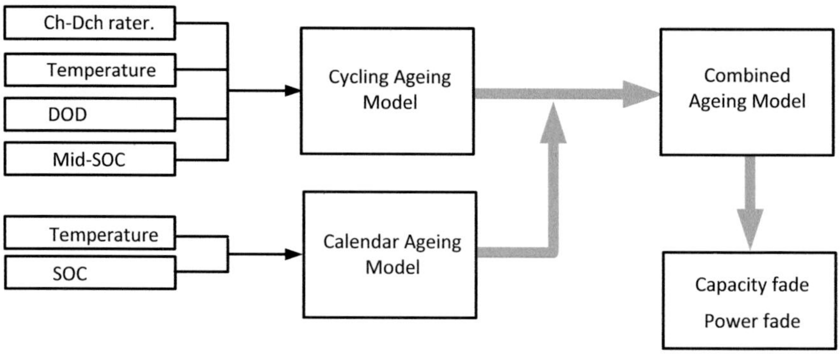

FIGURE 4.45 Schematic overview of the complete ageing model. (Adapted from Timmermans, J.-M. et al., Batteries 2020-Lithium-ion battery first and second life ageing, validated battery models, lifetime modelling and ageing assessment of thermal parameters, *EPE16-ECCE Europe, European Conference on Power Electronics and Applications*, Karlsruhe, Germany, September 6–8, 2016.)

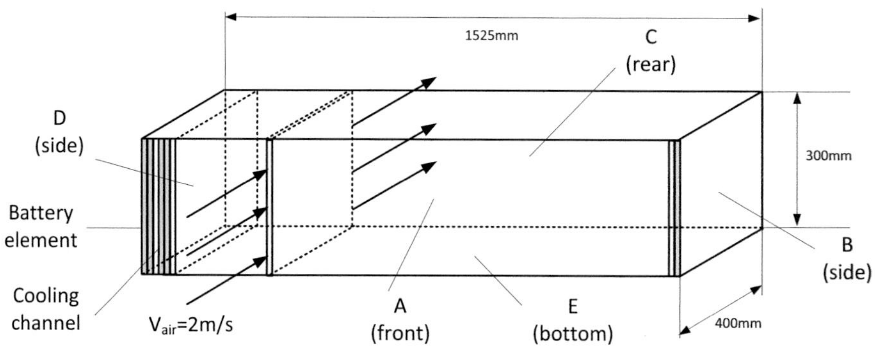

FIGURE 4.46 Battery pack.

TECHNICAL DATA

Battery: The battery is realized on the basis of 109 lithium polymer elements (see Exercise 1):

$U_b = 3.7$ V
Capacity: 50 Ah
$R_b = 0.7$ mOhm
Dimensions (mm) $W = 268, L = 265, T = 9.3$
Weight of the element: 1.260 kg

TASKS:

1. Calculate the average discharge current of the battery during the 1-hour run.
2. Calculate the battery internal losses during this run $\left(P_{loss1h}\right)$.

3. Calculate the thermal resistance of the battery housing (natural convection through the five walls A, B, C, D, E).

   For this calculation, a heat exchange coefficient *h* for the natural convection is given:

$$h = 4.5\ \mathrm{W}/\left(\mathrm{m}^2 * \mathrm{K}\right)$$

4. Calculate the steady-state end value of the temperature elevation $\Delta\theta$ due to a continuous dissipation at P_{loss1h}.
5. Calculate the thermal capacity of the whole battery on the basis of the mass of one element and its "entire specific heat capacity" (Table 4.3) of 927 J/kgK.
6. Determine the thermal time-constant of the battery thermal model.
7. Determine the value of the elevation of the battery temperature after the 1-hour run.
8. Represent the evolution of the battery temperature during the 1-hour run, together with the theoretical end-value and the thermal time constant.
9. For the fast charge of the battery, determine the internal power losses at 6C.
10. Determine the thermal resistance for the forced convection of one element (single-side forced convection, $V_{air} = 2\mathrm{m/s}$). For the forced convection at 2 m/s, a heat exchange coefficient h is given, $h_{fc} = 12\ \mathrm{W/m^2{*}K}$).
11. Determine the end-value of the elevation of the temperature of the battery if the fast charge dissipation would be maintained continuously.
12. With the thermal capacity of one element and the thermal resistance calculated under 9, calculate the thermal time constant of the thermal model for the forced convection
13. Calculate the value of the battery temperature reached after fast charge when the full capacity is reached (SoC 100%).
14. Represent the evolution of the battery temperature during the fast charge, together with the theoretical end-value and the thermal time constant. Take into account the initial temperature reached during the 1-hour run.

SOLUTION TO EXERCISE 3

1. Discharge during the 1-hour run

   The total battery energy capacity is $E_{sto} = 20kWh$
   Its discharge from 100% to 20% means

$$\Delta E_{sto} = 0.8 \cdot 20kWh = 16kWh \tag{4.60}$$

   This discharge within 1 hour means

$$P_{disch} = \frac{16kWh}{1h} = 16kW \tag{4.61}$$

   With a battery voltage of 109*3.7V = 403V

The battery discharge current is

$$I_{disch} = \frac{16kW}{403V} = 39A \quad (4.62)$$

2. The battery losses

The battery losses are calculated through

$$P_{loss\_1h} = N_{cell} \cdot R_{cell} \cdot I_{disch}^2 = 109 \cdot 0.7 \cdot 10^{-3}\Omega \cdot (39A)^2 = 116W \quad (4.63)$$

3. Thermal resistance

The thermal resistance of the complete battery pack is calculated with its five sides (the upper side of the case is the interface to the car and has a marginal contribution to the cooling).

Length of the case: 1525 mm

Width 400 mm

Height 300 mm

Surfaces

Front and rear walls	2 * 1.525 m * 0.3 m =	0.915 m^2
Side walls	2 * 0.4 m * 0.3 m =	0.24 m^2
Bottom	1.525 m * 0.4 m =	0.61 m^2
Total surface		1.765 m^2

The heat exchange coefficient is given as h = 4.5 W/m^2K, and the resultant heat conductivity becomes

$$H = S_{exch} \cdot h = 1.765m^2 \cdot 4.5W/m^2 \cdot K = 7.94W/K \quad (4.64)$$

The thermal resistance R_{th}

$$R_{th} = 1/H = 1/(7.94W/K) = 0.125K/W \quad (4.65)$$

4. Elevation of the temperature (steady-state)

The elevation of the temperature (steady-state, final value) is determined by the dissipated power and the thermal resistance

$$P_{loss\_1h} = 116W = H(\theta_b - \theta_{amb}) = \frac{\theta_b - \theta_{amb}}{R_{th}} = \frac{\Delta\vartheta}{R_{th}} \quad (4.66)$$

finally

$$\Delta\vartheta = 116W \cdot 0.125K/W = 14.6K \quad (4.67)$$

5. Thermal capacitance

The thermal capacitance is calculated first for one cell

$$C_{th} = m_{cell} \cdot c_{cell} = 1.260kg \cdot 927J/kgK = 1168J/K \quad (4.68)$$

for the whole battery

$$C_{th\_tot} = N \cdot C_{th} = 109 \cdot 1168 J/K = 127'312 J/K \tag{4.69}$$

6. Thermal time-constant

   The thermal time constant of the naturally cooled battery pack is

$$\begin{aligned} \tau_{th} &= R_{th} \cdot C_{th\_tot} = 0.125 \frac{K}{W} \cdot 127'312 \frac{W \cdot s}{K} = 15'914 s \\ &\triangleq 265 \min \\ &\triangleq 4.42 h \end{aligned} \tag{4.70}$$

7. Temperature elevation after 1 hour

$$\begin{aligned} \Delta\vartheta_{1h} &= \Delta\vartheta \cdot (1 - e^{-t/\tau_{th}}) = 14.6K \cdot (1 - e^{-(60\min/265\min)}) = \\ &= 14.6K \cdot (1 - 0.79) = 2.96K \end{aligned} \tag{4.71}$$

8. Evolution of the temperature during the 1-h run

   The evolution of the temperature is represented in Figure 4.47.

9. Power losses during fast charge

   The charging current during fast charge is indicated as 6C. This means

$$I_{fch} = 6 \cdot I_{nom} = 6 \cdot 50A = 300A \tag{4.72}$$

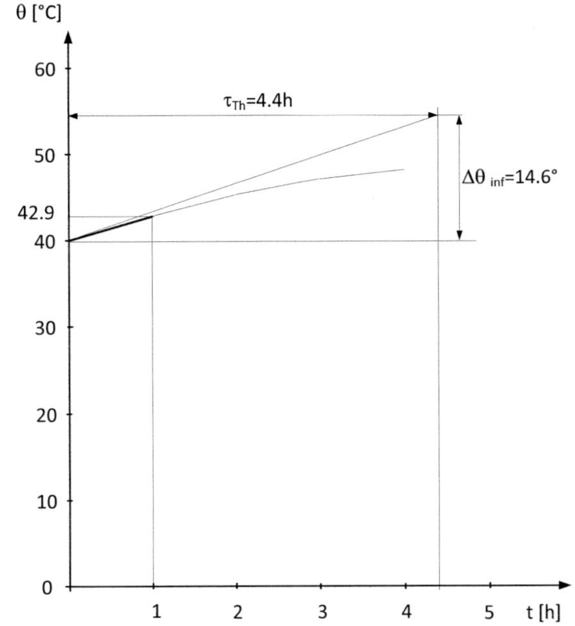

FIGURE 4.47 Elevation of the temperature after 1-h run.

Because the cooling during fast charge is a forced convection through cooling channels between the cells, the power dissipation, the thermal capacity and the thermal resistance will be calculated at the level of one cell.

The losses in one element are

$$P_{fch\_cell} = R_b \cdot (I_{fch})^2 = 0.7 \cdot 10^{-3} \cdot (300A)^2 = 63W \tag{4.73}$$

10. Thermal resistance of the forced cooling

    According Figure 4.46, the battery stack is realized with one cooling channel alternating with two battery elements. This corresponds to a single-side cooling of each element.

    The thermal resistance is then calculated with the surface of one side of the elements.

$$R_{th\_fcool} = \frac{1}{H_{fcool}} = \frac{1}{h_{fcool} \cdot S_{cell}} = \frac{1}{12\,{}^{W}\!/_{m^2 \cdot K} \cdot 0.268m \cdot 0.265m} = 1.17\,{}^{K}\!/_{W} \tag{4.74}$$

11. End value of $\Delta\theta$

    The theoretical end value of the temperature increase due to the fast charge current is calculated

$$\Delta\vartheta_{\inf} = P_{fch} \cdot R_{th\_fcool} = 63W \cdot 1.17\,{}^{K}\!/_{W} = 107.1° \tag{4.75}$$

    This value would be reached if the fast charge is maintained for an infinite time, which is not the case in this application.

12. Thermal time-constant of the forced cooling

    The thermal time-constant is calculated according to

$$\tau_{th\_fcool} = R_{th\_fcool} \cdot C_{th\_cell} = 1.17\,{}^{K}\!/_{W} \cdot 1168\,{}^{W \cdot s}\!/_{K} = 1366s \mathrel{\hat{=}} 22.7\,\text{min} \tag{4.76}$$

13. Battery temperature after the fast charge

    The charging time must first be calculated:

$$t_{fch} = \frac{\Delta E_{sto}}{P_{fc}} = \frac{\Delta E_{sto}}{I_{fc} \cdot U_b} = \frac{16kWh}{300A \cdot 403V} = 0.13h \mathrel{\hat{=}} 7.9\,\text{min} \tag{4.77}$$

    and finally

$$\Delta\vartheta_{fch} = \Delta\vartheta_{\inf} \cdot (1 - e^{-({}^{t_{fc}}\!/_{\tau_{fc}})}) = 107.1 \cdot (1 - e^{-(6.9\,\text{min}/22.7\,\text{min})}) = 31.5° \tag{4.78}$$

    The final temperature after a 1-hour run followed by a fast charge of 8 min is

$$\vartheta_{final} = \vartheta_{init} + \Delta\vartheta_{1h} + \Delta\vartheta_{fch} = 40°C + 2.96°C + 31.5°C = 73.7°C \tag{4.79}$$

14. The evolution of the battery temperature during the 1-hour run followed by a fast charge at a rate of 6C is indicated in Figure 4.48. In order to obtain a figure with acceptable dimensions and readability, the scale has been adapted and the high value of theoretical temperature increase characterizing a continuous regime of fast charge ($\Delta\theta_{inf} = 107°C$) has been reduced by a factor of 2. The same adaptation has been done for the thermal time constant of the fast charge process ($\tau_{th\_fcool}/2 = 11.35$ min). Consequently, the rate of rise of the temperature is represented correctly.

The rapid increase of the temperature during the fast charge is not strongly influenced by the forced cooling because of the short charging time (8 min) related to the corresponding time constant (22.7 min). The produced heat is absorbed by the heat capacity of the elements. However, if the forced cooling is maintained after the fast charge, the battery temperature can be decreased rapidly down to the ambient. As a consequence, the ageing phenomena due to temperature can be limited.

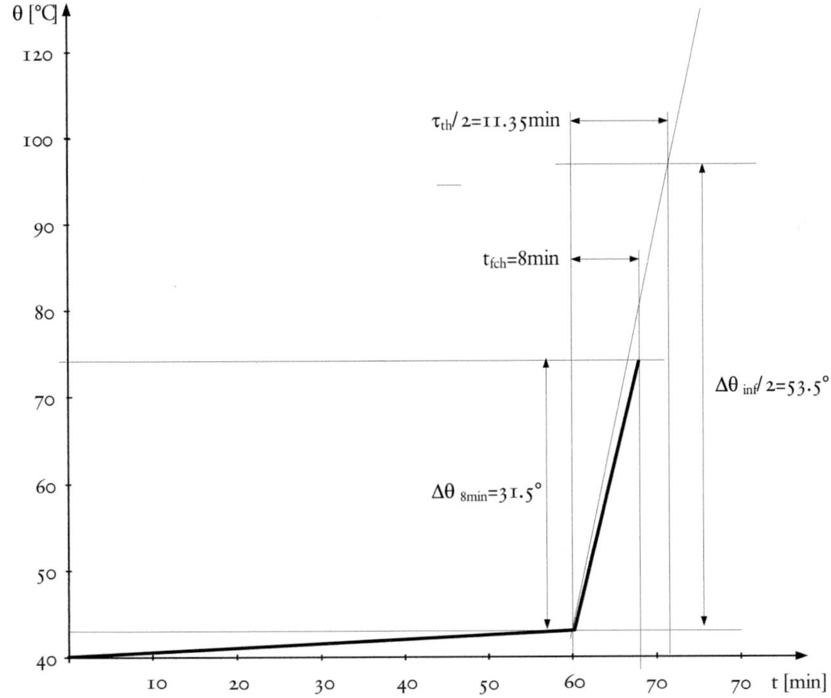

FIGURE 4.48 Evolution of the battery temperature during the 1-hour run, followed by a fast charge of 8 min.

REFERENCES

[1] Odru, P., *Le stockage de l'énergie*, 2nd edition, Dunod, Paris, France, 2013.

[2] Greiner, W., Neise, L., Stocker, H., *Thermodynamics and Statistical Mechanics*, Springer-Verlag, New York, 1995.

[3] Fusalba, F., Martinet, S., Electrochemical storage, cells and batteries, in *Energy Storage* (edited by Y. Brunet), John Wiley & Sons, Hoboken, NJ, 2011.

[4] Girault, H., *Analytical and Physical Electrochemistry*, EPFL Press, Lausanne, Switzerland, 2004.

[5] O'Connor, J. P., *OFF GRID Solar—a Handbook for Photovoltaics with Lead-Acid or Lithium-Ion Batteries*, OCon Energy Consulting, 2016, ISBN: 1537622617; ISBN-13:978-1537622613.

[6] Kawakami, N. et al., Development and field experiences of stabilization system using 34MW NAS batteries for a 51MW wind farm, *2010 IEEE International Symposium on Industrial Electronics*, Bari, Italy, 2010, pp. 2371–2376, https://doi.org/10.1109/ISIE.2010.5637487.

[7] Bartolozzi, M., Development of redox flow batteries. A historical bibliography, *Journal of Power Sources*, 27(3), 219–234, 1989.

[8] Skyllas-Kazacos, M., Menictas, C., The vanadium redox battery for emergency back-up applications, *Intelec 97*, Melbourne, Australia, 1997.

[9] Blanc, C., Rufer, A., Multiphysics and energetic modelling of a vanadium redox flow battery, *ICSET 2008: 2008 IEEE International Conference on Sustainable Energy Technologies*, Singapore, 2008.

[10] Blanc, C., Modeling of a vanadium redox flow battery electricity storage system, Thesis EPFL No. 4277, EPFL, Lausanne, Switzerland, 2009, https://infoscience.epfl.ch/record/129758/files/EPFL_TH4277.pdf. Accessed on September 22, 2017.

[11] Barsali, S., Ceraolo, M., Dynamic models of lead-acid batteries: Implementation issues, *IEEE Transactions on Energy Conversion*, 17(1), 16–34, March 2002.

[12] Enache, B.-A. et al., A LiFePO4 battery discharge simulator for EV applications—part 1: Determining the optimal circuit based battery model, *The Ninth International Symposium on Advanced Topics in Electrical Engineering*, Bucharest, Romania, May 7–9, 2015.

[13] Mousavi, S. M. G., Nikdel, M., Various battery models for various simulation studies and applications, *Renewable and Sustainable Energy Reviews*, 32, 477–485, 2014.

[14] Waheed, A., Eng, M., Buss, K., Modelling and validation of a three-dimensional thermo-electric model of a 50 Ah lithium-iron-phosphate battery cell at three different ambient temperatures in the new European driving cycle (NEDC), in *Automotive Meets Electronics*, VDE Verlag GMBH, Dortmund, Germany/Offenbach, Germany, March 1–2, 2016, ISBN 978-3-8002-4167-0.

[15] Buhler, H., *Electronique de puissance, Traité d'électricité*, vol. 15, Presses Polytechniques Universitaires Romandes (PPUR), Lausanne, Switzerland.

[16] Intersil, *Battery Management System Tutorial*. Intersil Corporation, Milpitas, CA [Online], www.intersil.com/content/dam/Intersil/whitepapers/battery-management/batterymanagement-system-tutorial.pdf. Accessed on July 25, 2017.

[17] *Multi-Cell Battery Management System Using MSP430F5529 and bq76PL536, Application Report, SLAA478*, Texas Instruments Incorporated, Dallas, TX, December 2010 [Online], www.ti.com/lit/an/slaa478/slaa478.pdf. Accessed on July 25, 2017.

[18] Barrade, P., Pittet, S., Rufer, A., Energy storage system using a series connection of supercapacitors with an active device for equalizing the voltages, *IPEC 2000, International Power Electronics Conference*, Tokyo, Japan, April 3–7, 2000.

[19] Caspar, M., Eiler, T., Hohmann, S., Comparison of active battery balancing systems, *2014 IEEE Vehicle Power and Propulsion Conference (VPPC)*, Coimbra, Portugal, 2014, pp. 1–8, https://doi.org/10.1109/VPPC.2014.7007027.

[20] Arioli, V. T. et al., Ageing evaluation of lead acid battery used in off-grid photovoltaic system, *Battcon 2016, International Stationary Battery Conference*, Boca Raton, FL, May 2016, pp. 17-1–17-4.

[21] Timmermans, J.-M. et al., Batteries 2020-Lithium-ion battery first and second life ageing, validated battery models, lifetime modelling and ageing assessment of thermal parameters, *EPE16-ECCE Europe, European Conference on Power Electronics and Applications*, Karlsruhe, Germany, September 6–8, 2016.

5 Energy Storage by Means of Supercapacitors

5.1 GENERAL CHARACTERISTICS OF SUPERCAPACITORS

5.1.1 Principles and Properties

In Chapter 3, supercapacitors were briefly introduced as new storage components with lower energy capacity (or energy density) than batteries, but they can be realized at high capacity scale of several thousand farads. In addition, they can be utilized with a significantly higher number of possible cycles or longer lifetime than conventional batteries. These new components, also called ultracapacitors, formerly electric double layer capacitors (EDLCs), can be charged or discharged much faster than batteries, defining a relatively high power density.

Equation 5.1 is a standard formula for the calculation of the capacity of conventional capacitors.

$$C = \varepsilon\, A/d \tag{5.1}$$

Accordingly, capacitance C is greatest in capacitors made from materials with a high permittivity ε, large electrode plate surface areas A and small distance between plates d.

As a result, double-layer capacitors have much higher capacitance values than conventional capacitors, arising from the extremely large surface area of activated carbon electrodes and the extremely thin double-layer distance in the order of a few ångströms (0.3–0.8 nm). The amount of charge stored per unit voltage in an electrochemical capacitor is primarily a function of the electrode size [1].

Figure 5.1 shows the principle and structure of both a conventional capacitor with plan electrodes and a double layer capacitor.

5.1.1.1 Lifetime

Since no chemical changes take place within the electrodes or electrolyte, charging and discharging electric double-layer capacitors is in principle unlimited. Real supercapacitor lifetimes are only limited by electrolyte evaporation effects.

Figure 5.2 shows the position of EDLC capacitors regarding the number of cycles, with an additional indication of their efficiency. This image represents the order of magnitude of the life cycle of the first available elements (around 2000); modern components are able to achieve more than 1 million cycles [2].

The energy and power density of supercapacitors are indicated in Figure 5.3 [2]. Finally, Figure 5.4 gives a comparison between several technologies of supercapacitors with an electrolytic capacitor and with a small lithium accumulator.

DOI: 10.1201/9781003736486-5

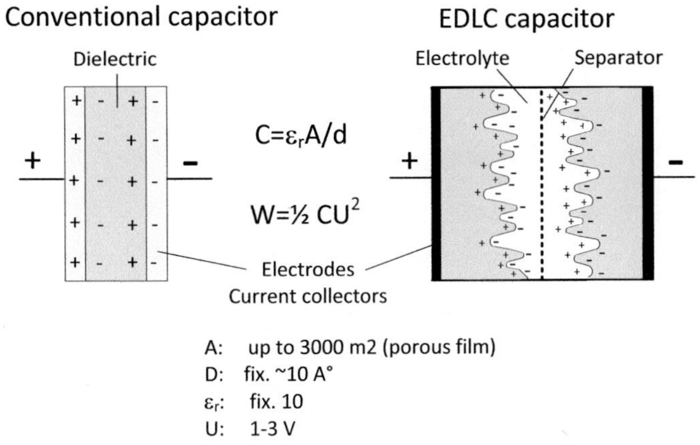

FIGURE 5.1 Principle and structure of an EDLC.

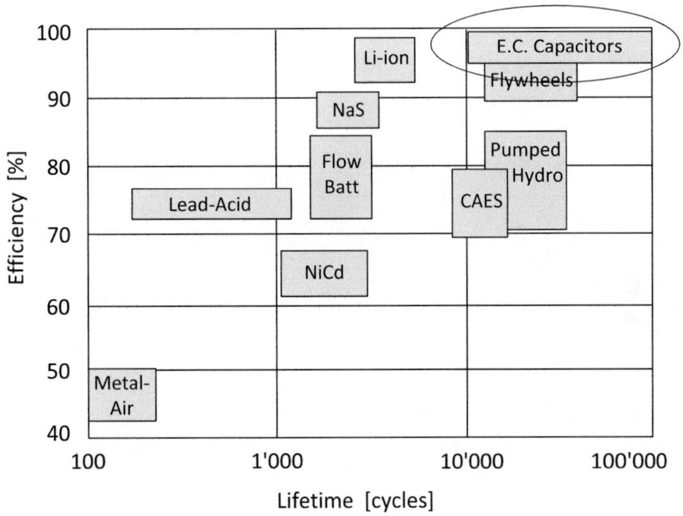

FIGURE 5.2 Lifetime and efficiency of storage components.

5.1.2 Phenomenology of Storage Devices

Supercapacitors, as new components in the sector of energy storage, have their own properties and advantages that must be highlighted properly. Table 5.1 shows the different phenomenology related to storage devices, where it appears that supercapacitors, even if they belong to the category of electrochemical devices, are based on an electrostatic phenomenology as classical capacitors. This is the main reason for their long life cycle, as represented in Figure 5.2.

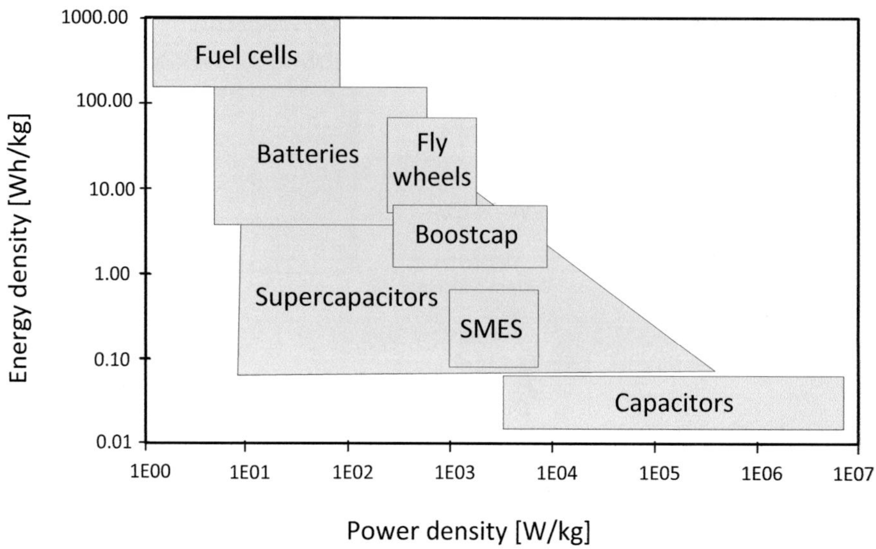

FIGURE 5.3 Ragone diagram showing the Energy Density versus Power Density of a Boostcap supercapacitor.

Electrolytic capacitor		Supercapacitor	Li-ion battery	Gas lighter
98 mF / 63 V	390 μF / 450 V	310 F / 2.7 V	1050 Ah / 2.7 V	7 ml Butan
W = 194 J	W = 39.4 J	W = 1129 J	W = 14'040 J	W = 205'800 J

FIGURE 5.4 Comparison of small storage elements.

TABLE 5.1
Phenomenology of storage devices

System/element		Phenomenology
Capacitors		
	Dielectric (film-foil, metallized film)	Electrostatic
	Oxide electrolytic (A, Ta)	Electrostatic
Electrochemical capacitor		
	Pseudocapacitor	Faradaic
	Double layer capacitor	Electrostatic
Battery		
	Pb, NiCa, NiMH, Li-ion, etc	Faradaic
Fuel-cell	H2, Methanol	Chemical transformation
Flywheel		
Superconducting Magnetic Energy Storage (SMES)		Magnetic

The operation mechanism and scale of nanometers of the internal structure as presented in section 5.1.1 are the reasons for significant power density and very high energy density in regard to other capacitive devices.

As defined in Chapter 1, the combined representation of energy density and power density can be done with the help of the Ragone diagram.

An example of a Ragone plot used for the comparison of the properties of supercapacitors is given in Figure 5.3.

For a comparison of real devices at the level of the used volume, several available products are represented in Figure 5.4. The large electrolytic capacitor (98 mF, 63 V, 194 J) has an amount of stored energy of one-fifth of the cylindric supercapacitor (310 F, 2.7 V, 1129 J), but its volume is significantly larger.

The comparison with a small Li-ion battery (14,000 J) gives a good relationship for the energy amount. The gas lighter at the right side of the figure is a good reference for the dimensions but also gives a reference value concerning the energy amount (205,000 J).

5.1.3 Materials and Construction

The materials used for the different components can play a significant role, depending on the application domain. In traction applications, for example, inflammable or toxic materials are generally forbidden. For the different components of a supercapacitor, the following materials are used:

- Electrolyte
    - Aqueous: high ionic conductance, high power density (1 V, 1.2 S*cm) KOH, H_2SO_4

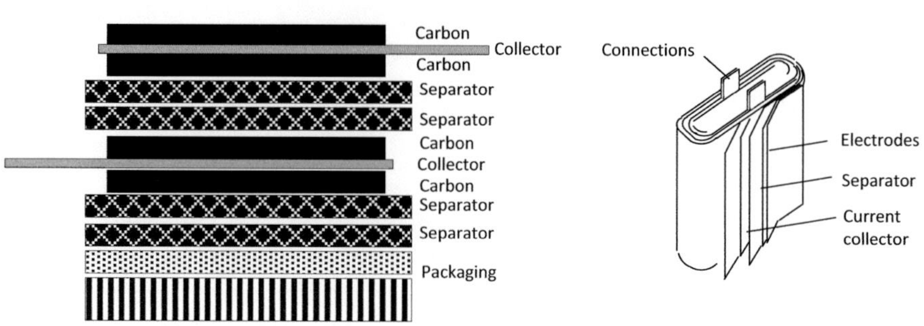

FIGURE 5.5 Construction of an EDLC.

    - Organic: "high" voltage, high energy density (2.5–3 V, 50 mS*cm) (Ex: 40 mS*cm)
- Electrode
    - Carbon: activated carbon
- Separator
    - Organic: polymer (ex: 25-μm microporous polyolefin), paper
    - Inorganic: glass fibers, ceramics
- Current collector
    - Metal foil (Al, Ti, Fe)

Figure 5.5 gives an idea of the physical construction of a supercapacitor, including the materials used.

5.1.4 Main Properties (Summary)

The long lifetime of supercapacitors is related to:

- Reversible physical electrostatic charging cycles
- No chemical redox reactions, unlike batteries
- Low voltage solicitation

The lifetime of supercapacitors is affected by the main ageing mechanisms:

- Type of aging
    - Electrical solicitations
    - Environmental (temperature, corrosive atmosphere)
    - Mechanical (vibrations, shocks)
- Electrical aging mechanisms
    - Ionic saturation of electrode surface
    - Increased contact resistance in the electrodes
    - Water diffusion
    - Redox reaction impurities
    - Electrolyte decomposition

5.1.5 Safety and Limitations

The safe operation of supercapacitors under accident or fire conditions is of high importance, especially in their use in transportation of persons. The following aspects must be considered by a given design

- Cell behavior under crushing conditions:
  - Strong resistance for axial crushing
  - Weak resistance for transverse crushing
- Temperature/fire
  - Opening of cells in 6 min (600°C without fire)
  - Opening of cells in 4 min in case of direct fire exposure
  - Flammable gas emission in case of opening cell if T > 500°C
- Gas emission in case of opening (HCn)
  - 90 cells 3000 F/for a 2 m^3 closed volume: risk of death

5.1.6 Example of Components on the Market

Figure 5.6 shows different forms of available supercapacitors. Cell components are available up to several thousands of Farad. For larger storage capacities, modules with series/parallel connections are also available.

CELL COMPONENTS:

- Maximum voltage: 2.7 V
- 1 Million charge/discharge cycles
- ~0.01 Euro/F

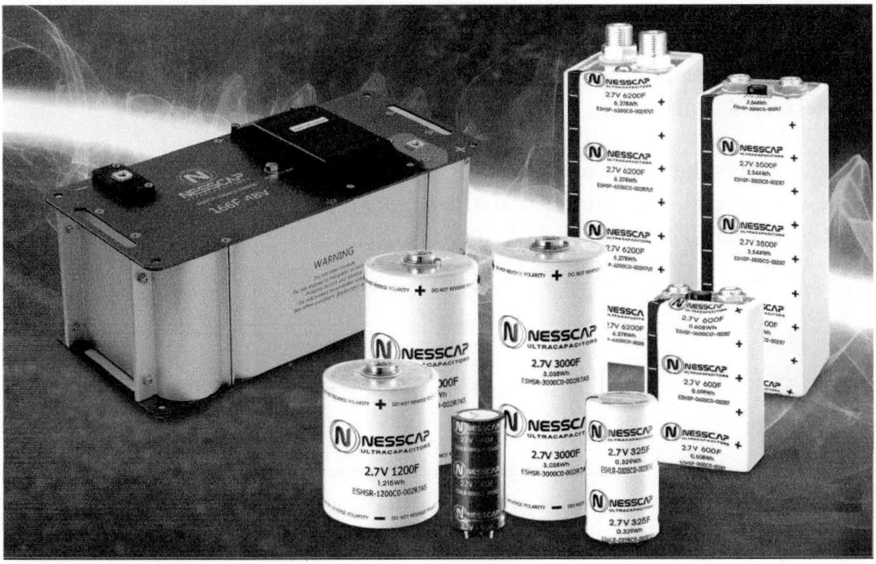

FIGURE 5.6 Examples of supercapacitors, elements and modules.

MODULES:

- From a few Farads to few thousands of Farads
- From the small module (tens of volts to larger modules with several hundreds of volts)

5.2 MODELING SUPERCAPACITORS

5.2.1 Basic Model

For a given application of supercapacitors, there will be some characteristics to consider for the design regarding the storage capacity and also for the definition of acceptable conditions of operation. The main parameters of a supercapacitor (from the application engineer's point of view) are:

- The energy that can be stored:
  - *Capacitance*: from 1 F to 3000 F (and more!)
  - *Maximum voltage*: typically 2.5 V–2.7 V (and more?)
- The energy efficiency of the charging/discharging process:
  - *Series resistor*: limitation of the charging/discharging current
- The self-discharge:
  - *Leakage resistor*: self-discharging of the component

5.2.2 Simple Equivalent Scheme

The classical model for a capacitor is given by its simple equivalent scheme. This scheme includes

an ideal capacitor C
a series resistor R_s
a leakage resistor R_p
a series inductance L_s

Figure 5.7 gives the simple equivalent scheme of a supercapacitor. As will be defined in the next section, the charging and discharging process is normally defined under constant current. The effect of L_S is then negligible.

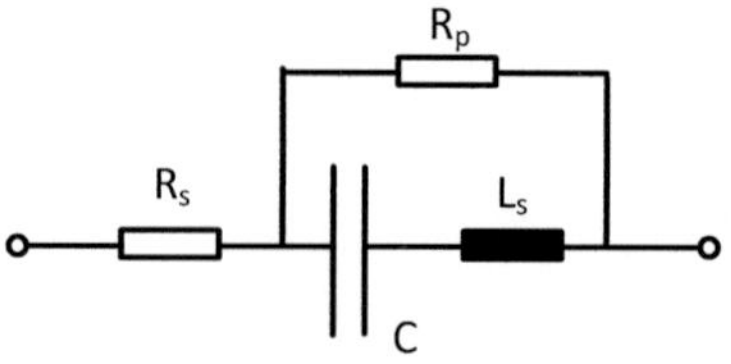

FIGURE 5.7 Simplified equivalent scheme of a supercapacitor.

The elements of the equivalent scheme are used for modeling defined physical phenomena or are related to the physical structure of the component. They can be estimated through the following conditions:

- The ideal capacitor
  - defined by the surface of electrodes, width of ions
- The series resistor
  - defined by the quality of carbon deposition on the aluminum current collectors
  - defined by the electrical conductivity of the carbon
  - defined by the ionic mobility of the electrolyte
- The leakage resistor
  - overcharge beyond the decomposition limit of the electrolyte
  - redox reaction impurities
  - redox reaction of functional groups on the edge of carbon particles
  - electronic conductance through the separator

The elements of the equivalent scheme can be constant or show some dependency on the operation conditions. As an example, Figure 5.8 shows the dependency of the capacitance and series resistor values from the operating frequency BCAP0010 (Montena Component SA).

5.2.3 Specific Behavior of Supercapacitors

Supercapacitors show some very specific behavior, such as the non-constant value of the capacitance and the relaxation phenomena.

5.2.3.1 Voltage-Dependent Capacitance

This phenomenon is related to the variation of the thickness of the double layer that appears under the action of the normal electrostatic force.

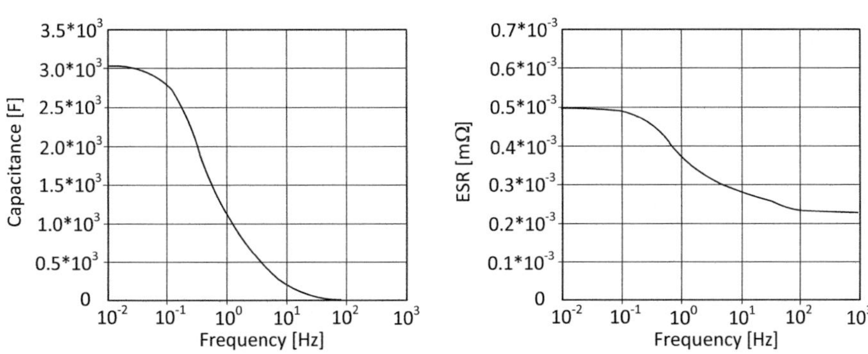

FIGURE 5.8 Capacitance and ESR versus frequency.

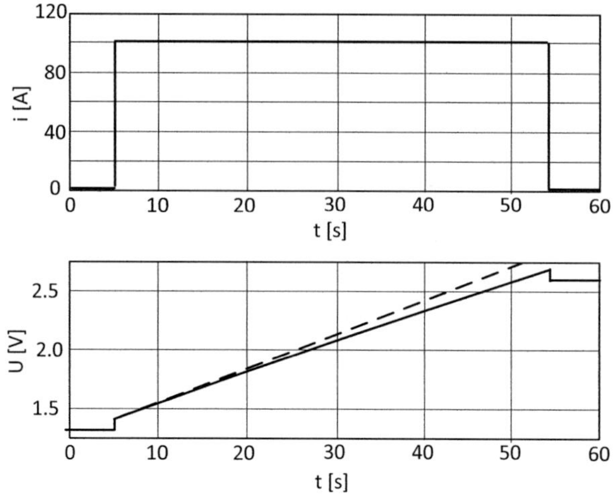

FIGURE 5.9 Effect of the voltage dependent capacitance.

As a consequence, the value of the capacitance increases with the capacitor voltage. In Figure 5.9, the voltage rate of rise is reduced when the voltage increases. The effect of the change of the value of the capacitance can be described through Equation 5.2.

$$C = C_0 + C_u = C_0 + K \cdot U \tag{5.2}$$

where C_u is the voltage-dependent part of the capacitance [3].

There is a first consequence in the calculation of the relation between the current and the voltage related to the charge variation:

$$Q = C \cdot U \Rightarrow i_c = \frac{dQ}{dt} \Rightarrow i_c = (C_0 + 2K \cdot U)\frac{dU}{dt_u} = C_i \cdot \frac{dU}{dt} \tag{5.3}$$

where C_i is called the "current capacitance".

The second effect of the variance of the capacitance is that it changes the energetic capacity of the element.

$$P = i_c \cdot U = (C_0 + K \cdot U)U\frac{dU}{dt} \Rightarrow W_c = \frac{1}{2}\left(C_0 + \frac{4}{3}KU\right)U^2 \tag{5.4}$$

where $C_W = C_0 + 4/3KU$ is called the "energetic capacitance".

The voltage-dependent capacitance and the related values of C_i and C_W are often documented for real components through C_0 and K (example BCAP0008, Montena Component SA) C = 1800 F, $C_o = 1800$ F and K = 150 F/V, U = 2.5 V.

These parameters are difficult to identify, as is the exact amount of energy really stored in a cell.

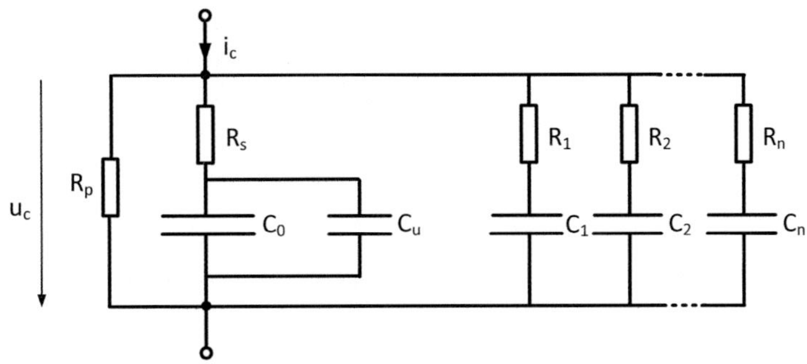

FIGURE 5.10 The completed equivalent scheme.

5.2.4 Completed Equivalent Scheme

Figure 5.10 gives the completed equivalent scheme of a supercapacitor where, in addition to C_0 the voltage-dependent capacitance C_U is added. On the right side of the figure, the equivalent scheme includes some additional RC circuits which represent the so-called relaxation phenomenon. This relaxation phenomenon is described in the next section.

5.2.5 Relaxation Phenomenon

The relaxation phenomenon is a dynamic variation of the properties of the supercapacitor due mainly to charge migration inside of the porous electrodes (misopores, mesopores, macropores). During a fast charge (discharge), ions will first enter (leave) into macropores, then mesopores. The diffusion of ions in misopores is characterized by longer timeconstants.

During the ageing process, the relaxation phenomenon is reinforced by impurities affecting the dimension of the pores.

After a fast charge (discharge), non-homogeneous repartition of charges appears on the electrodes.

The diffusion of the charges for reaching a homogenous distribution depends on the size of the pores and ions. The observable phenomena are:

- Voltage decrease (after charge), even if the current is set to zero.
- Voltage increase (after discharge), even if the current is set to zero.

5.3 DESIGN OF A SUPERCAPACITIVE BANK

5.3.1 Series and Parallel Connections of Elements for Greater Power and Higher Capacity

For storage applications of significant power, the relatively low voltage of supercapacitors gives only a limited power level and also a limited amount of stored energy. The example of a large element of 3000 F and 2.7 V illustrates this limitation when the power by a discharge at 200 A leads to only 540 W, the energy content being around

10 kJ. These limitations lead to the realization of storage banks with series and parallel connections. The design of a storage device must determine the total number of needed supercapacitors and their series/parallel arrangement. The operating voltage level will determine the number of series-connected elements in each branch.

5.3.2 Defining the Needed Energy Capacity

For a given application, the needed energy capacity of a storage device is evaluated for a basic cycle of the use of energy. Generally the known specification of an application is the needed power profile, defined over the complete cycle (time).

The needed energy capacity corresponds to the integral over time of the power according to Equation 5.5:

$$W_c = \int P_c dt \tag{5.5}$$

The design of the storage system is then based on the largest energy excursion in the same direction caused by the positive or negative power value of one cycle. Generally, the charge and discharge of energy are identical, defining a complete round-trip storage cycle.

As an example, the acceleration and positive consumption of energy (integral of power) of an elevator during ascension should be identical to the recovered power during the descent. For this example, the power curve is represented in Figure 5.11, together with the energy consumed and recovered by the elevator.

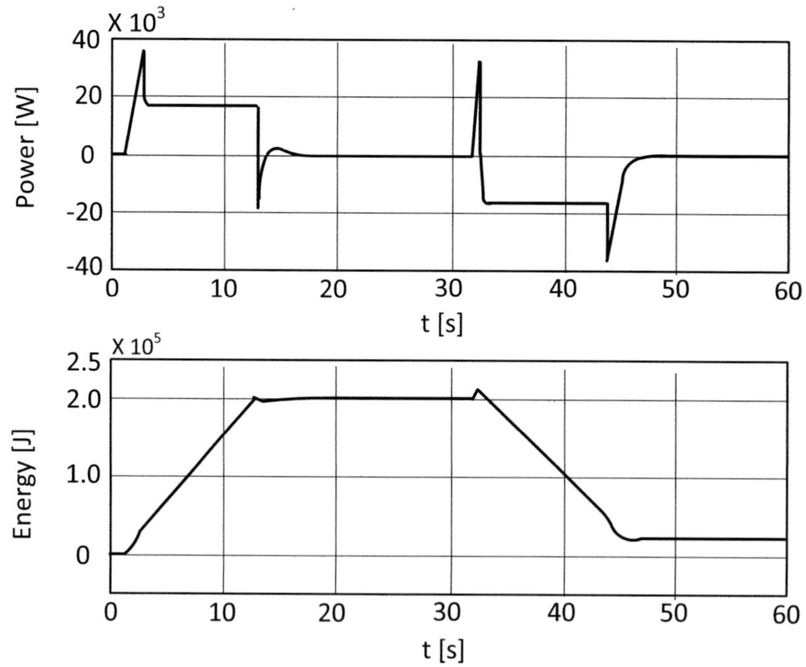

FIGURE 5.11 Power profile and associated energy excursion of an elevator. (Adapted from Rufer, A., Barrade, P., *IEEE Transactions on Industry Applications*, 38(5), 1151, 2002.)

The value of energy that must be retained for the design of the storage device corresponds to the maximum excursion reached during the cycle, 220 kJ in this example.

5.3.3 From the Needed Energy to the Number of Supercapacitors

The total amount of energy in a supercapacitor element is defined by Equation 5.6 (see Appendix B).

$$W_c = \frac{1}{2} C U_c^2 \tag{5.6}$$

To use the total stored energy, the voltage across the component should be decreased from its maximum allowed value U_M down to 0 V. But even if such a discharge is theoretically possible, it cannot fulfil the specified power demand, because the power curve follows the voltage curve, leading to nearly zero power by deep discharge. The compensation of the voltage decrease by elevating the current value would lead to enormous losses when the voltage is low.

As a consequence, a residual voltage level U_m must be respected during discharge, or, in other words, a voltage discharge ratio d must be defined, giving the minimum voltage U_m of a component after its discharge according Equation 5.7:

$$d = 100 \frac{U_m}{U_M} \tag{5.7}$$

According to Equation 5.6, the maximum energy to be stored in a capacitor with a voltage of U_M is

$$W_M = \frac{1}{2} C U_M^2$$

With the definition of the discharge ratio d, the usable energy W_u is only a part of W_M according Equation 5.8:

$$W_u = W_M \left(1 - \frac{d^2}{100^2}\right) \tag{5.8}$$

The ratio of W_u / W_M versus d is represented in Figure 5.12.

Then, the number of supercapacitors needed for a given amount of usable energy becomes:

$$N_s = \frac{2W_u}{C U_M^2 \left(1 - \frac{d^2}{100^2}\right)} \tag{5.9}$$

where

C is the capacity of the single element.

U_M is the maximum voltage of this single element.

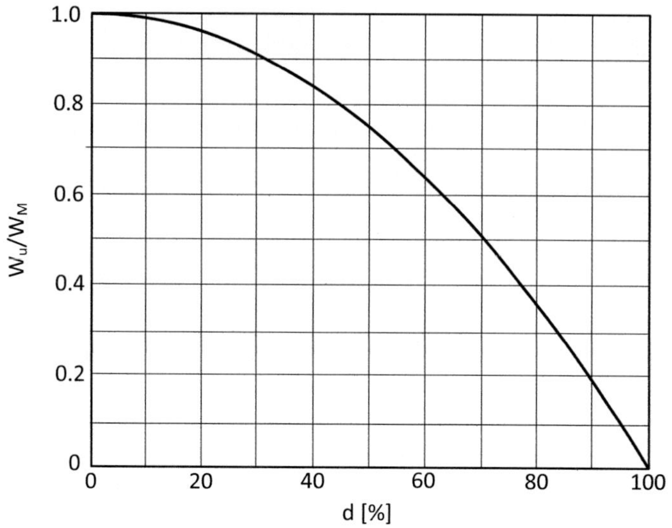

FIGURE 5.12 Usable energy versus discharge ratio.

TABLE 5.2
Number of Needed Supercapacitor Elements Depending on the Voltage Discharge Ratio

2600F, 2.5V		Total stored energy
d=50%	N=31	W=293.8kJ
d=60%	N=37	W=350.6kJ
d=70%	N=46	W=438kJ

Example of design:

- Using a 2600 F/2.7 V/0.4 mΩ supercapacitor
- Needed storage capacity: 220 kJ

In table 5.2, the number of supercapacitor elements is given depending on the value of the voltage discharge ratio *d*.

5.3.4 Power Capability

Due to the presence of the series resistor R_s of the supercapacitor, the energy efficiency during charge and discharge of the supercapacitors has to be taken into account. The internal dissipation is proportional to the square of the charging/discharging current. This will define the power capability of a supercapacitor bank.

Before defining these limits, the charging/discharging process will be described.

5.4 CHARGING AND DISCHARGING PROCEDURE OF SUPERCAPACITORS

5.4.1 Principles and Definitions

The state of charge of a supercapacitor is proportional to the square of the capacitor voltage according to Equation 5.6:

$$W_c = \frac{1}{2} C U_c^2 \tag{5.6}$$

Changing the capacitor voltage responds to the integration law

$$U_c = U_o + \frac{1}{C} \int_{t_0}^{t} I_c dt \tag{5.10}$$

Because of the low value of the series resistor of a supercapacitor, the charging procedure is preferably done with a controlled current source. Figure 5.13a shows the simple circuit with charging through a voltage source. Figure 5.13b shows the preferred circuit with charging through a current source. Figure 5.13c shows the principle of realizing a controlled current source with a static converter. The converter that is normally considered a voltage source is completed by a current measurement device and a current controller. Due to the nature of the modulation of the converter (for example, a PWM), there is a need for some smoothing/filtering elements. Typically, the inductor represented in Figure 5.12c is needed for the limitation of the current ripple. Further, it facilitates the control of the current.

a) Charging from a voltage source
b) Charging from a current source
c) Realization of a controlled current source

There are initial and final states to be defined for the charging and discharging process, according to the following definitions:

Initial state:

The voltage at the initial state is indirectly defined through the voltage discharge ratio d_i

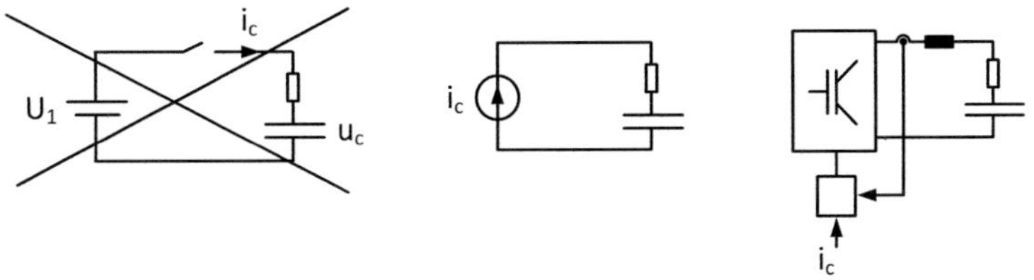

FIGURE 5.13 Principles of charging a supercapacitor.

$$d_i = 100\frac{U_{ini}}{U_M}$$

Final state:

The voltage at the final state is defined through the voltage discharge ratio d_f

$$d_f = 100\frac{U_{fin}}{U_M}$$

5.4.2 Properties of Charge/Discharge with Controlled Current

Table 5.3 gives the expressions related to the charge and discharge conditions under a controlled constant current. The charging time T_{ch} and discharging time T_{dis} are given, as well as the related dissipation on the internal resistor R_{s}: W_{r}. The stored and recovered energy amount ΔW_c is also given (Equation 5.11).

5.4.3 Efficiency for Charge/Discharge with Controlled Current

Corresponding to the same charging and discharging conditions under constant current, the resulting energy efficiency is given in Table 5.4. From Equation 5.12, one can see that for the charging process, the efficiency decreases when the current is increased, due to the fact that the dissipation term appears at the denominator of the expression. For the discharge, the efficiency also decreases when the current

TABLE 5.3
Properties of the charging/discharging of a supercapacitor under constant current

Charge	Discharge
$I_{cc}>0$ and $d_f>d$	$I_{cd}<0$ and $d_f<d_i$
Main equation	
$u_c = U_M\frac{d_i}{100}+\frac{1}{C}I_{cc}t$	$u_c = U_M\frac{d_i}{100}+\frac{1}{C}I_{cd}t$
Charge/discharge time	
$T_{ch} = C\frac{U_M}{I_{cc}}\frac{d_f-d_i}{100}$	$T_d = C\frac{U_M}{I_{cd}}\frac{d_f-d_i}{100}$
Energy loss	
$W_r = R_sCI_{cc}U_M\frac{d_f-d_i}{100}$	$W_r = R_sCI_{cd}U_M\frac{d_f-d_i}{100}$
Energy stored/recovered	

$$\Delta W_c = \frac{1}{2}CU_M^2\left[\left(\frac{d_f}{100}\right)^2-\left(\frac{d_i}{100}\right)^2\right] \quad (5.11)$$

TABLE 5.4
Efficiency of the charging/discharging of a supercapacitor under constant current

Charge		Discharge	
$I_{cc}>0$ and $d_f>d$		$I_{cd}<0$ and $d_f<d_i$	
Efficiency			
$\eta = \dfrac{\Delta W_c}{\Delta W_c + W_r}$		$\eta = \dfrac{\Delta W_c + W_r}{\Delta W_c}$	
$\eta = \dfrac{1}{1+2R_s\dfrac{I_{cc}}{U_M}\cdot\dfrac{100}{d_f+d_i}}$	(5.12)	$\eta = 1+2R_s\dfrac{I_{cd}}{U_M}\cdot\dfrac{100}{d_f+d_i}$	(5.13)
$\eta = \dfrac{T_{ch}}{T_{ch}+2R_sC\cdot\dfrac{d_f-d_i}{d_f+d_i}}$	(5.14)	$\eta = 1+2R_sC\cdot\dfrac{1}{T_{ch}}\dfrac{d_f-d_i}{d_f+d_i}$	(5.15)

increases (Equation 5.13) even if the dissipation term is now at the numerator. This is due to the negative sign of the current while discharging.

5.4.4 Energy Efficiency and Power Availability

In this section, the energy efficiency and the power availability of a supercapacitor will be shown with an example of a real component, namely a 2600 F/2.5 V/0.7 mΩ supercapacitor.

5.4.4.1 Charge

To reach 90% of energy efficiency, the current and the power for charging/discharging have to be limited. The efficiency conditions can also be reached through the setting of the charging/discharging time. From the representation given in Figure 5.14a, the boundary conditions for a given energy efficiency (for example, 90%) can be read as a condition on the current limitation ($I_c < 320$ A), on the power limitation ($P < 700$ W) or as a specification of the charging time ($T_{ch} > 10$ s). In Figure 5.14, there is not a big difference between the charging curve with constant current and the charging curve with constant power.

As a result, the power density of the device is only 806 W/kg (instead of 4300 W/kg as indicated in the data sheet of the component.). In this case, it appears clearly that the real value of the power density should be determined through the design and use of the supercapacitor.

5.4.4.2 Discharge

For the discharge process, similar boundaries can be represented for the same element. Figure 5.14b) indicates the energy efficiency for the discharge of the

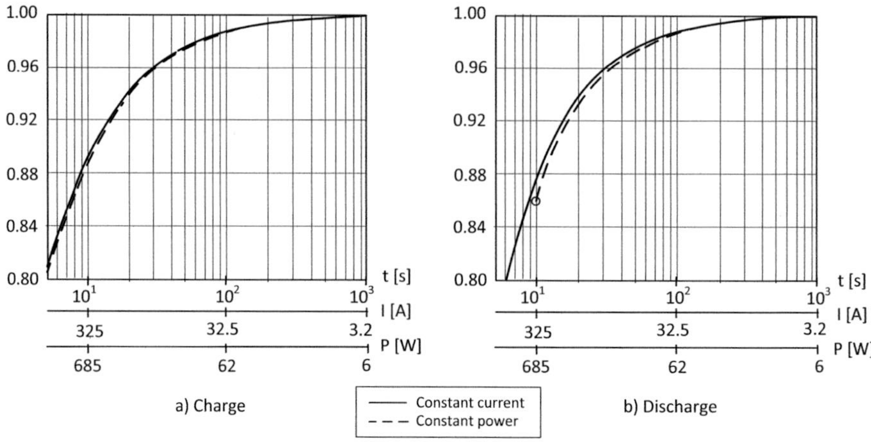

FIGURE 5.14 Efficiency of the charge (a) and discharge (b) process under constant current and constant power. (Adapted from Barrade, P., Supercondensateurs: Principes, dimensionnement, interfaces de puissance et applications, in *Problématiques du stockage d'énergie, Traite EGEM—Genie Electrique* (edited by Y. Brunet), Lavoisier, Paris, 2009, pp. 283–312.)

supercapacitor under constant current and under constant power. The diagram also indicates the dependency of the discharge time. As for the charging, the discharge must be made in a time higher than 10 seconds in order to respect an energy efficiency of 90%. The (constant) discharge current must be kept smaller than 320 A, or the (constant) power must be limited to 400 W. The power density is in this case also only 806 W/kg (instead of 4300 W/kg). It must be precised that the value of 806 W/kg corresponds to the highest value of the power density that appears during a discharge process under constant current. While the power of the transfer is given by the product of voltage and current, the maximum value of the power appears at the maximum of the voltage value. In other words, the maximum of the power appears at the beginning of the discharge or at the end of the charge.

At the left side of the figure, the curve of constant power discharge reaches a final value (small circle), corresponding to the state of full discharge. At this state, the total of initially available energy in the supercapacitor has been extracted or dissipated internally by the element.

5.5 DESIGN EXAMPLE OF A SUPERCAPACITIVE BANK

5.5.1 Analysis of Energy Efficiency and Power Availability

In the previous sections, the design aspects of a supercapacitor bank have been presented. One specific application example has been illustrated in Figure 5.11, where the power profile and the energy to be stored for an elevator are shown.

Based on a simple criterion regarding energy capacity only, the number of needed elements is calculated through Equation 5.9, and the results are shown in Table 5.2.

In this section, the design is completed by another criterion that is related to the energy efficiency.

The main parameters of the example are:

- Supercapacitor used (element): 2600 F/2.7 V/0.4 mΩ
- Specification of the storage: 220 kJ to store, +/–40 kW, with 90% energy efficiency

In Figure 5.15, the design according the energy criterion (Equation 5.8) is given though the continuous line curve. The values correspond to the numbers indicated in Table 5.2. The diagram is completed by curves related to volume and weight of the supercapacitor bank.

Based on additional considerations related to the energy efficiency, the design of the bank takes different values, as indicated through the dotted curves in Figure 5.15. In this case, and under the condition that the bank is discharged with constant power, the energy efficiency can be calculated through Equation 5.16. The number of needed elements to reach a specified value of the efficiency can then be calculated (5.17). More detailed design rules for the supercapacitive bank are given in Reference [4].

$$\eta = \frac{W_T}{W_u} = \frac{2\frac{R_s P}{N}\ln\frac{u_{(t=T_{ch})}}{u_{(t=0)}} - u_{(t=T_{dis})}^2 + u_{(t=0)}^2}{U_M^2\left(1-\left(\frac{d}{100}\right)^2\right)} \tag{5.16}$$

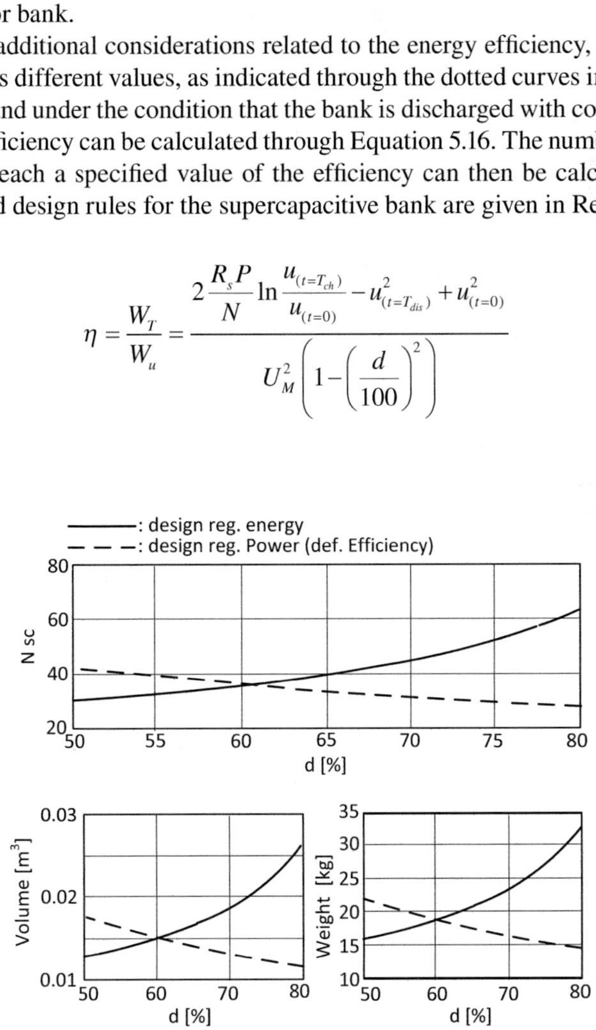

FIGURE 5.15 Number of needed capacitors, volume and weight.

with $u_{(t=0)} = \frac{U_M}{2} + \frac{1}{2}\sqrt{U_M^2 - 4\frac{R_s P}{N}}$ and $u_{(t=T_{dis})} = \frac{U_M}{2}\frac{d}{100} + \frac{1}{2}\sqrt{\left(U_M^2 \frac{d}{100}\right)^2 - 4\frac{R_s P}{N}}$

$$N = \frac{2R_s P \cdot \ln(u_{t=T_{dis}} / u_{t=0})}{U_M^2\left(1 - \left(\frac{d}{100}\right)^2\right) \cdot \eta_p + u_{t=T_{ch}}^2 - u_{t=0}^2} \tag{5.17}$$

From Figure 5.15, the design can be done with a number of 31 Scaps 2600 F/2.7 V if the voltage of the bank is kept above the limit of 50% of the nominal voltage. In this case, the energy efficiency is not taken into account.

However, if an additional design criterion regarding a value of 90% of energy efficiency is considered, the number of needed elements is equal to 36. For that design, the final arrangement will be chosen as a single series chain of series connected elements, leading to a maximum bank voltage of 97.2 V. For this application where the operation conditions are imposed by the elevator drive system running at a DC link voltage of several hundred volts, another choice for the supercapacitor element (smaller capacity) would lead to a higher number of elements and as a consequence to a higher nominal voltage nearer from the DC link voltage of the drive.

For the chosen elements, the realized design of Figure 5.15 corresponds to a volume of the bank equal to 12.9 L for a weight of 16.9 kg.

5.5.2 Thermal Aspects

The internal losses of a supercapacitor have to be transferred to the surrounding, avoiding a too-high temperature of the elements. The temperature of a supercapacitive tank must be monitored, and the temperature of each element should not reach too high a value.

Two main cooling systems can be chosen for the design, natural cooling based on thermal conduction and natural convection.

For highly solicited systems (high power level for a short time, for example) a forced cooling can be designed.

For the thermal design of a supercapacitive bank, thermal models can be made in different ways, from classical thermal impedance calculation and a thermo-electric equivalent scheme.

For a thermal model, the main parameters have to be identified:

- Thermal resistance R_{ca} case to ambient given in [°C/W]
  - Generally given by manufacturer but deduced from experimental protocol that gives quite vague data
- Thermal capacitance C_{th} given in [J/°C]
  - Also given by several manufacturers

An accurate method for the thermal design of a dedicated supercapacitor bank is described in Reference [5]. The simulation has included the internal dissipation due

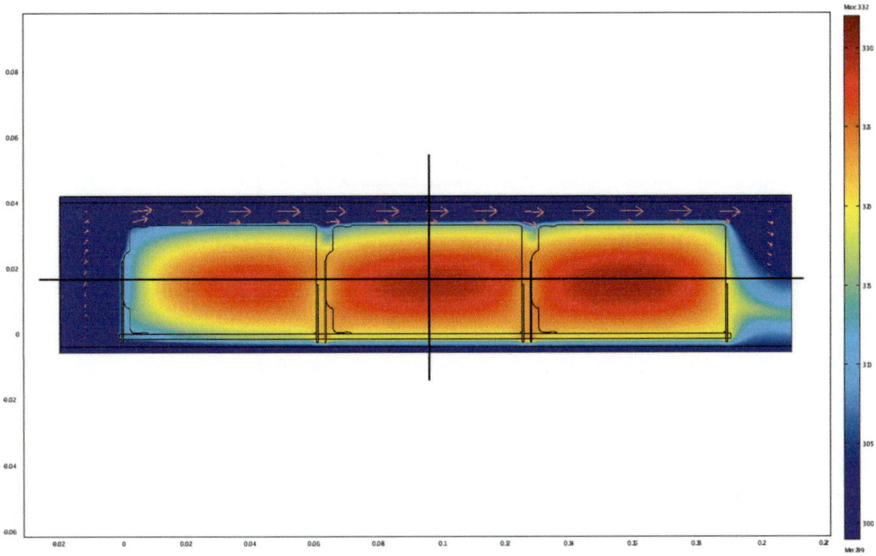

FIGURE 5.16 Modeling of the thermal behavior of a supercapacitor bank with the help of FEM. (adapted from Rufer, A. et al., Multiphysic modeling of a hybrid propulsion system for a racecar application, *EET-2008: European Ele-Drive Conference, International Advanced Mobility Forum*, Geneva, Switzerland, March 11–13, 2008.)

to the current during one cycle, and the thermal behavior has been implemented in a FEM tool. Figure 5.16 shows the final state after one cycle of high current solicitation. In this example, the air flow of the forced convection has also been simulated.

5.6 POWER ELECTRONICS CONVERTERS

5.6.1 Power Electronics Interfaces

Between the supercapacitor bank and the electric outer circuit, a power converter is used in order to allow a controllable power exchange. This power converter must fulfil the specification given by the storage bank.

During charging and discharging, the voltage across a supercapacitors bank varies:

$$W_c = \frac{1}{2} C U_c^2 \tag{5.18}$$

$$U_c = \frac{1}{C} \int i dt \tag{5.19}$$

A flexible power interface between the supercapacitor bank and the application circuit allows the compensation of the variation of the capacitor voltage (Figure 5.17).

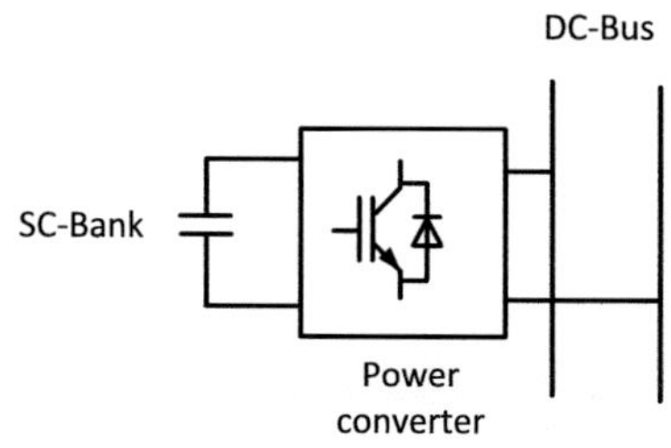

FIGURE 5.17 Power converter between the supercapacitors and the output circuit.

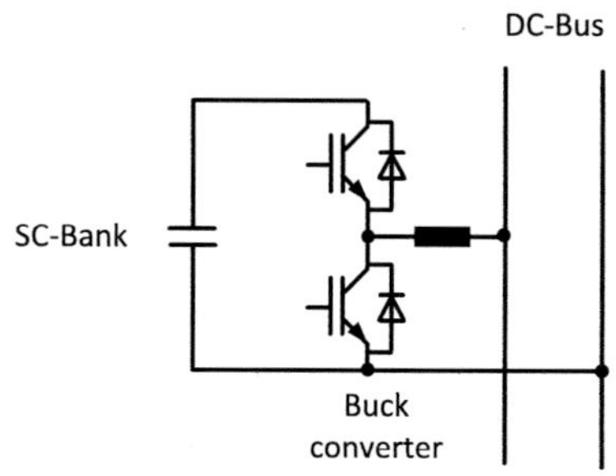

FIGURE 5.18 Buck converter.

The power converter must be able to provide a positive and a negative current corresponding to the charging and discharging processes. Two solutions can be chosen: using a Buck or a Boost topology (Appendix A). For the Buck topology (Figure 5.18), the supercapacitor voltage can vary from a given level that is higher than the DC bus voltage (usually the double of that value), down to a level that must be kept above the bus voltage. Regarding the current in the supercapacitors, it corresponds to a strongly discontinuous waveform that should eventually be smoothed by an additional filter. The inductor used at the output of the buck topology is smoothing the output current.

For the Boost topology (Figure 5.19), the supercapacitor voltage varies between the DC bus voltage level down to around half of this value. In this case, the supercapacitor current is non-discontinuous. The typical inductor used in the boost topology smooths the capacitor current.

5.6.1.1 Association of a Boost and a Buck Converter

This topology has to be used when the DC bus voltage varies strongly and can become lower than that of one of the Scaps (Figure 5.20). This system leads to higher losses due to double conversion.

A detailed analysis of the different topologies with a comparison of the total weight of many simple and complex conversion circuits is given in Reference [6].

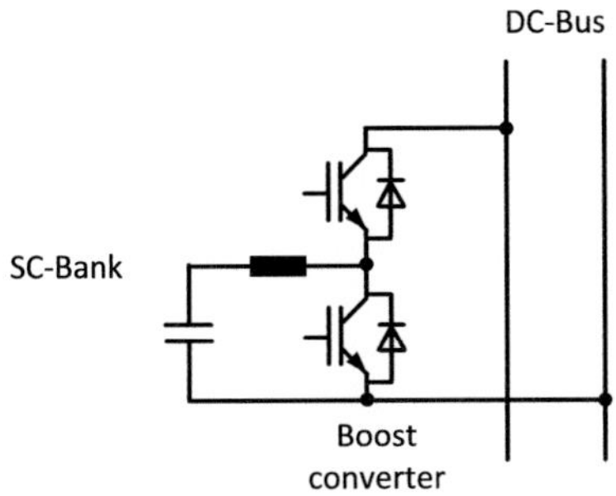

FIGURE 5.19 Boost converter.

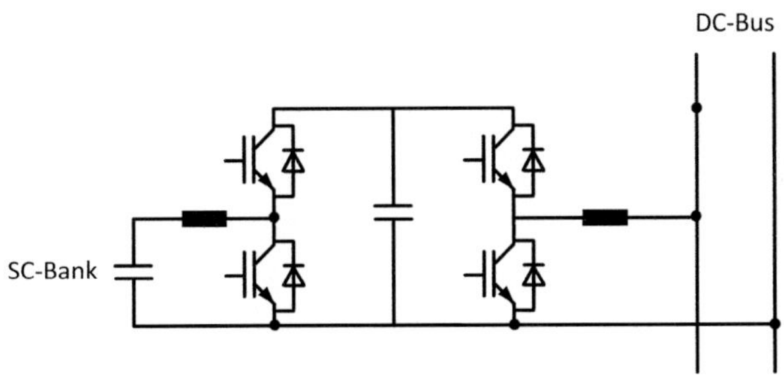

FIGURE 5.20 Association of a Boost and a Buck converter.

5.6.2 Series Connection of Supercapacitors

In order to reach a higher level of power, but also in order to reach good energetic efficiency, the operating voltage of the supercapacitor bank must be much higher than the on-state voltage of the power semiconductor switches used in the converter.

As a consequence, the supercapacitor bank should be realized with the highest possible number of series-connected elements. There is then another advantage, which is to obtain a reduction of the charging/discharging current.

5.6.3 Voltage Balancing

Due to the dispersion of the values of capacitances of series connected elements, imbalance of the voltages can occur when they are charged or discharged by the same current (series connection) [7, 8].

From Figure 5.21, it is evident that U_{c1} and U_{c2} will reach different values of voltage when charged by the same current and when the values of C_1 and C_2 are different.

The smaller capacitor will reach the limit voltage before the bigger, the first being fully charged while the second is not.

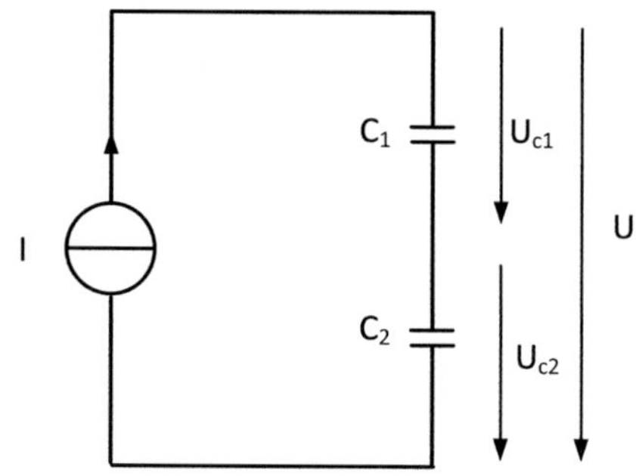

FIGURE 5.21 Series connection of supercapacitors.

TABLE 5.5
Influence of Voltage Imbalance on the energy capacity

	$C_1 = C_2$	C2 = 80% f C1 No balanncing	C2 = 80% of C1 With balancing
U [V]	5	4.5	5
U_{C1} [V]	2.5	2	2.5
U_{C2} [V]	2.5	2.5	2.5
E [J]	6250	4500	5625

5.6.4 Example of Voltage Unbalance and Consequences on the Energy Content

The voltage unbalance limits the amount of stored energy if the charging process is stopped when the element that presents the higher voltage has reached its limit. In Table 5.5, the consequence is shown on the total energy content if two supercapacitors are connected in series, with and without active balancing.

5.6.5 Voltage Equalization Circuits

Voltage balancing solutions can be based on switched resistors (Figure 22a) or using active power electronics circuits offering a higher efficiency (Figure.22b and c). Series connection of supercapacitors and equalizing devices are discussed in References [4, 5].

5.7 APPLICATION EXAMPLES

5.7.1 EnergyStorage in Transportation

For transportation systems with many starts and stops, kinetic energy can be recovered and stored during braking and re-used by the following acceleration [9, 10]

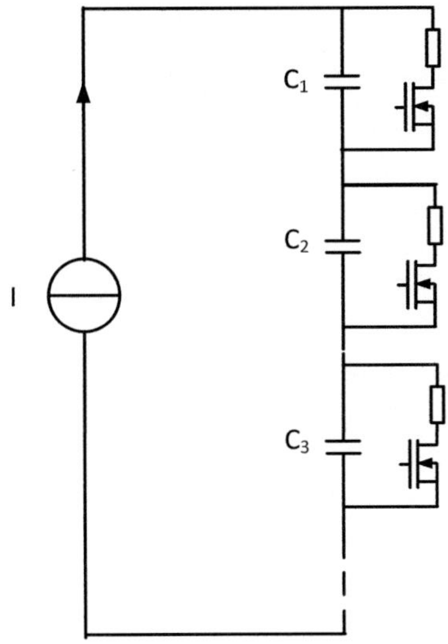

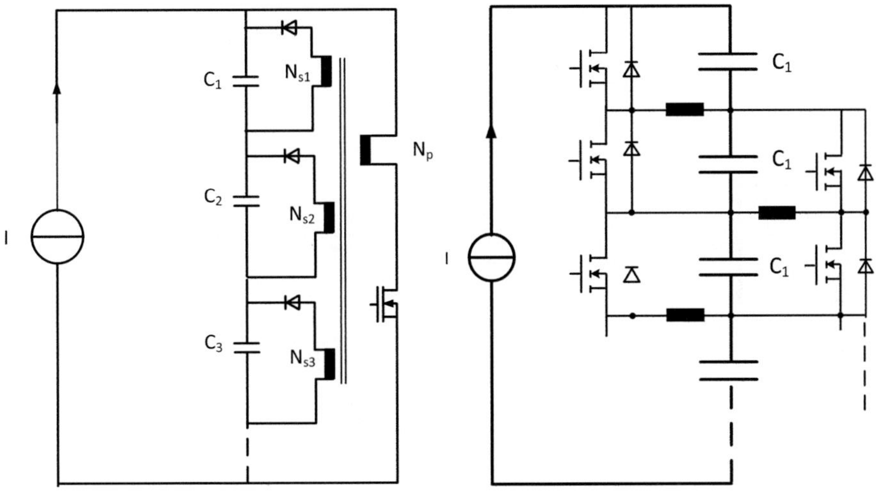

FIGURE 5.22 Active balancing circuits.

a) Balancing with switched resistors.

b) Active balancing using a flyback topology with distributed secondary windings.

c) Association of buck-boost dc-dc converters.

Figure 5.23 shows a modern diesel-electric train where the propulsion system is an electric motor fed from a power electronics converter. The electric drive is a power reversible one, allowing electric braking. In the classical system architecture, the braking energy is dissipated in resistive elements (Figure 5.24a). The same system is represented again in Figure 5.24b, where an energy recovery system based on supercapacitors replaces the resistive dissipation [9–11].

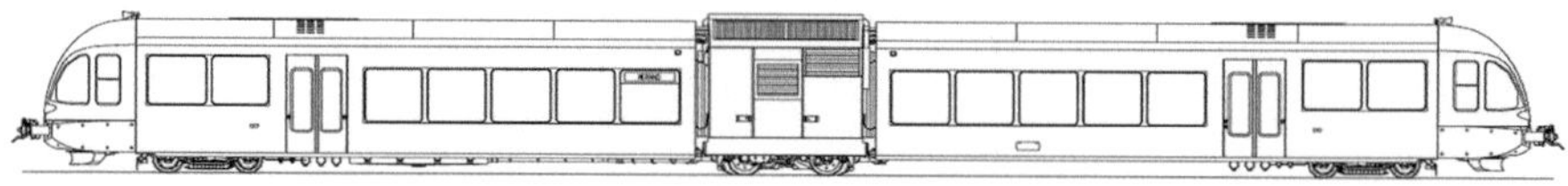

FIGURE 5.23 Diesel electric train. (Adapted from Destraz, B., Assistance énergétique à base de supercondensateurs pour vehicules à propulsion électrique et hybride, These EPFL No. 4083, Ecole Polytechnique Federale de Lausanne, Lausanne, Switzerland, 2008, http://infoscience.epfl.ch/record/118661/files/EPFL_TH4083.pdf. Accessed on September 22, 2017.)

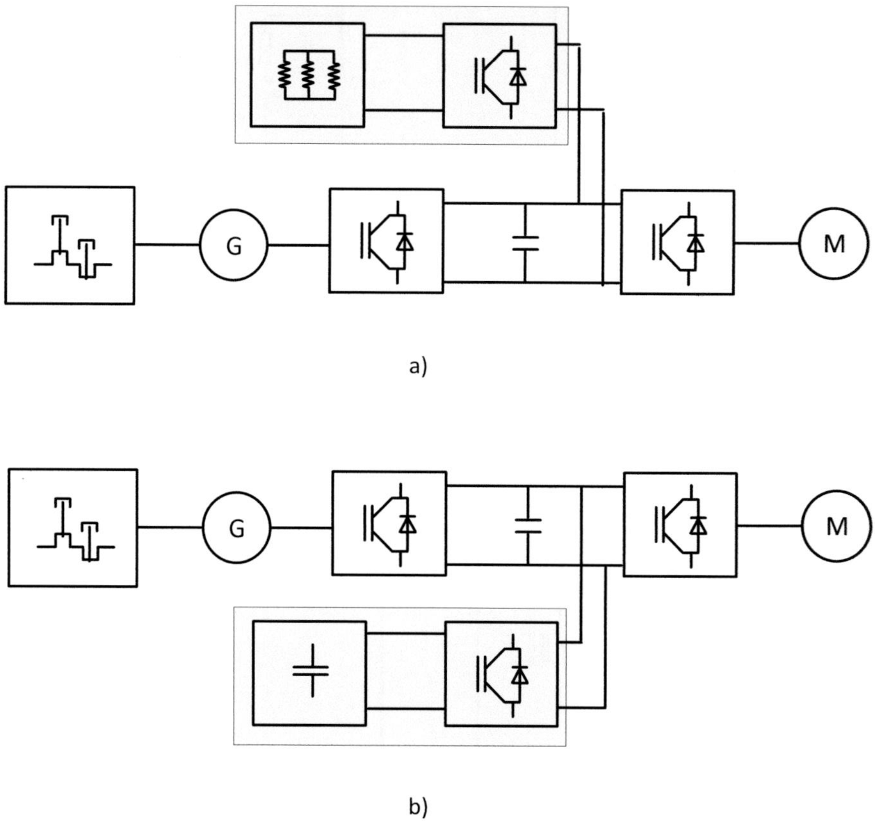

FIGURE 5.24 Propulsion technique with (a) resistive braking, (b) regenerative braking with storage.

In the case of Diesel-electric propulsion, the storage system presents a great advantage in energy consumption due to the possible re-use of the braking energy for the following acceleration. Additionally, there is another advantage: the possibility to downsize the internal combustion engine (ICE). As a consequence, this engine can operate at a higher mean value of the power, where the energy efficiency is better.

These advantages can be illustrated through a real example and for a defined track. Figure 5.25 shows in its upper part the specific track in North Italy (Malles-Meran) with the altitude of the track. The slopes are clearly visible.

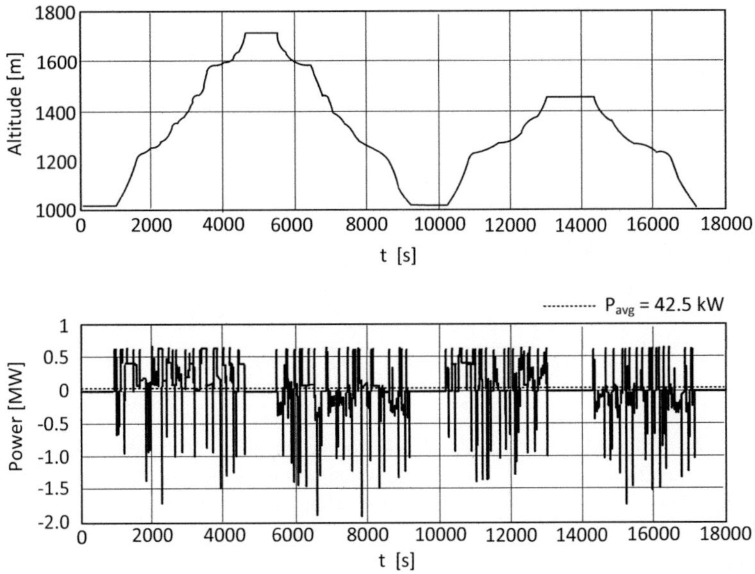

FIGURE 5.25 Characteristics of the Malles-Merano track. (Adapted from Destraz, B., Assistance énergétique à base de supercondensateurs pour vehicules à propulsion électrique et hybride, These EPFL No. 4083, Ecole Polytechnique Federale de Lausanne, Lausanne, Switzerland, 2008, http://infoscience.epfl.ch/record/118661/files/EPFL_TH4083.pdf, accessed on September 22, 2017.)

The lower part of Figure 5.25 indicates the power of the train, including the braking power. The mean value of the power (42 kW) is added to the curve, showing that the ICE has a strongly variable power, often being in idle mode.

Finally, Figure 5.26 indicates the downsizing potential of the ICE of the diesel electric train in function of the installed storage capacity. If the downsizing of the ICE down to the mean value of the power would be theoretically possible, the reality of the needed amount of storage capacity (550 MJ) makes such a design impossible.

However, a better adapted design with around 50 MJ would make it possible to divide the ICE power by two.

A significant reduction of the dieselpower is obtained with only a small amount of storage-capacity (Figure 5.26)!

5.7.2 Elevators

Elevators with counter-weights are systems with relatively high inertia. The counter-weight compensates only the steady state force (speed = const.) (Figure 5.27). For dynamic phenomena, the inertia of the cabin and of the counter-weight must be summed. Different configurations are possible for handling the braking energy or minimizing the power at the point of coupling to the grid [12, 133].

Figure 5.28 shows the speed, position, power and energy profiles of an elevator:

The technical data: car weight: 720 kg, counter weight: 1440 kg, load: 1400 kg

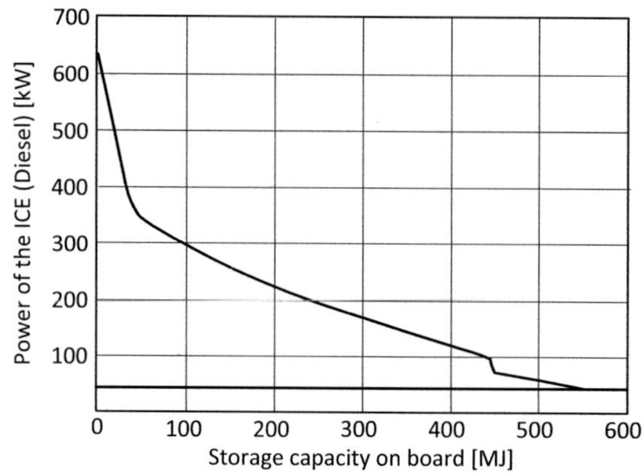

FIGURE 5.26 Needed motor power versus storage capacity. (Adapted from Destraz, B., Assistance énergétique à base de supercondensateurs pour vehicules à propulsion électrique et hybride, These EPFL No. 4083, Ecole Polytechnique Federale de Lausanne, Lausanne, Switzerland, 2008, http://infoscience.epfl.ch/record/118661/files/EPFL_TH4083.pdf, accessed on September 22, 2017.)

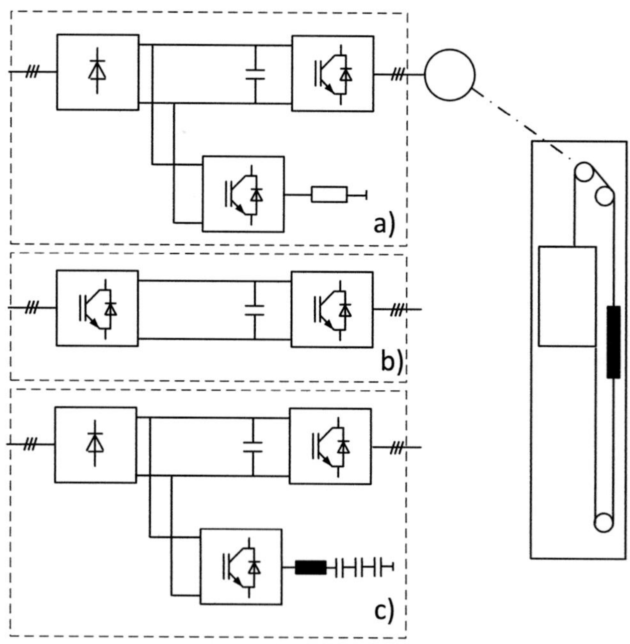

FIGURE 5.27 Drive techniques for elevators with braking capability. (a) Dissipative braking, (b) regenerative front-end, (c) storage with supercapacitors from the DC link. (Adapted from Rufer, A., Barrade, P., A supercapacitor-based energy storage system for elevators with soft commutated interface, *IEEE Transactions on Industry Applications*, 38(5), 1151, 2002.)

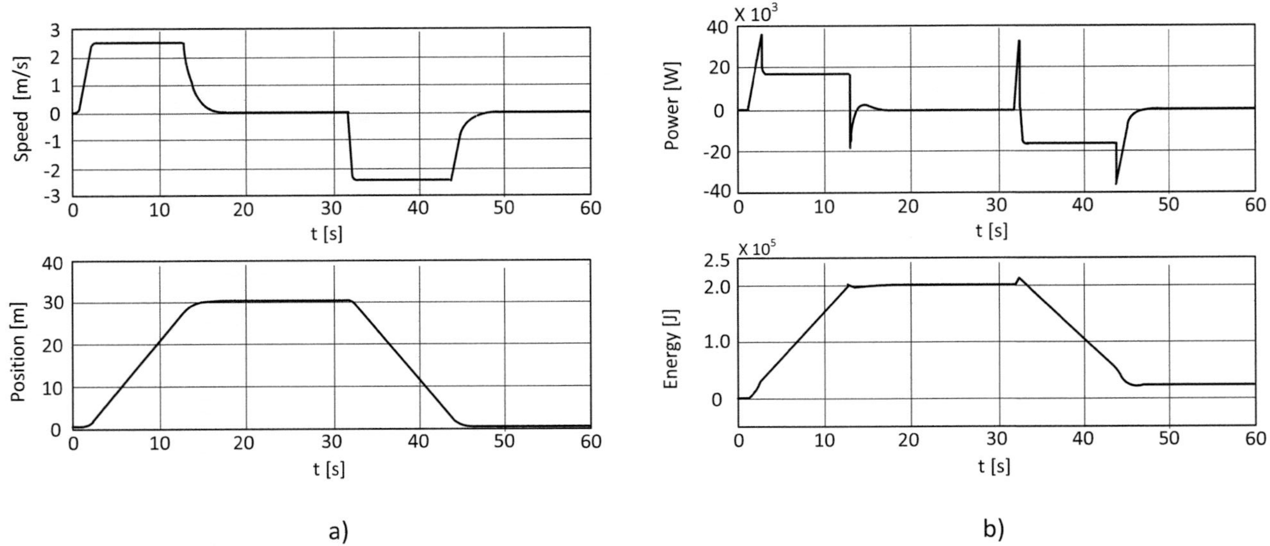

FIGURE 5.28 Elevator, speed, power, position, energy. (a) Speed, position for a 10-floor up/down run; (b) power and energy for a 10-floor up/down run. (Adapted from Rufer, A., Barrade, P., A supercapacitor-based energy storage system for elevators with soft commutated interface, *IEEE Transactions on Industry Applications*, 38(5), 1151, 2002.)

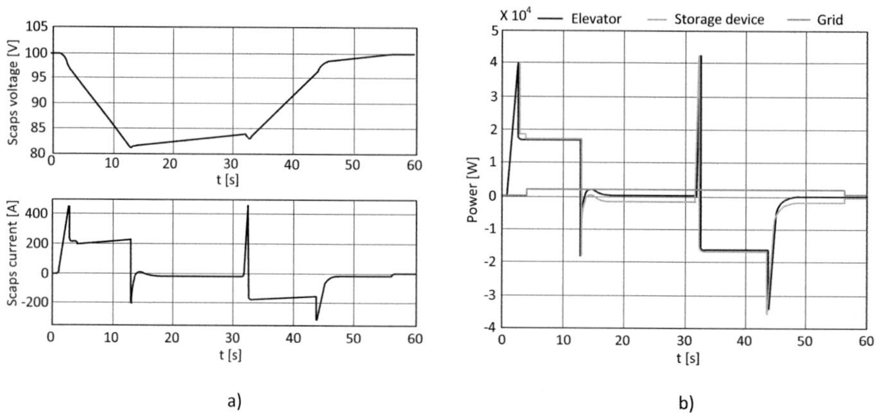

FIGURE 5.29 Power and energy profile. (a) Scaps voltage/current (10-floor up/down run); (b) powers for a 10-floor up/down run. (Adapted from Rufer, A., Barrade, P., A supercapacitor-based energy storage system for elevators with soft commutated interface, *IEEE Transactions on Industry Applications*, 38(5), 1151, 2002.)

The elevator reaches 10 floors.

As presented in section 5.3.2, the needed energy storage capacity is equal to 220 kJ for this example.

Figure 5.29 represent the different curves as the supercapacitors bank voltage together with its current (Figure 5.29a). Figure 5.29b shows the values of the power at different levels. The black curve corresponds to the power demanded by the elevator, the light grey curve shows the power delivered by the supercapacitors and the dark grey curve shows the current exchanged with the input circuit (the grid). There is a clear reduction of the current exchanged with the grid (from around 40 kW down to around 2.5 kW).

5.7.3 Innovative Concept: Sequential Supply

Electrical transportation systems in dense cities often suffer from the need of a catenary. The concept of Sequential supply can solve this problem. The trains (or busses) can be equipped by storage means for a short autonomy and therefore have to be recharged periodically at the stations [14] (Figure 5.30).

The idea of recharging buses at their stops appeared in the mid-1950s with the Oerlikon Electrogyro [6]. As its name indicates, the electrically propelled bus was equipped with a steel flywheel, serving as a short-range energy storage device. The flywheel was recharged at all stops by means of a three-phase pantograph. The high level of the recharging power, however, demanded a strong distribution grid.

The solution of locally providing high power for a fast recharge has been proposed through the use of intermediary storage facilities placed at bus stops [14]. These stationary buffers based on supercapacitors can easily provide the high power needed for the fast energy transfer to the bus and can be recharged slowly during the time separating the passage of two buses. The recharging power of the local buffers does not require a special distribution grid. It can be of a low-voltage type.

The fast recharging of the storage device that is onboard the vehicles leads to a high transfer power in comparison to the normal driving power level (Figure 5.31).

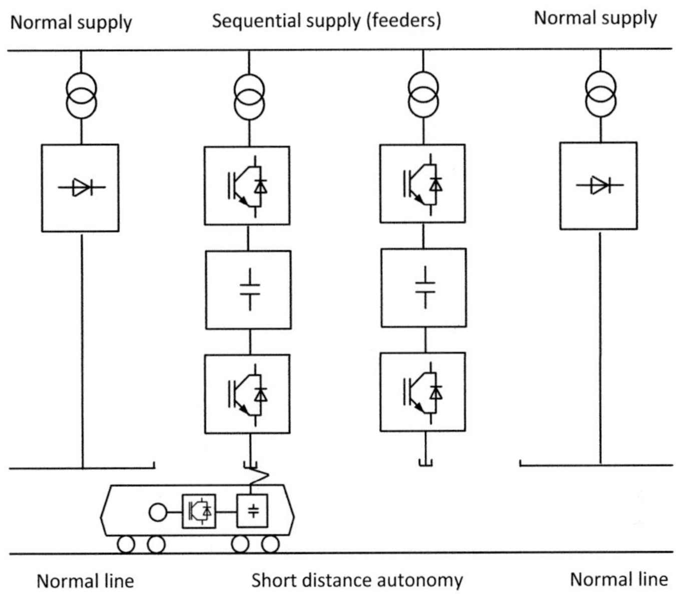

FIGURE 5.30 Sequential supply.

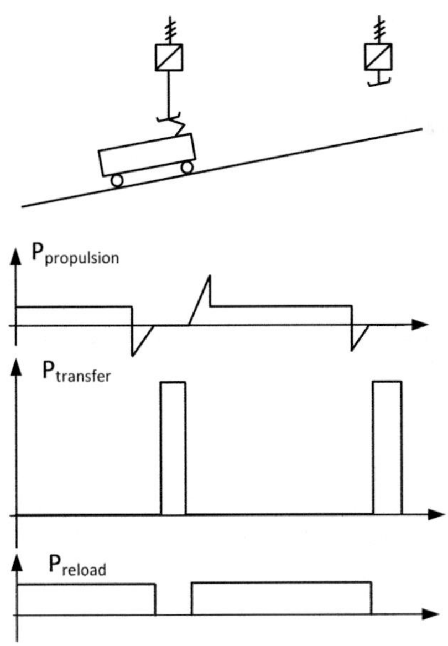

FIGURE 5.31 Fast reload at stations (up). (Adapted from Rufer, A. et al., Sequential supply for electrical transportation vehicles: Properties of the fast energy transfer between supercapacitive tanks. *Journal of Circuits, Systems and Computers, World Scientific*, 13(4), 941–955, 2004.)

The concept can be extended to the regeneration (Figure 5.32). In this case, the excess power is evacuated from the bus to the local stations and, if needed, further back to the distribution grid.

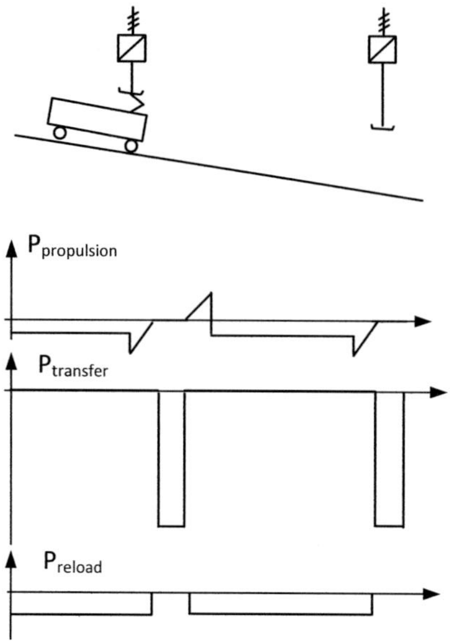

FIGURE 5.32 Fast recovery at stations (down). (Adapted from Rufer, A. et al., Sequential supply for electrical transportation vehicles: Properties of the fast energy transfer between supercapacitive tanks. *Journal of Circuits, Systems and Computers, World Scientific*, 13(4), 941–955, 2004.)

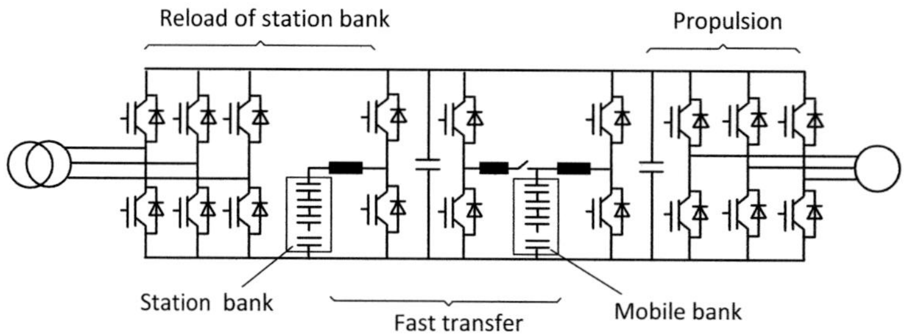

FIGURE 5.33 power electronics interface circuits. (Adapted from Rufer, A. et al., Sequential supply for electrical transportation vehicles: Properties of the fast energy transfer between supercapacitive tanks. *Journal of Circuits, Systems and Computers, World Scientific*, 13(4), 941–955, 2004.)

For the concept of the sequential supply, dedicated power electronics circuits are needed. Figure 5.33 shows the complete system as proposed in Reference [14] with stationary and on-board equipment.

A more elaborate concept similar to the sequential supply has been proposed recently as a real product. This system that is in fact an electric bus is based on the use of high power accumulators [15].

Exercise 4: Energy Storage with Supercapacitors, Recovery of Kinetic Energy of a Train

A given train has a total mass of around 40 tons and is running at 75 km/h (Figure 5.34). The train has to be braked with a deceleration of 1.2 m/s^2 down to zero [10].

The propulsion system of the train consists of an asynchronous motor fed by a voltage source converter connected to a constant voltage DC link.

During braking, the kinetic energy of the train has to be recovered within a supercapacitor bank.

The goal of the exercise is to design and to analyze the properties of the energy storage device on the basis of the use of 3000 F supercapacitors. The datasheet of the supercapacitors is given in Table 5.6.

Energetic properties and evolution of the voltage of the supercapacitors should be shown.

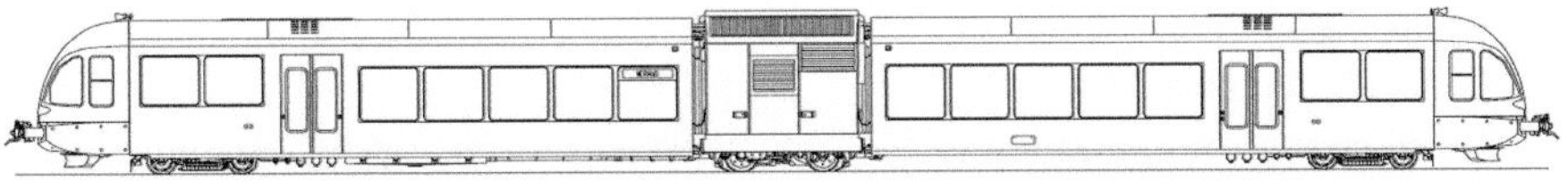

FIGURE 5.34 GTW train. (Adapted from Destraz, B., Assistance energetique a base de supercondensateurs pour vehicules a propulsion electrique et hybride, These EPFL No. 4083, Ecole Polytechnique Federale de Lausanne, Lausanne, Switzerland, 2008, http://infoscience.epfl.ch/record/118661/files/EPFL_TH4083.pdf. Accessed on September 22, 2017.)

TABLE 5.6
Technical data of supercapacitors (courtesy Maxwell)

Rated Capacitance	3000 F
Minimum Capacitance	3000 F
Maximum Capacitance	3600 F
Maximum ESR	0.29 mΩ
Test current for Capacitance and ESR	100 A
Rated Voltage	2.70 V
Absolute Maximum Voltage	2.85 V
Absolute Maximum Current	1900 A
Leakage Current (25°C)	5.2 mA

Tasks

First Approach

1. Draw the approximated speed profile while braking with the defined deceleration
2. Calculate the braking force
3. Draw the approximated power profile during deceleration with constant torque. The constant braking torque is assumed for the drive through its control.
4. Draw the electric scheme of the converter that allows the interface of the supercapacitor bank with the intermediary DC circuit of the traction chain. The converter must be of a bidirectional type, because the stored energy will be reused for the next acceleration.

   The converter will be idealized as one single converter (even if in practice some multichannel solution could be preferred).

   The voltage of the DC link is given as 800 V.
5. Calculate the number of needed cells in order to recover the total amount of kinetic energy of the train. The criterion is only on the basis of the energy amount.
6. The design of the supercapacitor bank has to be done for a discharge ratio of the elements equal to 0.5.
7. Define one configuration of the supercapacitive storage device (number of series and parallel elements of the bank).
8. Determine the maximum voltage of the supercapacitor bank at its full charged status (zero current).
9. Represent an equivalent scheme for the storage bank and calculate the values of the equivalent capacity and the equivalent resistance.
10. Determine the operating point where the current inside the storage elements is at its maximum value.
11. Calculate the loss and the resultant "local" energy efficiency at this operating point.
12. Calculate the energy efficiency for a second operating point where the internal ideal voltage of the supercapacitor bank has reached the level of 75% of its nominal value.
13. Represent the evolution of the supercapacitor's voltage during the braking.

Second Approach

Propose one method for a more accurate estimation of the resulting global round trip efficiency for the recovery system (whole deceleration and acceleration sequence).

Give some comments about the power profile. What can be done in order to increase the energy efficiency of the storage system?

SOLUTION TO EXERCISE 4

First Approach

1. Speed profile.

   The speed profile of the train during braking is given in Figure 5.35 together with the corresponding power.
2. The deceleration force is

$$F_{decel} = m \cdot a = 40000kg \cdot 1.2m/s^2 = 48 \cdot 10^3 kgm/s^2 = 48kN \quad (5.20)$$

3. The braking power (maximum value)

$$P_{\max} = F_{decel} \cdot V_{\max} = 48 \cdot 10^3 N \cdot \frac{75 \cdot 10^3 m}{3600s} = 1000 \cdot 10^3 Nm/s = 1MW \quad (5.21)$$

   The braking time being

$$t_{brake} = V_{\max} / a = \frac{75 \cdot 10^3 m}{3600s} / 1.2m/s^2 = \frac{20.8m/s}{1.2m/s^2} = 17.36s \quad (5.22)$$

4. The supercapacitive storage device is interfaced to the propulsion system at the level of the DC bus, the electrical scheme is given in Figure 5.36.

   For the right operation of the system, the maximum voltage of the supercapacitor cells connected in series must be smaller than the DC bus voltage

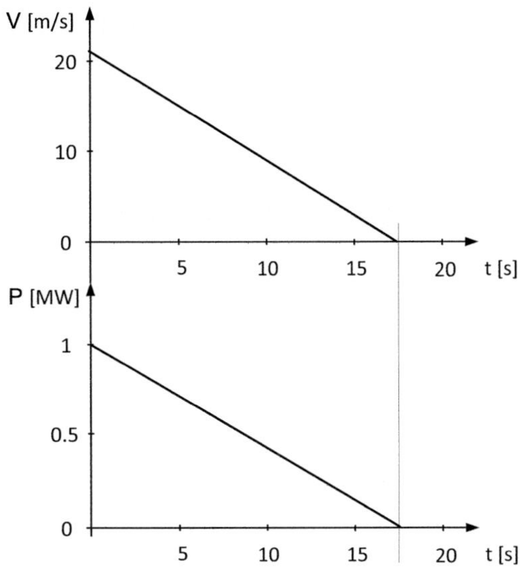

FIGURE 5.35 Speed and power profiles.

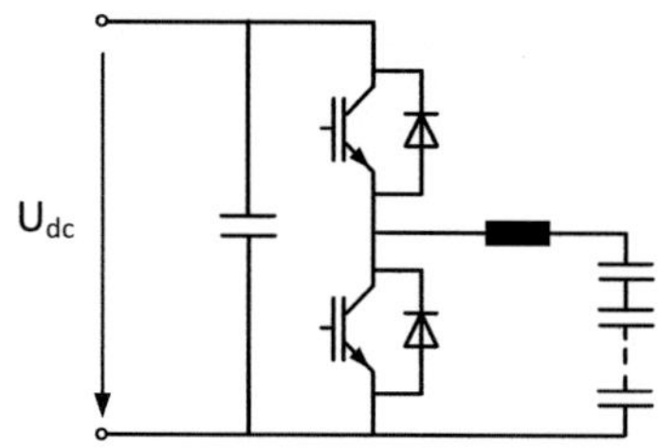

FIGURE 5.36 Interface converter for the supercapacitive storage bank.

(800 V). Details on the operation of this type of converter are given in section A.3.3.

5. The energy recovered (idealized case).

   It is supposed that the recovery system can store the full kinetic energy of the train, down to speed zero.

$$E_{decel} = \int_{0}^{t_{decel}} P(t) \cdot dt = \frac{1MW \cdot 17.36s}{2} = 8.6MJ \tag{5.23}$$

6. The number of needed supercapacitor cells depends on E_{decel} and on the voltage discharge ratio (see Equation 5.9 in Section 5.3.3)

$$N = \frac{2E_{decel}}{CU^2\left[1-(d/100)^2\right]} = \frac{2 \cdot 8.68MJ}{3000F \cdot (2.7V)^2(1-0.5^2)} = 1058 \tag{5.24}$$

7. The total number of elements of the whole storage bank is

$$N = n_s \cdot n_p \tag{5.25}$$

   n_s is the number of elements connected in series, and n_p is the number of parallel connected branches.

   The sum of the voltages of the cells connected in series must be smaller than the voltage of the DC link

$$\begin{aligned} & n_s \cdot 2.7V < 800V \\ & => n_s < \frac{800V}{2.7V} \\ & => n_s < 296 \left(n_s \text{must be integer}\right) \end{aligned} \tag{5.26}$$

   so the smaller value for n_p becomes

$$n_p > \frac{1058}{296} => n_p = 4 \tag{5.27}$$

All branches of the supercapacitive bank must have the same number of series-connected elements. Consequently, the total number N must be increased to a multiple of 4.

$$N => 1060 \tag{5.28}$$

and finally

$$n_s = \frac{1060}{4} = 265 \tag{5.29}$$

The detailed scheme of the supercapacitive storage bank is represented in Figure 5.37.

8. Maximum voltage of the supercapacitive bank

$$U_{sc0\max} = 2.7V \cdot 265 = 715.5V \tag{5.30}$$

U_{sc0} means the sum of the supercapacitor internal voltage (no-load voltage). This value represents the state of charge of the device as

$$E = \frac{1}{2} \cdot C \cdot U^2_{sc0\max} \tag{5.31}$$

The terminal voltage of the bank is affected by the ohmic voltage drop on the internal resistance when the current is flowing.

$$U_{sc} = U_{sc0} + R \cdot I_{sc} \tag{5.32}$$

9. Equivalent scheme of the storage bank.

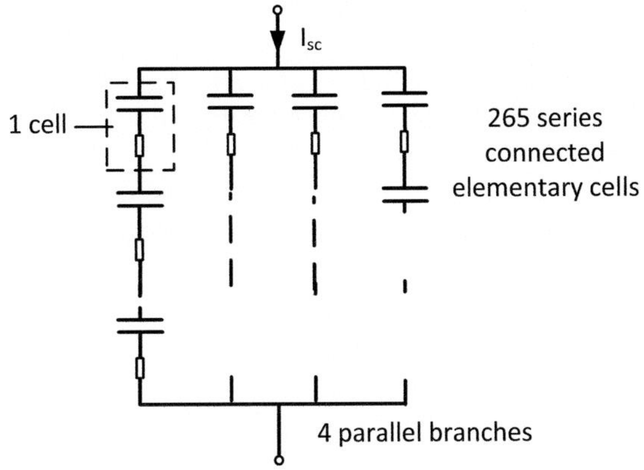

FIGURE 5.37 Scheme of the supercapacitive storage bank.

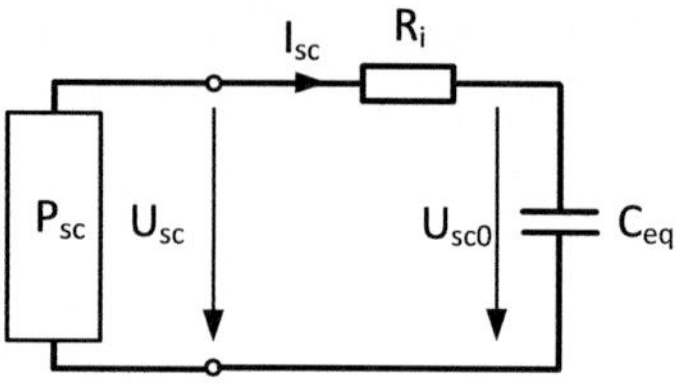

FIGURE 5.38 Equivalent scheme of the storage bank.

The equivalent scheme of the supercapacitive bank is shown at the right side of Figure 5.38. The power electronics converter feeding the storage bank is represented through the power source P_{sc}
The equivalent capacity of the bank is given by

$$C_{eq} = \frac{C}{n_s} n_p = \frac{3000F}{265} \cdot 4 = 45.2F \tag{5.33}$$

and the internal resistance R_i

$$R_i = \frac{R \cdot n_s}{n_p} = \frac{0.29 \cdot 10^{-3}\Omega \cdot 265}{4} = 0.0192\Omega \tag{5.34}$$

10. Operating point at the highest current.
    The operating point with the highest current corresponds to the situation where the storage device has its lowest voltage, namely the voltage with the lowest discharge ratio d = 0.5. It corresponds to the beginning of the braking phase (t_0), where logically the recovering storage capacity is the highest.
    At this point, the value of the internal voltage of the supercapacitor bank is

$$U_{sc0} = d \cdot U_{sc0\max} = 0.5 \cdot 715V = 357.5V \tag{5.35}$$

11. Losses and efficiency.
    According Figure 5.36, the value of the power at t_0 is equal to 1MW, and taking into account the converter efficiency, the injected power into the storage device is

$$P_{sc1} = P_{rec} \cdot \eta_{conv} = 1MW \cdot 0.98 = 980kW \tag{5.36}$$

    The current can be calculated from

$$I_{sc1} = \frac{P_{sc1}}{U_{sc0} + R_i \cdot I_{sc1}} \tag{5.37}$$

The quadratic equation (5.38) gives two solutions for the value of the current (5.40)

$$I_{sc1}^2 \cdot R_i + I_{sc1} \cdot U_{sc0} - P_{sc1} = 0 \tag{5.38}$$

$$\begin{aligned} I_{sc1} &= \frac{-U_{sc0} \pm \sqrt{U_{sc0}^2 - 4 \cdot R_i \cdot (-P_{sc1})}}{2 \cdot R_i} = \\ &= \frac{-357 \pm \sqrt{(357)^2 + 4 \cdot 0.0192\Omega \cdot (980 \cdot 10^3 W)}}{2 \cdot 0.0192\Omega} = 2428A \end{aligned} \tag{5.39}$$

Only the solution related to the positive sign is considered; the other one (–21015A!), even if it corresponds to a physical case, is not realistic here, because it is negative and would cause the discharge and not the charge of the supercapacitors.

The current in one element (one branch) is equal to

$$I_{cell1} = \frac{I_{sc1}}{n_p} = \frac{2428A}{4} = 607A \tag{5.40}$$

The current through the internal resistor R_i (Figure 5.38) causes an ohmic voltage drop of

$$\Delta U_1 = R_i \cdot I_{sc1} = 0.0192\Omega \cdot 2428A = 46.6V \tag{5.41}$$

The energy efficiency of the charging process at this operating point is

$$\eta_1 = \frac{P_{sc1} - R_i \cdot I_{sc1}^2}{P_{sc1}} = \frac{980 \cdot 10^3 W - 0.0192\Omega \cdot (2428A)^2}{980 \cdot 10^3 W} = 0.88 \tag{5.42}$$

12. Operating point at 75% of the voltage.

The value of 75% for the internal voltage of the supercapacitors (d = 0.75) is

$$U_{sco} = 0.75 \cdot U_{sco\max} = 0.75 \cdot 715.5V = 536.5V \tag{5.43}$$

Hypothesis: In order to simplify the evaluation, it is hypothesized that the energy recovery by braking occurs without losses. This would correspond to the case where the stored energy in the supercapacitors is identical to the recovered kinetic energy of the train.

So the energy transferred to the supercapacitor bank is calculated as

$$\begin{aligned} \Delta E_{sc} &= E_{0.75} - E_{0.5} = \frac{1}{2} \cdot C \cdot U_{sc_{0.75}}^2 - \frac{1}{2} \cdot C \cdot U_{sc_{0.5}}^2 = \\ &= \frac{1}{2} \cdot 45.2F \cdot (536.5V)^2 - \frac{1}{2} \cdot 45.2F \cdot (357V)^2 = 3.625MJ \end{aligned} \tag{5.44}$$

The kinetic energy of the train has been reduced by ΔE_{sc} and reaches the value of

$$E_{train_{0.75}} = E_{dec\max} - \Delta E_{sc} = 8.6MJ - 3.6MJ = 5MJ \tag{5.45}$$

The velocity of the train can be calculated from the basic definition of the kinetic energy

$$E_{kin\_train} = \frac{1}{2} \cdot m \cdot v^2 \tag{5.46}$$

$$v_{0.75} = \sqrt{\frac{2 \cdot E_{kin\_train}}{m}} = \sqrt{\frac{2 \cdot 5000 \cdot 10^3}{40 \cdot 10^3}} = 15.8m/s \tag{5.47}$$

The braking power at the second operating point (0.75) is calculated through

$$P_{0.75} = P_{\max} \cdot \frac{v_{0.75}}{v_{init}} = 1MW \cdot \frac{15.8m/s}{20.8m/s} = 0.759MW \tag{5.48}$$

Equations (5.39) and (5.40) can be reused for the calculation of the current at the second operating point

$$I_{sc2}^2 \cdot R_i + I_{sc2} \cdot U_{sc0} - P_{sc2} = 0 \tag{5.38bis}$$

$$\begin{aligned} I_{sc2} &= \frac{-U_{sc0} \pm \sqrt{U_{sc0}^2 - 4 \cdot R_i \cdot (-P_{sc2})}}{2 \cdot R_i} = \\ &= \frac{-536.5 \pm \sqrt{(536.5)^2 + 4 \cdot 0.0192\Omega \cdot (759 \cdot 10^3 W) \cdot 0.98}}{2 \cdot 0.0192\Omega} = 1324A \end{aligned} \tag{5.39bis}$$

Each branch of the supercapacitor array has a current of

$$I_{cell} = I_{sc2}/4 = 1324A/4 = 331A \tag{5.49}$$

The voltage drop on R_i at the second operating point is (5.42bis)

$$\Delta U_2 = R_i \cdot I_{sc2} = 0.0192\Omega \cdot 1324A = 25.4V \tag{5.41bis}$$

Finally, the efficiency at the second operating point becomes (5.43bis)

$$\eta_2 = \frac{P_{sc2} - R_i \cdot I_{sc2}^2}{P_{sc2}} = \frac{0.98 \cdot 759 \cdot 10^3 W - 0.0192\Omega \cdot (1324A)^2}{0.98 \cdot 759 \cdot 10^3 W} = 0.954 \tag{5.42bis}$$

13. Voltage of the supercapacitors.

The evolution of the supercapacitor's voltage is given in Figure 5.39.

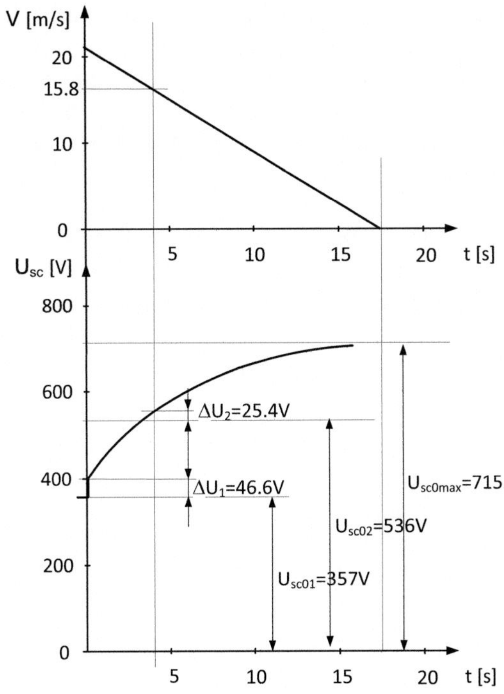

FIGURE 5.39 Velocity of the train and voltage of the supercapacitive storage device.

Conclusions

From the calculations here, it appears that the design of the storage bank based on an energy capacity criterion leads to a high current (607 A) in the cells at the highest speed of 20.8 m/s. This value corresponds to 31.6% of the absolute maximum value of the current specified by the manufacturer (datasheet).

The efficiency of the bank is at this point of operation equal to U_0.

The current value of 607 A is further equal to 2.89 times the maximum continuous current $(\Delta T = 40°C)$, but it is strongly decreasing during the deceleration process. The deceleration time being short (17 s), the corresponding solicitation of the cells is admissible.

The current decrease is due to two superposed effects:

- The braking power is decreasing when the speed decreases (constant torque braking).
- The current in the supercapacitors is reduced through the converter's duty cycle (see Section A.3) because the operating voltage of the elements increases while the supercapacitors are charged.

Second Approach

An accurate estimation of the energetic performance of the braking energy storage system is only possible by continuous simulation of the whole process.

The simulation must calculate the exact current profile during braking.

The dissipation and the temperature increase of the supercapacitors must be calculated during one (and further successive) braking using the thermal characteristics from the datasheet, namely the thermal resistance R_{thca} (case-to-ambient) and the thermal capacitance C_{th}. The thermal models given in Section 4.3.4 for battery elements can be used in this case.

Simulation Model for the Braking Process

Figure 5.40 shows a simulation diagram (Simulink) of the deceleration process, where the mechanical timeconstant of the train is represented by the integrator 1.

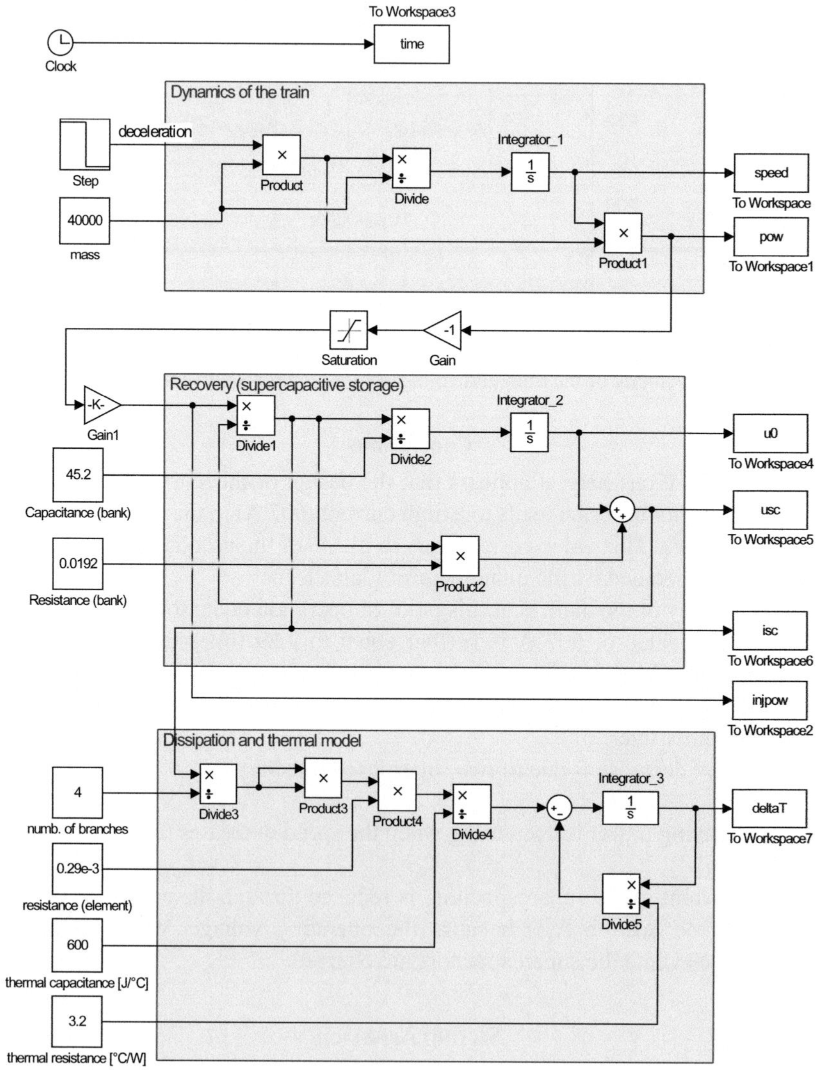

FIGURE 5.40 Simulation diagram of the braking process.

The output variable is the speed of the train. Then, with a constant deceleration, the braking force is calculated. Multiplying the speed by the braking force gives the deceleration power.

The current injected in the storage bank is calculated from the braking power divided by the supercapacitor bank voltage according to Equation 5.37. The internal voltage of the supercapacitor bank is calculated with the integrator 2.

Finally, with the current in one supercapacitor, its temperature elevation is computed with the integrator 3 representing the thermal capacity of one element (600 J/°C). The influence of the thermal resistance (3.2°C/W) is introduced according to the thermal model given in Figure 4.17 (Section 4.3.4).

The simulated curve for the recovered power is given in Figure 5.41a and the current injected in the supercapacitive storage bank in Figure 5.41b.

The evolution of the terminal voltage U_{sc} is represented in Figure 5.42a together with the internal voltage of the supercapacitors U_0.

Finally, the temperature increase of the supercapacitors is shown in Figure 5.42b. The low value of the temperature elevation is due to the low internal dissipation (max. around 100 W) and to the high thermal capacity of the elements (600 J/°C).

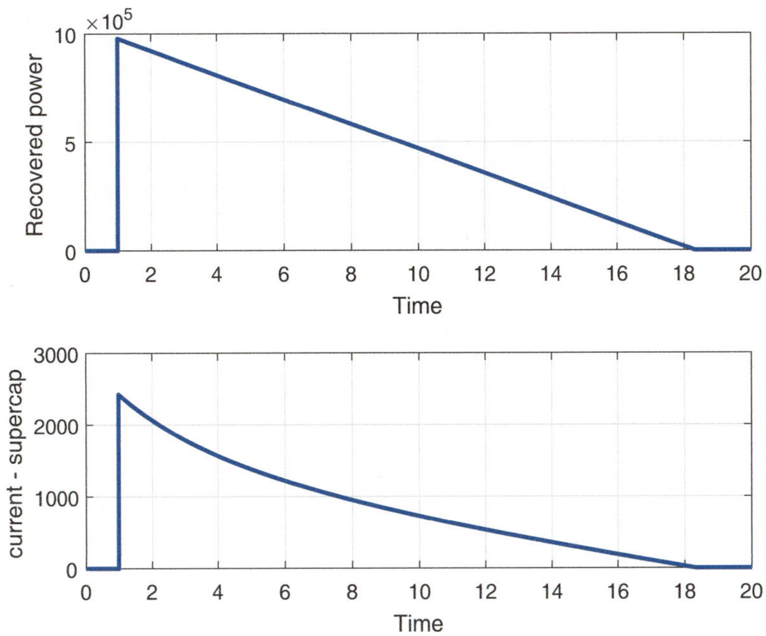

FIGURE 5.41 Recovered power during braking (a) and current injected in the supercapacitive storage device (b).

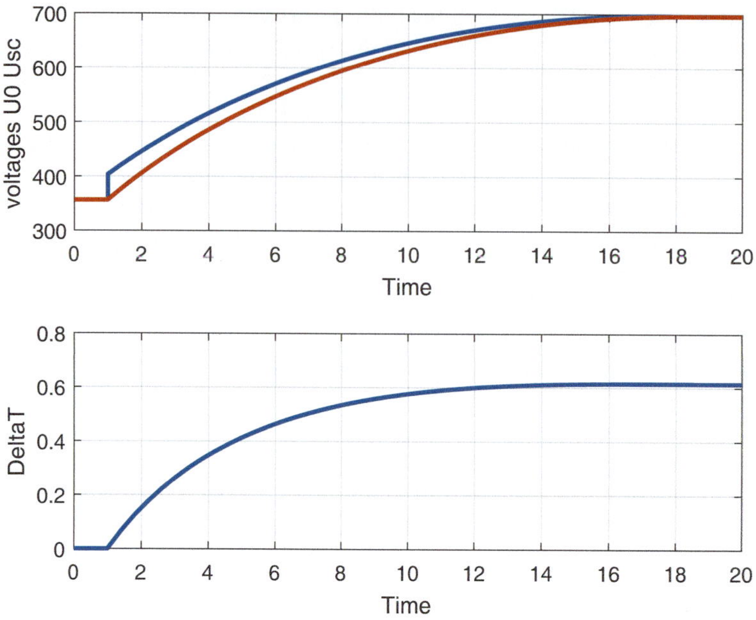

FIGURE 5.42 Evolution of the terminal and internal voltage of the supercapacitor bank (a), Elevation of temperature of the elements (b).

REFERENCES

[1] Conway, B. E., *Electrochemical Supercapacitors, Scientific Fundamentals and Technological Applications*, Springer US, New York, 1999.

[2] Schneuwly, A., Gallay, R., Properties and applications of supercapacitors, from the state-of-the-art to future trends, *Proceedings of PCIM 2000, Conference on Power Conversion and Intelligent Motion*, Nueremberg, GermanyMay 2000, pp. 1–10.

[3] Barrade, P., Supercondensateurs: Principes, dimensionnement, interfaces de puissance et applications, in *Problématiques du stockage d'énergie*Traite EGEM—Genie Electrique (edited by Y. Brunet), Lavoisier, Paris, 2009, pp. 283–312.

[4] Barrade, P., Rufer, A., Current capability and power density of supercapacitors: Considerations on energy efficiency, *EPE2003*, Toulouse, France, 2003.

[5] Rufer, A., Barrade, P., Correvon, M., Weber, J.-F., Multiphysic modeling of a hybrid propulsion system for a racecar application, *EET-2008: European Ele-Drive Conference, International Advanced Mobility Forum*, Geneva, Switzerland, March 11–13, 2008.

[6] The Oerlikon Electrogyro, its development and applications for Omnibus services, *Automobile Engineer*, 559–566, December 1955.

[7] Barrade, P., Pittet, S., Rufer, A., Energy storage system using a series connection of supercapacitors ith an active device for equalizing the voltages, *IPEC 2000, International Power Electronics Conference*, Tokyo, Japan, April 3–7, 2000.

[8] Barrade, P., Pittet, S., Rufer, A., Series connection of supercapacitors, with an active device for equalizing the voltages, *PCIM 2000: International Conference on Power Electronics, Intelligent Motion and Power Quality*, Nuremberg, Germany, June 6–8, 2000.

[9] Destraz, B., Assistance energetique a base de supercondensateurs pour vehicules a propulsion electrique et hybride, Thesis EPFL No. 4083, Ecole Polytechnique Federale de Lausanne, Lausanne, Switzerland, 2008, http://infoscience.epfl.ch/record/118661/files/EPFL_TH4083.pdf. Accessed on September 22, 2017.

[10] Destraz, B., Barrade, P., Rufer, A., Power assistance for diesel—electric locomotives with supercapacitive energy storage, *IEEE-PESC 04: Power Electronics Specialist Conference*, Aachen, Germany, June 20–25, 2004.

[11] Destraz, B., Barrade, P., Rufer, A., Klohr, M., Study and simulation of the energy balance of an urban transportation network, *EPE 2007: 12th European Conference on Power Electronics and Applications*, Aalborg, Denmark, September 2–5, 2007.

[12] Rufer, A., Barrade, P., A supercapacitor-based energy-storage system for elevators with soft commutated interface, *IEEE Transactions on Industry Applications*, 38(5), 1151–1159, 2002.

[13] Bilbao, E., Barrade, P., Ion, E.-O., Rufer, A., Luri, S., Gil, I., Optimal energy management strategy of an improved elevator with energy storage capacity based on dynamic programming, *IEEE Transactions on Industry Applications*, 50(2), 1233–1244, 2014, https://doi.org/10.1109/TIA.2013.2276015.

[14] Rufer, A. et al., Sequential supply for electrical transportation vehicles: Properties of the fast energy transfer between supercapacitive tanks, *Journal of Circuits, Systems and Computers, World Scientific*, 13(4), 941–955, 2004.

[15] Patey, T., Fluckiger, R., Poland, J., Segers, D., Wicky, S., Charged in a flash—optimization of batteries for a flash-charged bus, *ABB Review*, 4(16), 8–12, 2016.

6 Energy Storage Systems Based on Compressed Air

6.1 GENERALITIES ON COMPRESSED AIR ENERGY STORAGE SYSTEMS

6.1.1 System Description

Compressed air energy storage (CAES) is based on the compression and expansion of the air. Such systems use generally a compression stage comprising an electric motor driving a compression machine. Then the compressed air is stored in a reservoir. For the recovery of the stored energy, an expansion system is also provided. This expansion machine is composed of a volumetric expander driving an electric generator. The general scheme of CAES is represented in Figure 6.1.

The amount of energy stored in a reservoir of volume V_1, pressurized at a pressure level P_1 and stabilized at the same temperature as the surroundings, can be calculated through Equation (6.1) [1, 2].

$$E = P_1 \cdot V_1 \left[\ln\left(\frac{P_1}{P_a}\right) - 1 + \frac{P_a}{P_1} \right] \tag{6.1}$$

P_a is the pressure of the surrounding (atmosphere). The expression gives the value of the maximum amount of energy that can be recovered from the reservoir that corresponds to a full expansion under isothermal conditions.

As introduced briefly in chapter 3, CAES systems allow storage of a high amount of energy at a high power level. CAES systems belong to the category of systems with long lifecycles, and the round-trip efficiency that can be reached is in the range of 70–80% (Sections 3.1 and 3.7).

6.1.2 History

CAES systems have been built for decades, dedicated to diverse applications in industry, house-delivered energy and transportation [3] (Figure 6.2). The first utility CAES system was developed in Germany as a 290 MW plant in Huntorf [4]. Other systems have been realized, such as the McIntosh facility (Alabama) [5–7].

In specific applications like compressed air cars [8], more recent investigations and developments are conducted at the university level but also as new industry [9–12].

DOI: 10.1201/9781003736486-6

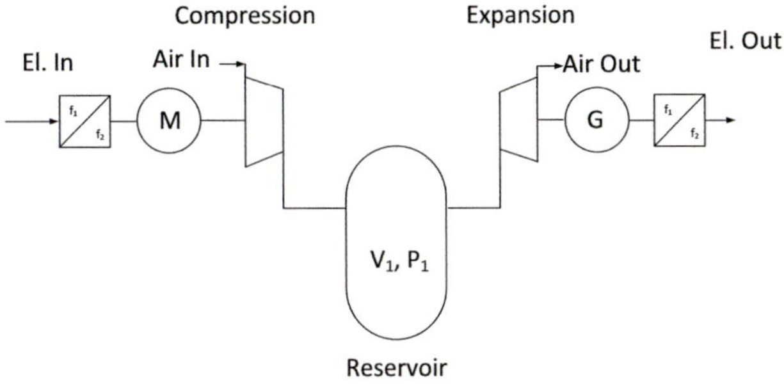

FIGURE 6.1 General scheme of CAES.

FIGURE 6.2 Tram Mekarski (~1875).

Figure 6.3 shows the CAES plant in Huntorf, Germany. The Huntorf plant stores energy in an underground cavern, using the compressor stage of a gas turbine ((7) and (9) in Figure 6.3) for loading, and its turbine in the discharge sequence ((3) and (5) in Figure 6.3). In order to compensate the decrease of the air temperature during the discharge sequence, the burners of the gas turbine are maintained ((2) and (4)).

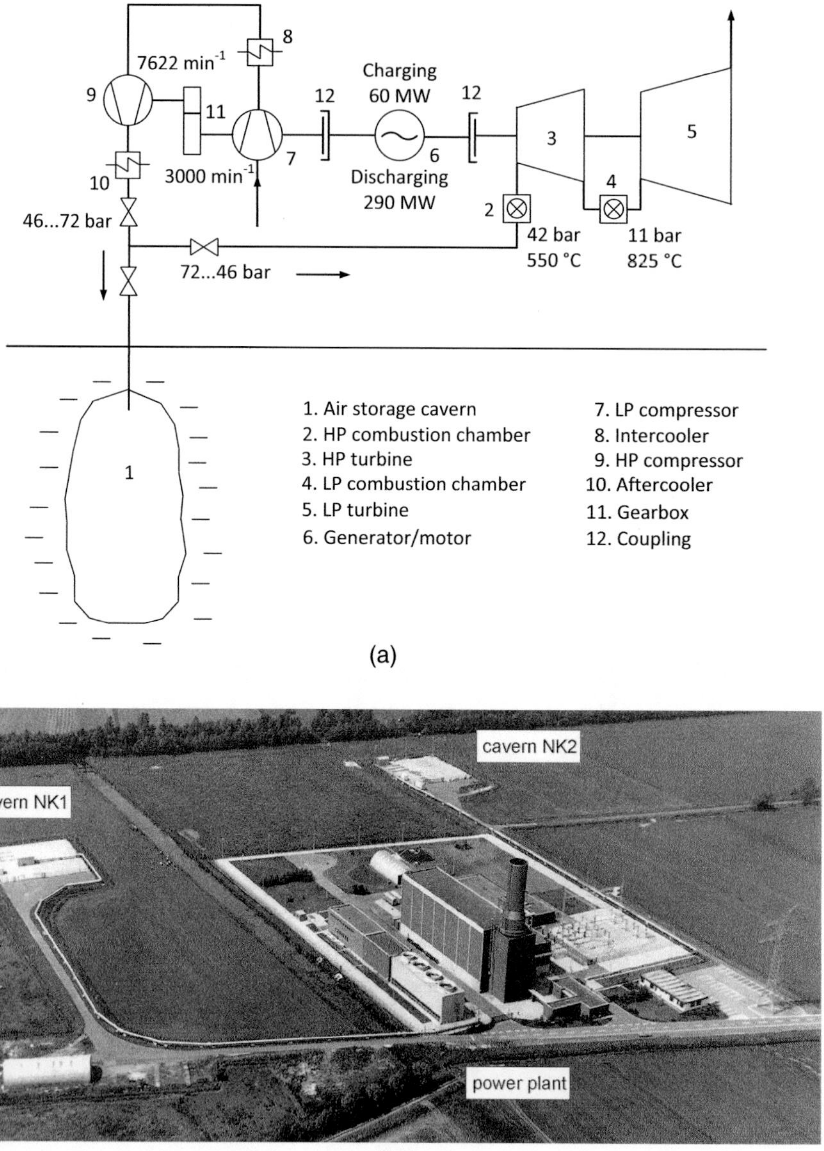

FIGURE 6.3 Classical CAES Huntorf (Brown-Boveri, Mannheim, Germany/Alstom). (a) Schematic representation of the plant and (b) air view of the plant.

6.1.3 Motivation

Electrochemical batteries represent the most-used electrical energy storage devices. However, they suffer from a limited number of cycles due to non-reversible aging phenomena. In addition, they have to face the problem of recycling sensitive materials. As a possible alternative, the development of CAES systems goes in the direction of reversible physics where longer life-cycles can be reached, and no materials are used which are difficult to recycle. Most CAES systems are composed of subsystems or components that are individually repairable or exchangeable.

6.2 ELEMENTARY PROPERTIES AND SUBSYSTEMS

6.2.1 Physical Principles

In the general scheme of CAES represented in Figure 6.1, two main types of components are used, compression/expansion machines where mechanical work is the main input vector and storage vessels where the mechanical work is equal to zero. The two types of components are represented in Figure 6.4a and b.

Both components can be considered separated control volumes and respond to the rate form of the First Law of thermodynamics.

$$\dot{W}_i = \dot{Q}_e + \dot{m}_e \cdot h_e - \dot{m}_i \cdot h_i + \frac{dU}{dt} \tag{6.2}$$

where

$\dot{W}_i$ and $\dot{Q}_e$ are the work and heat flows transferred to the gas from the external
$\dot{m}_i$ and $\dot{m}_e$ are the input and exit mass flows
h_e, h_i are the corresponding specific enthalpies [J/kg]

The compression and expansion machine's main function is to change the thermodynamic state of the gas inside the control volume and further to maintain the input and exit flowrates.

The reservoir itself is characterized by the absence of work transferred to the gas ($\dot{W} = 0$).

The description of a compression machine is based on the assumption of the air being considered an ideal gas $\left(PV = mRT\right)$ and further on the basic relations for the work and the heat

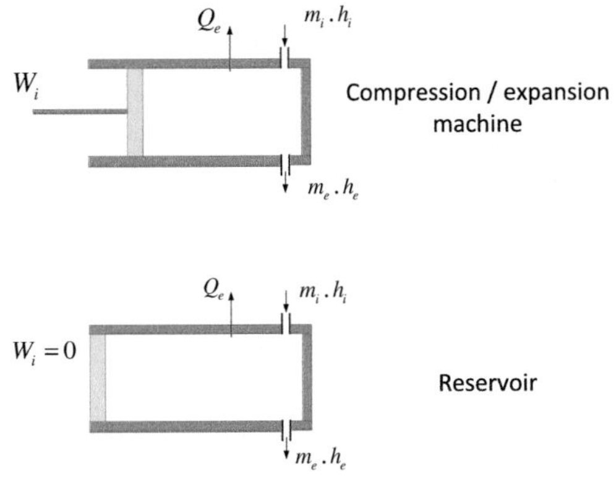

FIGURE 6.4 Main components of a CAES.

$$W = \int p dV \tag{6.3}$$

$$\dot{Q} = H_c A_c (T - T_w) \tag{6.4}$$

where

H_c is the heat transfer coefficient
A_c is the cylinder surface area exposed to convection
T_w is the temperature of the surface area
T is the instantaneous gas temperature

For a compressor, assuming steady state conditions where no energy is accumulated in the device, the following relation can be written as a combination of the first and second laws of thermodynamics [2, 13].

$$\dot{W} = (1 - \frac{T_0}{T})\dot{Q}^- + \dot{m} \cdot \psi \tag{6.5}$$

where Ψ is the flow energy defined by

$$\psi = (h - h_0) - T_0(s - s_0) \tag{6.6}$$

where h and s are the specific enthalpy and entropy, respectively, and the subscript 0 indicates that the properties are taken at reference temperature and pressure $\left(T_0 = 20°\text{C}, p_o = \ 1\text{bar}\right)$.

The exergy flow (usable energy) of the produced air stream is then expressed as

$$\dot{X} = \dot{m}[h - h_0 - T_0(s - s_0)] \tag{6.7}$$

In the case of an ideal gas flow

$$h - h_0 = c_p(T - T_0) \tag{6.8}$$

$$s - s_0 = c_p \ln\frac{T}{T_0} - R\ln\frac{p}{p_0} \tag{6.9}$$

The air stream exergy can be split into two parts, the pneumatic part and the thermal part, as follows

$$\dot{X} = \dot{X}_{(pn)} + \dot{X}_{(th)} \tag{6.10}$$

where

$$\dot{X}_{(pn)} = \dot{m}RT_0 \ln\frac{p}{p_0} \tag{6.11}$$

$$\dot{X}_{(th)} = \dot{m}c_p\left(T - T_0 - T_0 \ln\frac{T}{T_0}\right) \tag{6.12}$$

The compression work can lead to various ratios between pneumatic and thermal parts, according to various processes such as isothermal, adiabatic or polytropic ones.

The compressed air is stored in the reservoir, where the thermal exergy can be further dissipated by cooling down, leading to a decrease of the pressure [2].

The final pressure in the reservoir becomes

$$p_2 = p\frac{T_0}{T} \tag{6.13}$$

The exergy of the compressed air in the reservoir after reaching the thermal equilibrium with ambient can be obtained by integrating the recoverable exergy of the air stream after cool down

$$\dot{X}_{(pn)2} = \dot{m}RT_0 \ln\frac{p}{p_0}\left(\frac{T_0}{T}\right) \tag{6.14}$$

as

$$\begin{aligned} X_2 &= \int_{p=p_0}^{p=p_2} \dot{X}_{(pn)2}dp = \int_{p=p_0}^{p=p_2} \dot{m}RT_0 \ln\frac{p}{p_0}\left(\frac{T_0}{T}\right)dp = \\ &p_2V\left(\frac{1}{p_2}+\ln(p_2)-1\right) = m_1RT_0\left[\ln\left(\frac{p_2}{p_0}\right)+\frac{p_0}{p_2}-1\right] \end{aligned} \tag{6.15}$$

Isothermal compression and expansion represent the best conditions for storage, because the thermal energy exportation towards the external atmosphere allows, in addition to the high efficiency, a greater energy density inside of the storage vessel.

As shown in Figure 6.5, the phenomenon can be easily understood from the example of a hand pump with the orifice closed [1]. By adding a supplementary pressure of 1.5 bar, from a stabilized initial condition of 1 bar of pressure in a unity volume at 15°C, the piston will go down, reducing the volume to a new value of 0.5, while the temperature increases up to 101°C in totally adiabatic conditions (first step).

When in the second step the thermal energy corresponding to the added mechanical work is exported through the external ambient air by new temperature stabilization on 15°C, the volume will be reduced again down to 0.4. The pressure is kept on the previous level of 2.5 bar. Then, by removing the additional pressure (third step), the piston comes up again, bringing the volume up to 0.77. During this expansion, adiabatic conditions would decrease the temperature down to –51°C.

The temperature equalization (fourth step) will then increase the volume from 0.77 up to its initial value of 1. The match of the initial pressure and temperature conditions indicates that the energy added by isothermal compression can be recovered by isothermal expansion. The energy efficiency is then ideally equal to 1.

During isothermal compression, the internal energy of the cylinder has not been changed. This means that in reality, the energy is stored in the form of heat in the surrounding during compression and restored during expansion. The quality of heat transfer is therefore a key element for the performances of the storage process [1, 2].

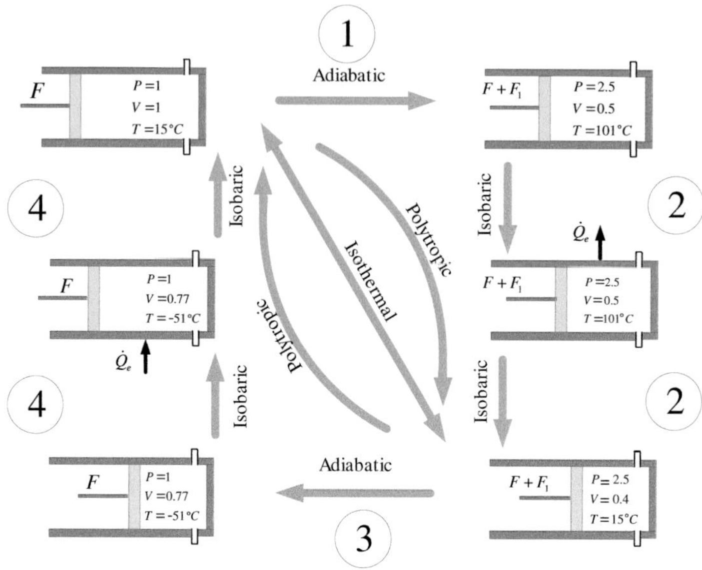

FIGURE 6.5 The basic compression and expansion cycle.

6.2.2 Adiabatic, Polytropic and Isothermal Compression and Expansion

Adiabatic, Polytropic and Isothermal compression/expansion have different energetic properties and can be described by the elementary formulations as

$$\begin{aligned} &\text{Adiabatic} && PV^k = \text{constant} \\ &\text{Polytropic} && PV^n = \text{constant} \\ &\text{Isothermal} && PV = \text{constant} \end{aligned} \tag{6.16}$$

where $n = k = c_p / c_v$

6.2.2.1 Compression

Significant energy is accumulated through increasing the gas pressure. In Figure 6.6, the dark grey surface represents the spent energy for bringing the thermodynamic state of the gas from its initial volume V_i to the final one V_f when the compression is realized in isothermal conditions. The surfaces under the mid-grey (polytropic) and light grey (adiabatic) curves are larger, and therefore the compression work is higher than under isothermal conditions.

The adiabatically reached point (2) from state (1) corresponds to a state with increased temperature.

After such a compression, the cooling down to the initial temperature (as at point 1) brings the system to point (3), where the stored energy is identical to the energy accumulated by isothermal compression.

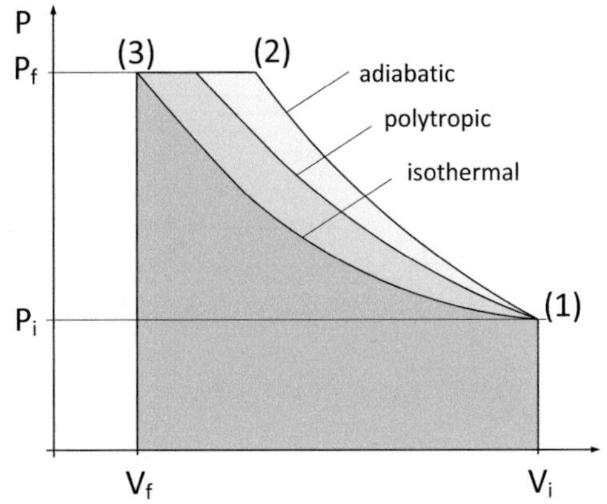

FIGURE 6.6 Compression in adiabatic, polytropic and isothermal conditions.

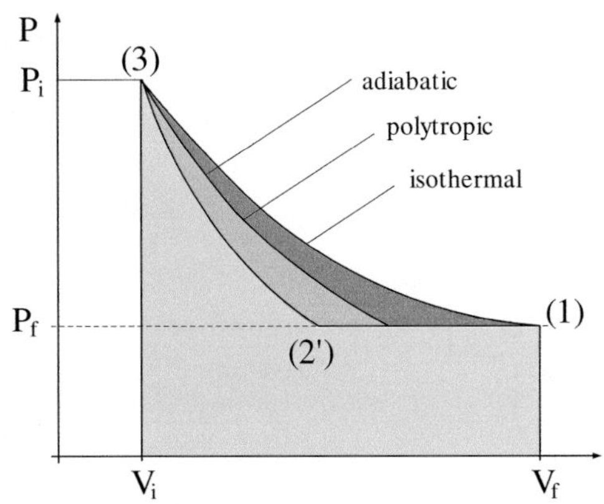

FIGURE 6.7 Expansion in adiabatic, polytropic and isothermal conditions.

6.2.2.2 Expansion

In Figure 6.7, the dark grey surface represents the recovered energy from bringing the thermodynamic state from V_i to V_f when the expansion is realized in isothermal conditions. The surface under the mid-grey (polytropic) and under the light grey (adiabatic) curves is smaller, and therefore the expansion work is smaller than under isothermal conditions.

The adiabatically reached point (2') from state (3) corresponds to a state with decreased temperature. After such an expansion, the heating up of the gas to the initial temperature (as at point 3) brings the system to point (1), where the remaining energy is identical to that resulting from an isothermal expansion [2].

6.2.3 From Single Stage to Multistage

In Figure 6.8, the same thermodynamic cycle is represented as in the previous paragraph. The cycle is realized with only one compression stage.

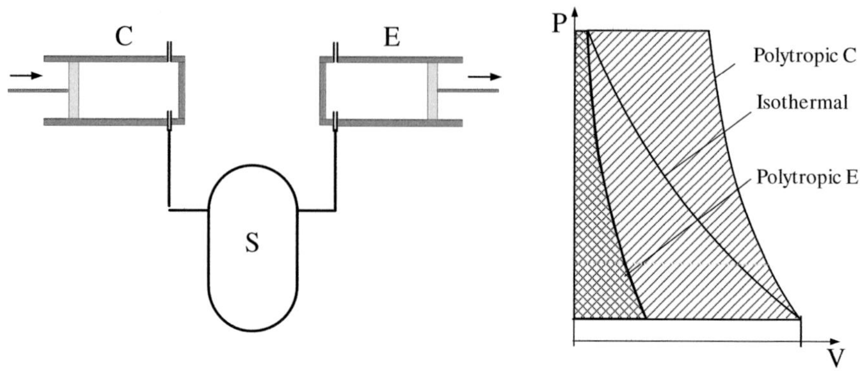

FIGURE 6.8 Compression and expansion with one stage.

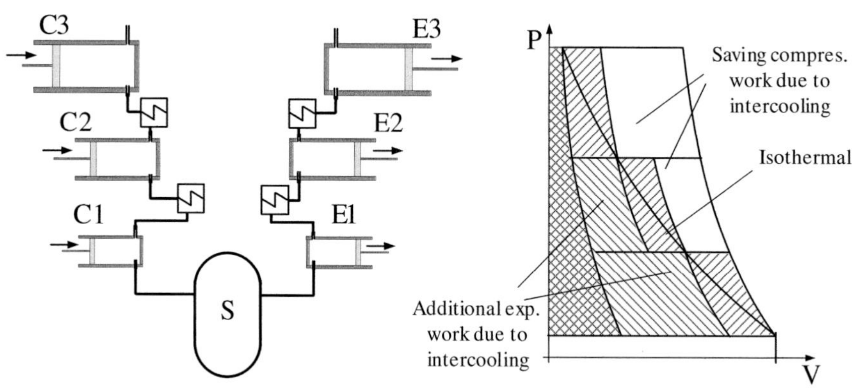

FIGURE 6.9 Compression and expansion with a multiple-stage machine and intercoolers.

In Figure 6.9, the same cycle is run again but using a three stage compression machine with intercoolers between the stages. This allows it to save some work during compression and to recover more work during expansion.

6.2.4 Higher Storage Pressure for Increased Energy Density

For a given storage device, it is interesting to reach a sufficiently high value of energy within a given space. Section 3.2 defined the energy density. In Figure 6.10, the volumetric energy density of a CAES device is represented as a function of the pressure in the reservoir [1].

The parameter n indicates the conditions of compression and expansion, namely if they are isothermal, polytropic or adiabatic conditions.

As a reference, the value of 50 kWh/m^3 is chosen (typical lead acid batteries), and the diagram shows that a storage pressure of 300 bar should be chosen.

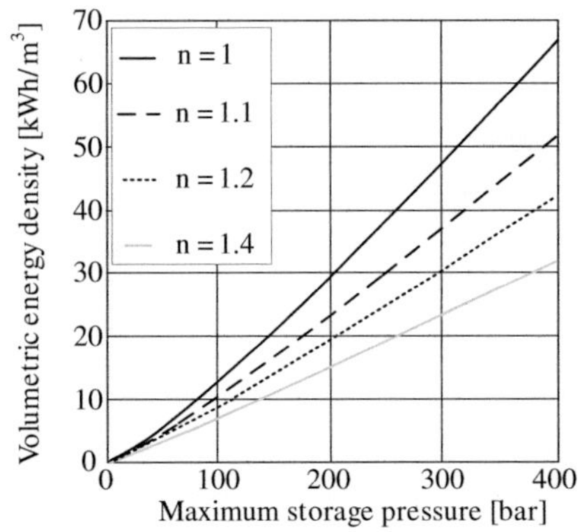

FIGURE 6.10 Energy density versus storage pressure.

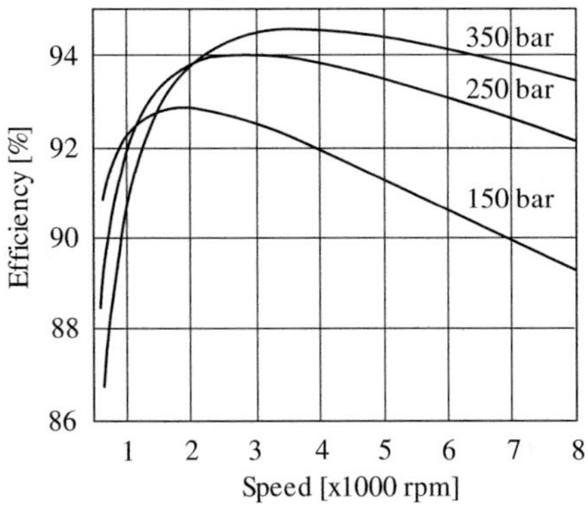

FIGURE 6.11 Efficiency curve of a hydraulic pump. (Adapted from Lemofouet, S. and Rufer, A., Hybrid energy storage systems based on compressed air and supercapacitors with maximum efficiency point tracking, *2005 European Conference on Power Electronics and Applications*, Dresden, Germany, September 11–14, 2005, p. 10, DOI: 10.1109/EPE.2005.219203.)

The polytropic factor *n* also indicates that the lost heat influences the final energy density as defined in Chapter 3.

6.2.5 Concept of the Liquid Piston: From "Pneumatic" to "Hydropneumatic"

In section 6.2.4, the dependency of the storage pressure was shown. According to the curves represented in Figure 6.10, pressure levels of hundreds of bars should be chosen for acceptable volumetric energy density.

Pneumatic compressors, even if they are realized with plural stages, have limited performance with regard to isothermal performance. Then, as an alternative solution for a compression machine, a hydraulic pump is proposed for achieving a pressure increase under very good efficiency conditions. Figure 6.11 shows the efficiency performance of a conventional hydraulic pump [14].

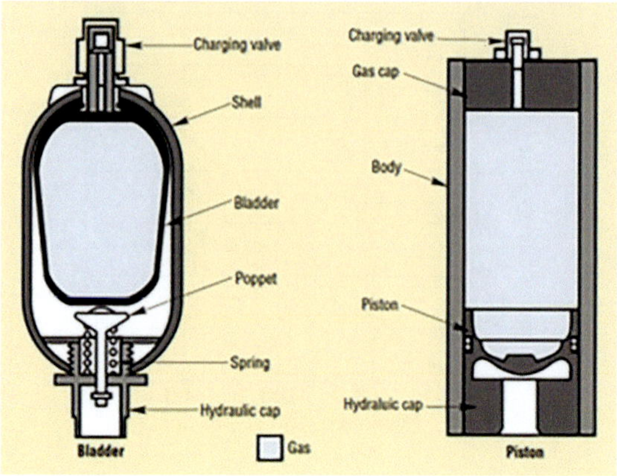

FIGURE 6.12 Hydraulic accumulators: (a) bladder and (b) cylinder and piston.

In fact, the idea of the liquid piston comes from the classical hydraulic accumulator (Figure 6.12) where a pre-charged volume of nitrogen is compressed by a piston or within a bladder and where the input power comes from a high-pressure hydraulic circuit. Hydraulic accumulators have no heat exchange equipment and are suited to gas temperature variations.

The first limitation of the hydraulic accumulator comes from the closed and limited volume of compressed gas, this volume reaching its smallest value at the highest pressure at the fully charged state. In order to elevate the energy density, the gas (air) circuit must be opened, taking the air from atmospheric pressure and compressing it into a vessel of constant volume. The principle of such a storage system is discussed in the next section.

6.2.6 Static and Dynamic Reservoir Systems

CAES systems use high-pressure vessels as energy carriers, which are of very different sizes and shapes. Normally the reservoirs have a constant volume, but it can also be realized as flexible under water envelopes, as will be described later in Section 6.5. One of the main advantages of UWCAES is that they are operated under constant pressure conditions given through the depth of immersion. Between the constant volume reservoir and the variable volume/constant pressure reservoir, an alternative solution of a kind of hybrid storage has been proposed [15]. Such a system, called a “dynamic reservoir”, combines accumulation in a variable air volume and in an integrated mechanical spring compressed by a moving piston. In opposition to the dynamic reservoir, a constant volume accumulator is called a “static reservoir”.

In a dynamic reservoir system, the air injected from the compressor moves the piston and compresses the spring where an additional amount of energy is stored. Near the solution of the static reservoir represented in Figure 6.13a, a schematic representation of a dynamic reservoir is given in Figure 6.13b. Figure 6.13c shows the same system in the full state of charge where the space occupied by the spring in its

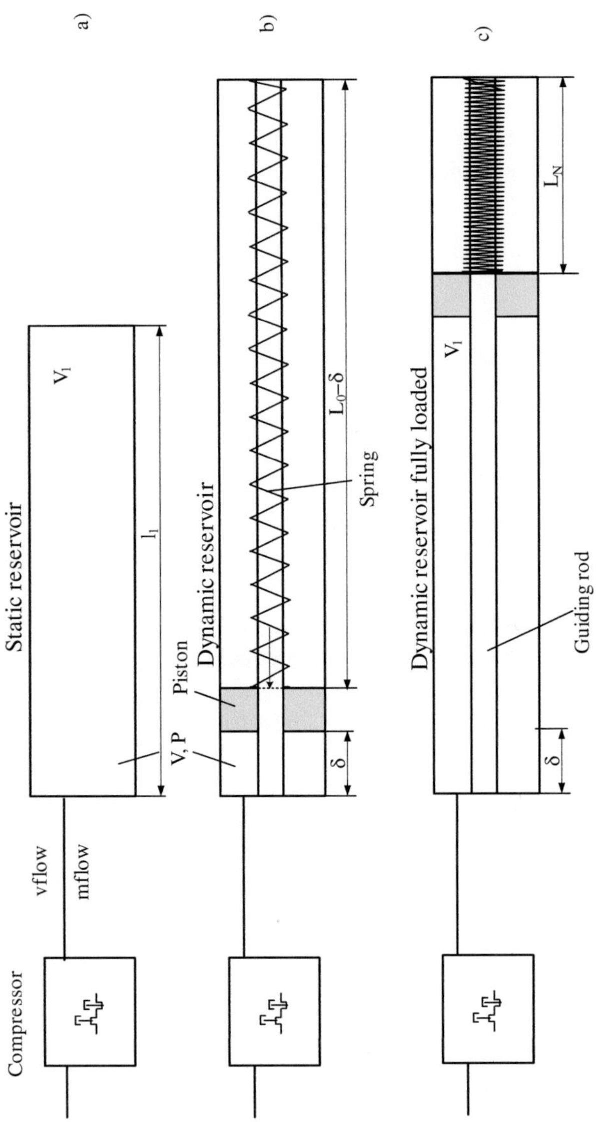

FIGURE 6.13 Static (a) and dynamic (b) compressed air reservoirs; (c) dynamic reservoir fully loaded.

compressed state becomes evident. The two cylindric volumes have the same section. The volume of the dynamic reservoir is equal to the volume of the static reservoir when the position of the piston d reaches the value of the length l_1 of the static reservoir. When the piston is at this position, the spring exerts a force equal to the pressure P multiplied by the piston's surface A.

$$F_{spring} = P \cdot A = k \cdot l_1 \tag{6.17}$$

The pressure P appears in the volume of the cylinder due to the injection of air by the compressor. The pressure depends on the mass of air inside the volume, and its value is given by Equation 6.18:

$$P = m \cdot \frac{RT}{V} \tag{6.18}$$

where
m is the time integral of the massflow imposed by the compressor

$$m = \int_0^t mflow \cdot dt \tag{6.19}$$

6.2.6.1 The Model of the Compressor

The compressor is considered a source of mass in this modeling and is characterized by its nominal volume flowrate $\dot{V}_{nom}$. The nominal volume flow rate $\dot{V}_{nom}$ is expressed in m^3/s from its given value in L/min. The value of 190 l/min is chosen in the considered example and corresponds to the characteristics of the example described in [15].

$$\dot{V}_{nom} = 190l / \min \cdot 10^{-3} / 60 = 0.00316 m^3/s \tag{6.20}$$

The real volume flowrate $\dot{V}_{real}$ is of a different value due to the effect of the expansion of the dead volume after the end of the discharge when the piston goes in its return stroke. The real volume flowrate is calculated with the help of a so-called volumetric yield η_V defined as

$$\eta_V = 1 - V_r \cdot C_r \tag{6.21}$$

where
V_r is the ratio of the top volume to the bottom volume of the compression cylinder

$$V_r = V_{top} / V_{bot} \tag{6.22}$$

And C_r the compression ratio or the ratio of the discharge pressure to the suction pressure and varies from 1 to $P_{\max} / P_{atm}$ during a full charging process.

$$C_r = P_{dis} / P_{suc} \tag{6.23}$$

The real volume flowrate becomes

$$\dot{V}_{real} = \dot{V}_{nom} \cdot \eta_v = \dot{V}_{nom}(1 - V_r \cdot C_r) \tag{6.24}$$

Then, the mass flow rate is calculated as

$$\dot{m} = \dot{V}_{real} \cdot \rho_0 \tag{6.25}$$

Numerically, the nominal mass flowrate of the example is calculated

$$\dot{m}_{nom} = \dot{V}_{nom} \cdot \rho_0 = \frac{190l/\min \cdot 0.001\,\mathrm{m}^3/l}{60s/\min} \cdot 1.2kg/m^3 = 0.00316m^3/s \cdot 1.2kg/m^3 = $$
$$= 0.0038kg/s \tag{6.26}$$

The effect of the expansion of the dead volume is illustrated by the P-V diagram of Figure 6.14. The piston of the compressor starts to move from the position with the largest volume ($V_{cyl} = V_{bott}$), and the pressure is equal to P_{suc} (1), (1'). Then, when the pressure reaches the discharge value (P_{dis}, P'_{dis}) (2), (2'), the discharge valve opens and the air is exhausted, the volume decreases down to $V_{top} = V_d$. During the return stroke the suction valve does not open until the pressure of the remaining air in the dead volume decreases under the value of the suction pressure P_{suc}. The effective suction volume becomes

$$V_{cyl} - V_d * C_r,$$

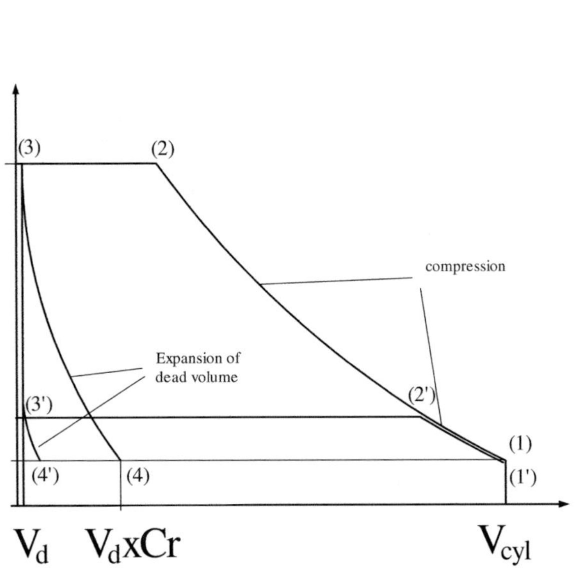

FIGURE 6.14 Effect of the expansion of the dead volume.

Leading to the definition of the volumetric yield η_V

$$\eta_V = 1 - V_r \cdot C_r \tag{6.27}$$

6.2.6.2 Modeling the Two Reservoirs

6.2.6.2.1 Parameters of the Reservoirs

This section will first describe the behavior of static and dynamic reservoirs as described in [15]. The parameters of these examples are listed in Table 6.1.

6.2.6.3 Structural Diagram of the Static Reservoir System

Figure 6.15 gives the structural diagram of the static reservoir system. The real mass flowrate provided by the compressor is equal to the product of the real volume flow multiplied by the density of the fluid. Then the mass of air accumulated in the reservoir corresponds to the time integral (block 1/s) of the mass flowrate. The pressure in the reservoir is then calculated from the accumulated mass of air with an ideal gas law considering the volume of the reservoir and the temperature of the air. The pressure P_{dis} is fed back to the calculation of the compression ratio Cr used for the volumetric yield.

TABLE 6.1
Parameters of the reservoirs

Static reservoir	
Volume	$0.1m^3$
Dynamic reservoir	
Variable volume	$0–0.1m^3$
Stroke of the piston	0.9m
Surface of the piston	$0.111m^2$
Constant of the spring	61.7kN/m
Moving mass	10kg

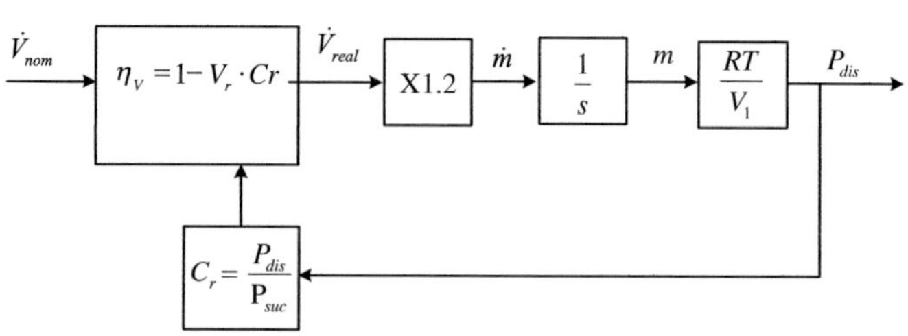

FIGURE 6.15 Structural diagram of the static reservoir system.

6.2.6.4 Structural Diagram of the Dynamic Reservoir System

In the dynamic reservoir system, the pressure exerted on the mobile piston by the injected air produces a displacement of it and a compression of the spring. The system comprises two different energy accumulation elements. First, energy is stored in the form of compressed air in the variable volume of the cylinder and second in the tension of the spring. The calculation of the piston's position and of the air accumulation volume is realized according to a dynamic model for the association of the mass of the piston and of the spring. The energy accumulated in the form of compressed air is calculated with the same model as for the static piston system described in the previous section, namely via the accumulated mass of air in the chamber and an associated state equation. In this case the accumulation volume varies with the position of the piston.

6.2.6.4.1 Model of the Spring-Mass System

The description of the movement of the piston and its impact on the compression of the spring can only be made by using a model based on integral causality [15, 16]. So the piston and the storage spring are modeled as a classical damped mechanical oscillator as represented in Figure 6.16. The spring characteristic and the corresponding differential equation can be written as:

$$F_{spring} = Kx \tag{6.28}$$

$$m\ddot{x} = F_{therm} - Kx - \mu\dot{x} \tag{6.29}$$

m is the mass of the mobile piston.

The force caused by a thermodynamic effect F_{therm} is the difference of the force caused by the pressure in the cylinder's chamber (absolute value) on the front side of the piston, minus the effect of the atmospheric pressure on the rear side of the piston. K is the spring constant and μ the velocity-dependent friction coefficient [17].

6.2.6.4.2 Structural Diagram

The structural diagram of the dynamic reservoir system is given in Figure 6.17. The compressor model is identical to the model used previously and uses the principle of the volumetric yield. The state equation considers the variable volume depending on the piston's position. The diagram includes the damped mass-spring model.

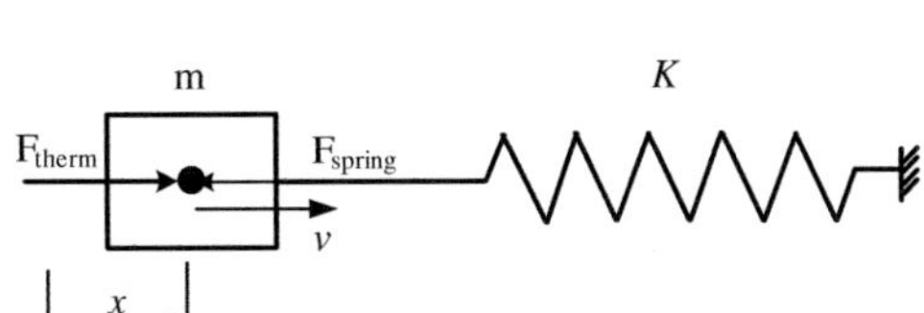

FIGURE 6.16 Schematic representation of the mass-spring system.

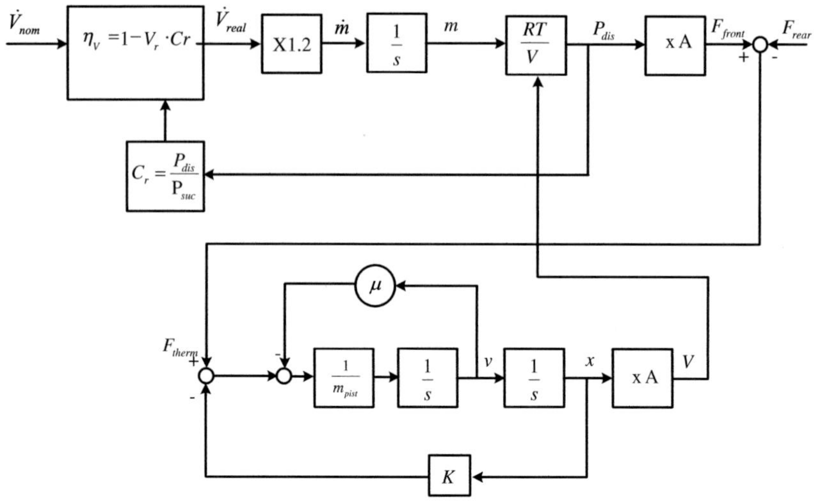

FIGURE 6.17 Structural diagram of the dynamic reservoir system.

6.2.6.5 Energies Accumulated

6.2.6.5.1 Volume of Air

The energy content of the volume of air depends on its volume, pressure and temperature. In the present system where the piston moves slowly, the thermodynamic phenomenon can be considered to evolve according to an isothermal characteristic. In the full state of charge, the pressure is 6 bar in a volume of 0.1 m^3. The corresponding stored energy is calculated as

$$E_{stat} = P \cdot V \left(\ln \frac{P}{P_{atm}} - 1 + \frac{P_{atm}}{P} \right) = 6 \cdot 10^5 \cdot 0.1 \left(\ln \frac{6}{1} - 1 + \frac{1}{6} \right) = \quad (6.30)$$

$$= 6 \cdot 10^5 \cdot 0.1 \cdot 0.958 = 57505J$$

This value is in the air volume in both cases, constant volume and variable volume when it reaches its maximum (0.1 m^3) and when the pressure is identical. The simulation of the charging process presented in the next section will show that this state of the variable volume is reached for t = 340 s (Figure 6.12). It is also the time where the pressure of 6 bar is reached. For the constant volume system, the pressure of 6 bar is reached after a time of 254 s. This confirms that the spring model and parameters are right.

6.2.6.5.2 Energy Accumulated in the Spring

$$E_{spring} = \int P \cdot dt = \int F \cdot \dot{x} \cdot dt = \int k \cdot x \cdot \dot{x} \cdot dt \quad (6.31)$$

$$\text{With } (uv)' = u'v + uv' \quad u = v \quad (6.32)$$

$$\int (u \cdot u)' = u^2 = \int u'u + uu' = 2\int uu' \quad (6.33)$$

$$\text{Then } E_{spring} = k\int x \cdot \dot{x} \cdot dt = \frac{kx^2}{2} \tag{6.34}$$

When $k = 61.7\text{kN/m}$ and $\text{x} = 0.9$

$$E_{spring} = \frac{kx^2}{2} = \frac{61.7 \cdot 10^3 N/m \cdot (0.9m)^2}{2} = 24.9kJ \tag{6.35}$$

6.2.6.6 Simulation Results

The structural diagrams of the static and dynamic reservoir systems have been implemented in a numeric simulation (Simulink) and produce the evolution of the different variables as functions of the time. In the static reservoir the accumulated mass of air and the pressure in the reservoir are relevant. For the dynamic reservoir additional variables such as the force exerted on the piston, its displacement and the variable volume are considered.

6.2.6.7 Evolution of the Pressures and Masses of Air

The evolution of the pressure in the static and dynamic reservoirs is presented in Figure 6.18. The blue curve shows the evolution of the pressure in the constant volume system and the red one the pressure in the variable volume with spring.

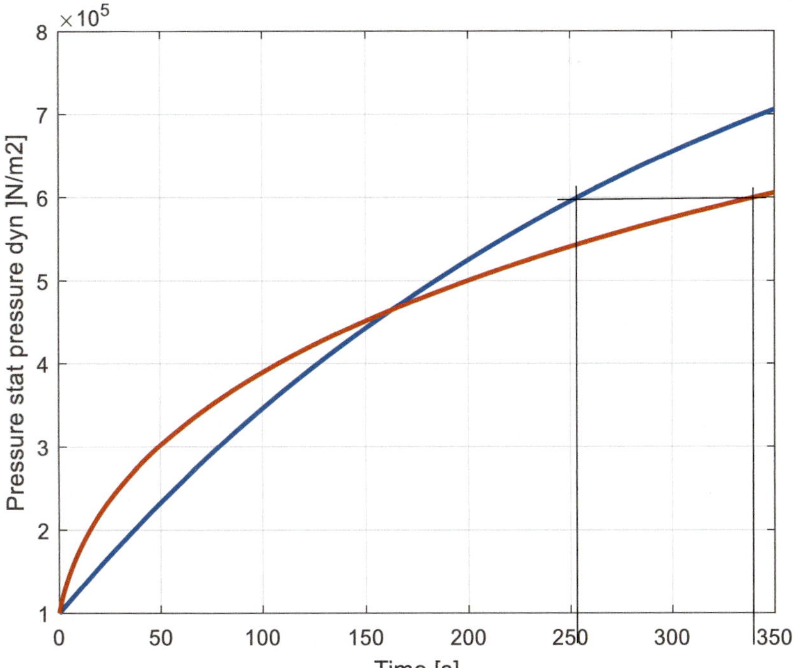

FIGURE 6.18 Pressure in the reservoirs (constant volume: blue, variable volume: red).

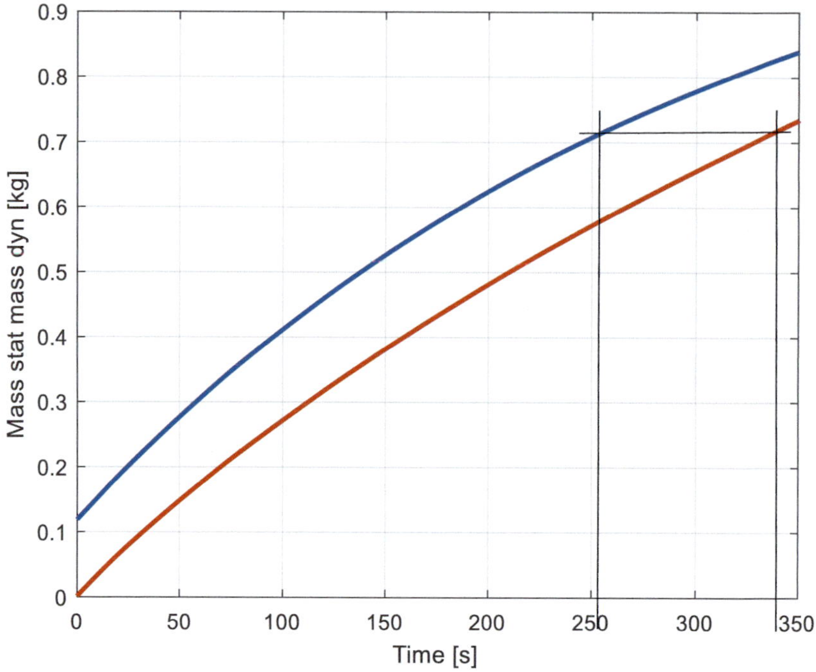

FIGURE 6.19 Mass of air accumulated in the two reservoirs (blue: constant volume, red: variable volume.

The evolution of the pressures is the result of the accumulated amount of air in the volumes (Figure 6.19) due to the injection by the compressor of the corresponding massflows (Figure 6.20). The blue color is again associated with the constant volume system and the red with the variable volume. The same masses of air are accumulated at the corresponding instants where the pressures are identical (254 s and 340 s).

6.2.6.8 Evolution of the Mechanical Variables

In opposition to the static reservoir where only the pressure changes, the geometry of the dynamic reservoir evolves according to the change of the pressure. The elevation of the pressure in the variable volume cylinder produces a directly proportional force. The evolution of the force as a function of time is represented in Figure. 6.21. From this force the characteristic of the spring produces a displacement of the piston which further increases the active volume of the cylinder. The variation of the position of the piston is also directly proportional to the force as the variation of the volume. The corresponding curves are given in Figure 6.22 and 6.23. In these figures one can see that the values reached at the time where the pressure is equal to 6 bar (340s) correspond to the correct design of the system (0.9 m and 0.1 m^3)

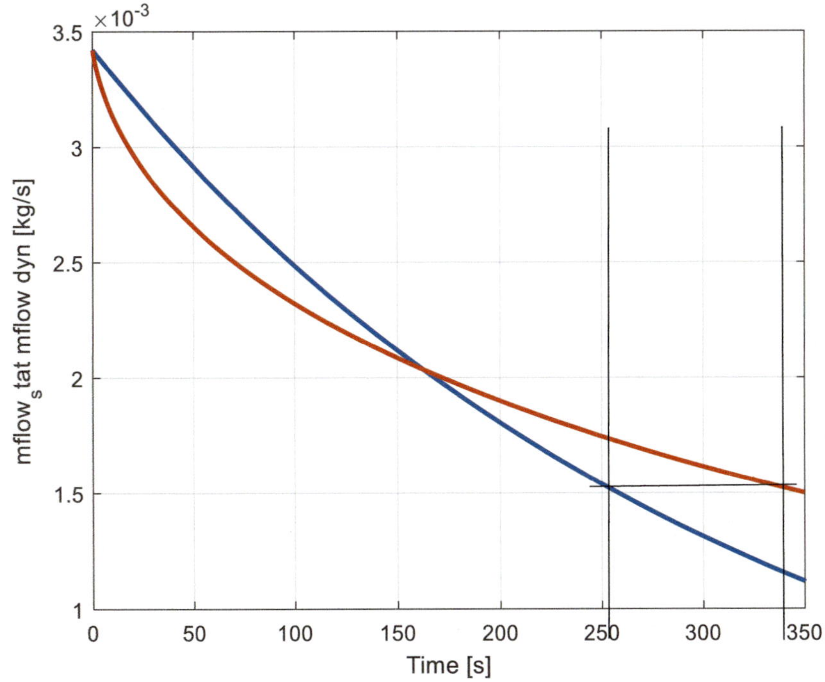

FIGURE 6.20 Massflows to the reservoirs (blue: constant volume red: variable volume).

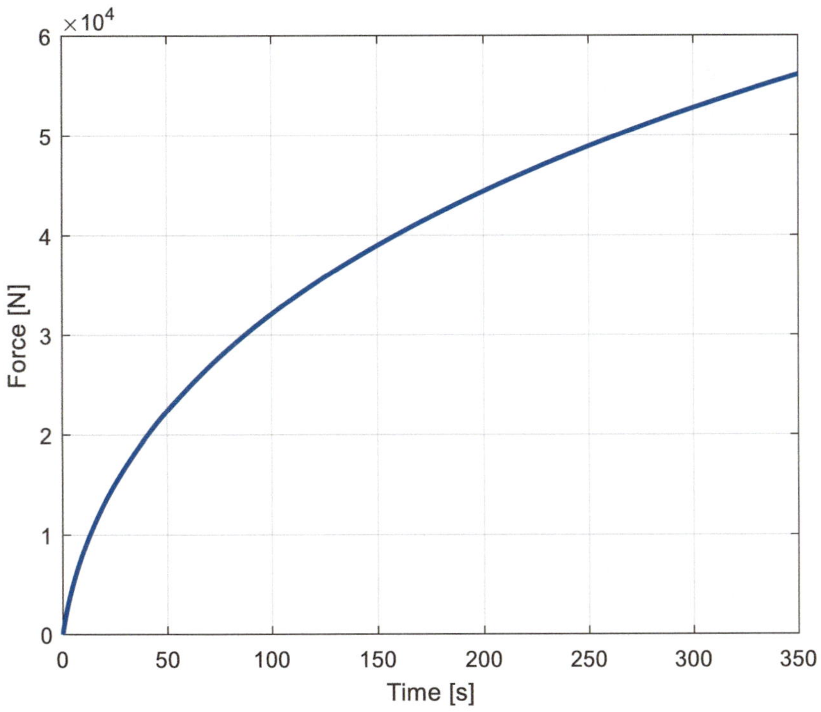

FIGURE 6.21 Evolution of the force as a function of time.

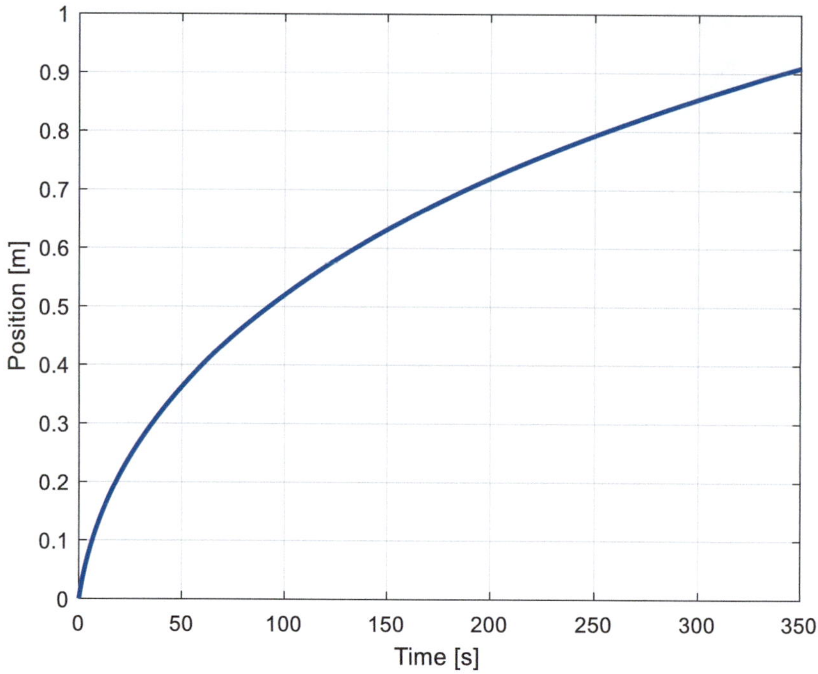

FIGURE 6.22 Position of the piston.

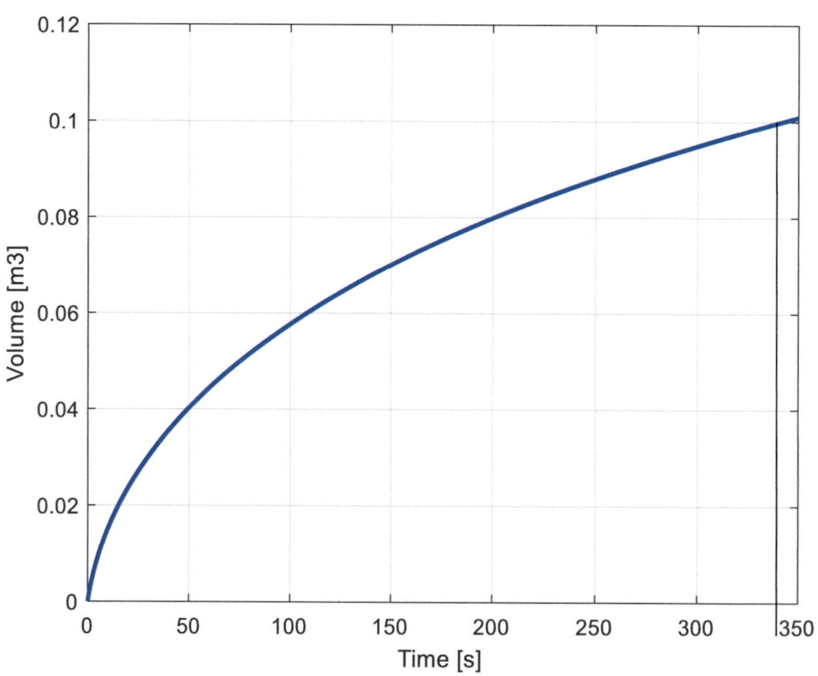

FIGURE 6.23 Volume of the dynamic reservoir.

6.2.6.9 Evolution of the Energies Stored in the Systems

In the static reservoir, energy is stored only in the volume of air. Concerning the dynamic reservoir, energy is stored in the volume of compressed air and in addition in the loaded spring. The different values of stored energy are indicated in Figure 6.24. The content in the static reservoir with constant volume is indicated in blue. The value of 57,505 J (Equation 6.30) is reached at the time t = 242 s. The same value is reached for the variable volume of the dynamic reservoir at the time t = 340 s (red curve). The energy stored in the spring reaches the value of 24.9 kJ at the same time (black curve). Finally, the total amount of stored energy in the dynamic reservoir (air and spring) is represented by the yellow curve. At the full state of charge the dynamic reservoir has accumulated the value of

$$E_{dyn} = E_{air} + E_{spring} = 57.5kJ + 24.9kJ = 82.4kJ \tag{6.36}$$

This value represents an increase of 43% referred to the static reservoir.

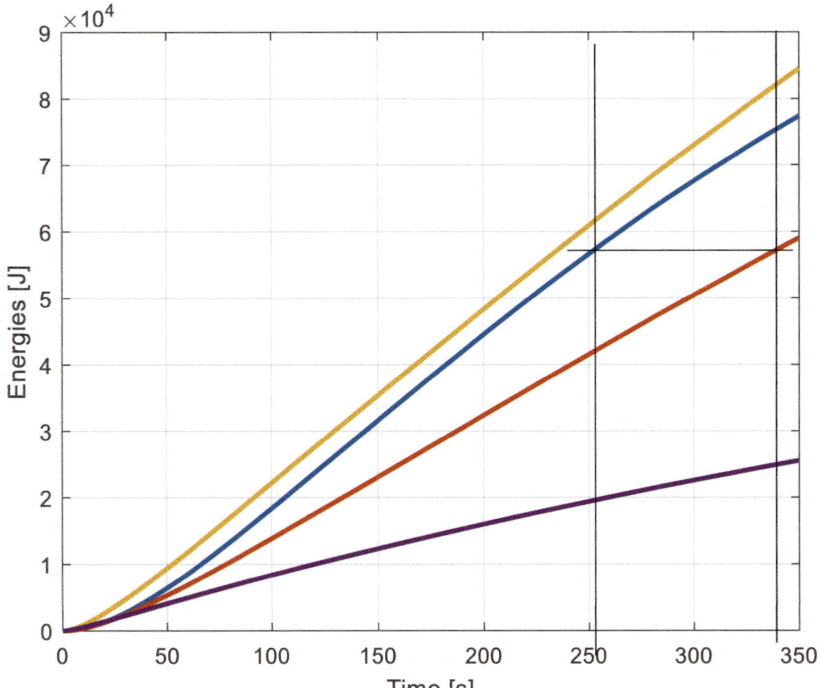

FIGURE 6.24 Energies accumulated in the static and dynamic reservoirs. Compressed air energy in the fixed reservoir (blue curve); compressed air energy in the variable volume (red curve); Energy accumulated in the spring (black curve); total energy of the dynamic reservoir (yellow curve).

6.2.7 Completed State Equation Models for the Simulation of Compression and Expansion

For nearly all compression and expansion process, simplified models have been proposed on the basis of relations valid for ideal gases. This assumption can be used for a gas like air, but the question of the validity of the models for other real gases like hydrogen is open. The reflections of this section anticipate the study of Chapter 9 dedicated to hydrogen storage, especially in the form of a pressurized gas. For the modeling of compression and expansion, the behavior of a piston-cylinder assembly is analyzed with different equations of state such as the Van der Waals model or other multiple parameter polynomial models for the evaluation of the resultant pressure and temperature. Sophisticated models available in powerful computer programs give accurate results regarding the thermodynamic states of different gases [18], but simple formalisms like the classical Van der Waals equations can be simply integrated in a simulation model where the mechanical variables of compression/expansion cylinders and of the coupled driving mechanisms are of major interest. The thermal behavior of the gas can be modeled using a complementary thermal model inserted in the loop of the state equation model evaluating the gas temperature.

6.2.7.1 Models of Compression

For the simulations of the compression, a specific piston-cylinder is used. The comparison study is based on an implementation of the physical elements and of the different thermodynamic state equations in a dynamic time-based representation where the velocity of the piston is imposed (Simulink). The dimensions and other parameters are overtaken from a previous study dedicated to hydrogen compression [19]. The simple piston-cylinder assembly is represented in Figure 6.25, together with the inlet pressure reservoir at 15 bar and the output receiver at 160 bar. Anti-return valves are placed between the reservoirs and the compression cylinder. *P*, *V* and *T* are the variables; *d* the inner diameter of the cylinder (27.5 mm); and *l* is the stroke (90 mm). The volume of the cylinder varies from the maximum value V_{cyl} at bottom dead center (BDC) down to the lowest value at the top dead center (TDC) with a velocity of vel = 0.18 m/s. The evolution of the volume in function of time is represented in Figure 6.26. A lower limit of the volume is introduced in order to avoid problems in the calculation of the volume ratio (no division by 0).

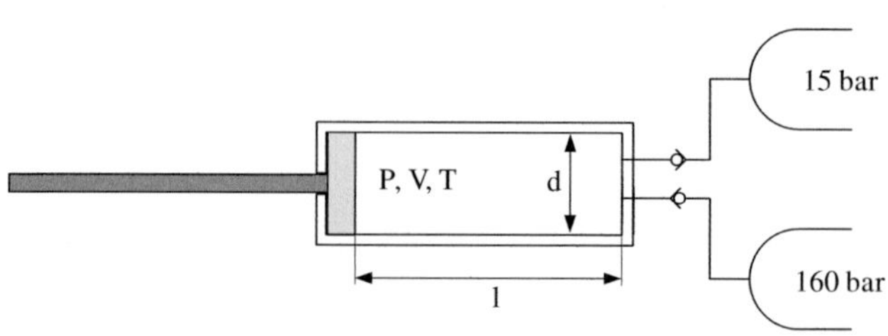

FIGURE 6.25 Compression piston-cylinder system.

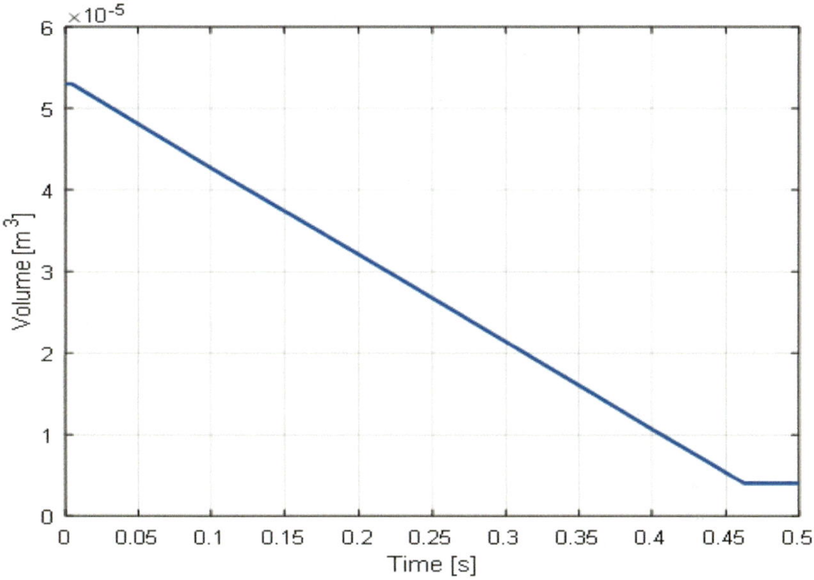

FIGURE 6.26 Evolution of the volume of the compression cylinder.

$$V_{cyl} = \pi \cdot d^2 / 4 \cdot l = \pi \cdot (27.5)^2 / 4 \cdot 90 = 53456 mm^3 \quad or \quad 53.5 \cdot 10^{-6} m^3 \tag{6.37}$$

$$vel = l / 0.5\text{s} = 0.090 m / 0.5 s = 0.18 m / s \tag{6.38}$$

A classical expression for the evolution of the pressure of an ideal gas depending on the variation of the volume is given in Equation 6.39. This expression has been used for the compression of air in adiabatic $(\gamma = 1.4)$ and in isothermal conditions $(\gamma = 1)$. The two curves are represented in Figure 6.27. According to the experimental set-up shown in Figure 6.25, the pressure in the cylinder varies from the intake level of 15 bar to the pressure in the output reservoir of 160 bar.

$$P_{gas} = P_{in\_gas} \left(\frac{V_{comp\_\max}}{V_{compr\_\text{var}}} \right)^{\gamma} \tag{6.39}$$

For a non-ideal gas like hydrogen, different equations of state can be used. A first relation directly deducted from the ideal gas law is used (Equation 6.40), and the corresponding curve is compared with the curve of the isothermal compression. The ideal gas law is a representation of the compression phenomenon supposing an inverse proportionality between pressure and volume.

$$P = \frac{R \cdot T}{v} = \frac{m \cdot R \cdot T}{V} \tag{6.40}$$

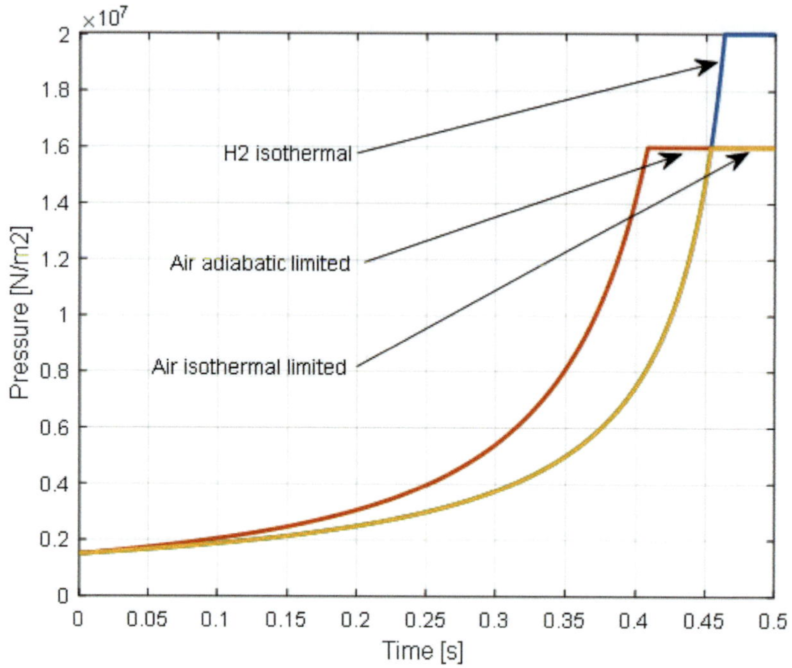

FIGURE 6.27 Comparison of compression curves.

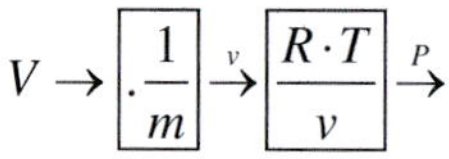

FIGURE 6.28 Functional diagram of the ideal gas law model.

In this expression for the pressure P, the mass volume v or the real volume V can be used. The functional diagram of Figure 6.28 is implemented for the representation of the evolution of the pressure.

In opposition to Equation 6.39 where only the initial pressure and the geometric (volumetric) conditions are considered, the ideal gas law uses the mass m of gas and the related gas constant R. The temperature is supposed to be known.

The mass of hydrogen sucked from the primary reservoir (15 bar) is calculated with the same relation (6.40) and with the parameters of the cylinder

$$m = \frac{P \cdot V}{R \cdot T} = \frac{15 \cdot 10^5 N/m^2 \cdot 53.5 \cdot 10^{-6} m^3}{4.124 kJ/(kg \cdot K) \cdot 293 K} = 0.664 \cdot 10^{-4} kg \tag{6.41}$$

the hydrogen-related constant R being:

$$R = 4.124 kJ/(kg \cdot K) \tag{6.42}$$

A compression curve of hydrogen is calculated according to Equation 6.40 under isothermal conditions and is represented in Figure 6.27. This curve (blue curve) is identical to the curve of the isothermal compression of air calculated with Equation 6.39 (yellow curve). The upper limit of the yellow curve is imposed as an upper pressure limit (160 bar). The blue curve represented reaches a different-upper limit. This value is related to the lowest value of the volume, as represented in Figure 6.26.

Two other equations of state are considered which are supposed to bring a more accurate representation of the evolution of the pressure. First the Van der Waals state equation is considered, and further a still more complex model is given by the Beattie–Bridgeman state equation [20].

The Van der Waals state equation is an improvement of the ideal gas equation where the attraction force between molecules is introduced $(\text{term}\, a / v^2)$. The volume occupied by the molecules (term b) is also considered. The expression for the pressure gives:

$$P = \frac{R \cdot T}{v - b} - \frac{a}{v^2} \tag{6.43}$$

The mass volume v is calculated from the system variable V (the real volume)

$$v = \frac{V}{m} \tag{6.44}$$

and the factors a and b are given through:

$$a = \frac{27 R^2 \cdot T_{cr}^2}{64 P_{cr}} \quad \text{and} \quad b = \frac{R \cdot T_{cr}}{8 P_{cr}} \tag{6.45}$$

a and b are calculated with the parameters of Table 6.2 [20].

The corresponding functional diagram of the Van de Waals model is represented in Figure 6.29.

TABLE 6.2
Constants of Hydrogen for the Van der Waals Model

	Formula	Molar Mass M Kg/kmol	Constant R kJ/kg*K	T_{cr} K	P_{cr} MPa	Vol m³/kmol
Hydrogen (normal)	H_2	2.016	4.124	33.3	1.3	0.0649

$$V \rightarrow \boxed{\cdot \frac{1}{m}} \xrightarrow{v} \boxed{\frac{R \cdot T}{v - b} - \frac{a}{v^2}} \xrightarrow{P}$$

FIGURE 6.29 Functional diagram of the Van der Waals model.

The state equation of Beattie–Bridgeman is based on five constants, which are determined experimentally. The expression of the pressure is given by Equation 6.46

$$P = \frac{R_u \cdot T}{\overline{v}^2}\left(1 - \frac{c}{\overline{v} \cdot T^3}\right)\left(\overline{v} + B\right) - \frac{A}{\overline{v}^2} \tag{6.46}$$

where $$A = A_0\left(1 - \frac{a}{\overline{v}}\right) \quad \text{and} \quad B = B_0\left(1 - \frac{b}{\overline{v}}\right) \tag{6.47}$$

The input variable of Equation 6.46 is the molar volume $\overline{v}$, which is calculated from the mass volume v according to

$$\overline{v} = M \cdot v = M \cdot V / m \tag{6.48}$$

where M is the molar mass

The universal gas constant Ru is recalculated from the specific gas constant R

$$R_u = M \cdot R \tag{6.49}$$

The parameters for hydrogen are given in Table 6.3 [20].

Finally, the implementation in the simulation is done according to the functional diagram of Figure 6.30.

In Figure 6.31, three compression curves for hydrogen are represented: an ideal gas law curve (red curve), a Van der Waals model curve (yellow curve) and a Beattie–Bridgeman model curve (blue curve). In this simulation, the pressure limits are given through the lower volume limit (see Figure 6.26). For pressure values under 50 bar, the difference between models is limited. Only for higher values of the pressure (above 50 to100 bar) does the difference become significant.

6.2.7.2 Calculation of the Temperature

For the calculation of the evolution of the temperature during compression, Equation 6.39 is completed by another analytic equation considering the variation of the volume (Equation 6.50):

TABLE 6.3
Constants for the Beattie–Bridgeman State Equation

	A_0	a	B_0	b	c
Hydrogen H_2	20.017	-0.00506	0.02096	-0.04359	504

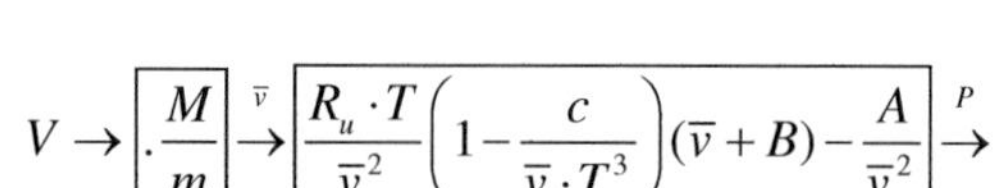

FIGURE 6.30 Functional diagram of the Beattie–Bridgeman model.

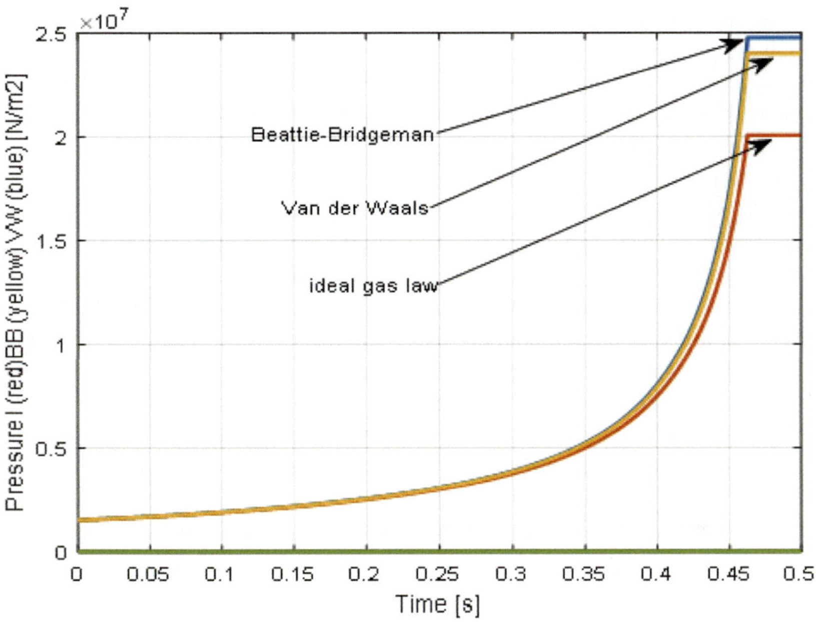

FIGURE 6.31 Different compression curves.

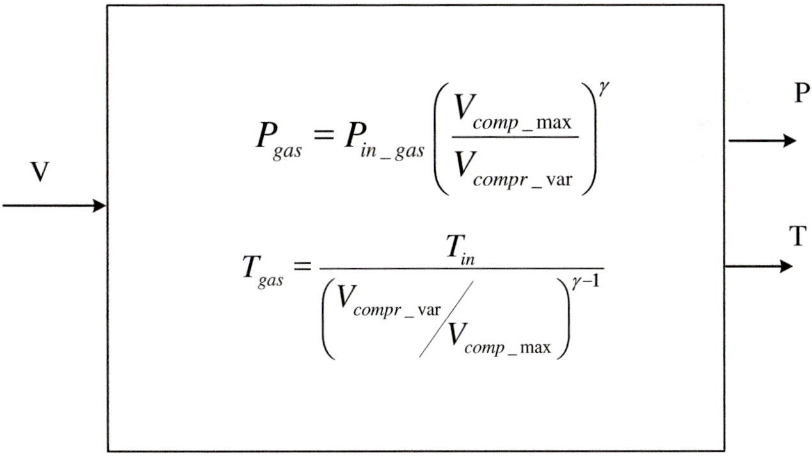

FIGURE 6.32 The analytic model for pressure and temperature.

$$T_{gas} = \frac{T_{in}}{\left(V_{compr\_var}/V_{comp\_\max}\right)^{\gamma-1}} \tag{6.50}$$

So the pressure and the temperature of the gas can be calculated with the global model, as represented in Figure 6.32, where the two analytic expressions are used together. In this model, the thermal conditions of the evolution are represented through the polytropic exponent γ.

Another method we call the Thermal Integrated Model (TIM) will be particularly useful in relation to the use of the two complex state equations (Van de Waals and Beattie–Bridgeman). This method includes a thermal model based on a thermal equivalent circuit where the thermal capacity of the mass of gas C_{th} and a thermal resistor R_{th} are used. R_{th} represents the heat flow from the compressed gas towards the surrounding. The thermal model is used "in the loop" together with the state equation of the gas. This new TIM representation allows calculation of the simultaneous and coupled evolutions of the pressure and of the temperature during compression.

The evaluation of the temperature using the thermal equivalent circuit supposes that the input variable of this circuit (the thermal power) is known. This variable is calculated as the mechanical power transmitted to the cylinder by the external force F as long as no gas is transferred from the cylinder to the reservoir. The force is calculated as the product of the pressure multiplied by the piston's surface S. Further, the product of the force by the imposed velocity of the piston gives the mechanical power Pow. (Figure 6.33). In this figure, a binary variable X intervenes in the functional block between the mechanical power *P*ow and the thermal power *P*th. This variable sets the condition that all the mechanical power is transmitted to the mass of gas as long as the exhaust valve remains closed. As soon as this valve opens, the mechanical power no longer contributes to the elevation of the temperature of the gas but "moves" the compressed gas to the exhaust reservoir at a constant value of the pressure (160 bar).

6.2.7.2.1 Comparison of Results

First, the evolution of the pressure during compression which is calculated with the analytic model is compared to the evolution calculated with the TIM. The compressed

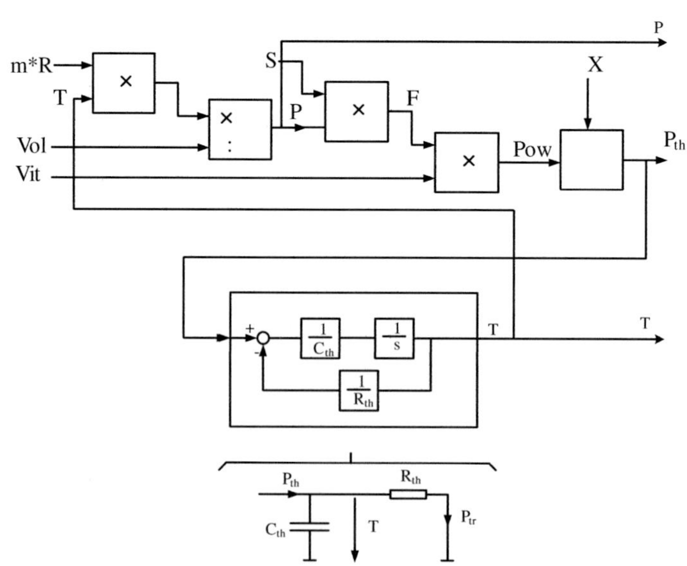

FIGURE 6.33 TIM with the thermal functional diagram "in the loop".

gas is in this first case air. The TIM uses the ideal gas law (Figure 6.33). The adiabatic compression is modeled with an exponent $\gamma = 1.4$ in the analytic model and with an infinite value of the thermal resistance R_{th} in the TIM. The feedback $1/R_{th}$ in the diagram is disconnected in this case.

For the thermal capacity C_{th}, Equation 6.52 is used with a mass of air equal to:

$$m = \frac{p \cdot V}{R \cdot T} = \frac{15 \cdot 10^5 \text{ N/m}^2 \cdot 53.5 \cdot 10^{-6} \text{ m}^3}{0.287 \text{ kJ/(kg} \cdot \text{K)} \cdot 293 \text{ K}} = 9.54 \cdot 10^{-4} \text{ kg} \quad (6.51)$$

$$C_{th} = m \cdot c_v = 9.54 \cdot 10^{-4} \text{ kg} \cdot 0.718 \text{ kJ/(kg} \cdot \text{K)} = 0.682 \text{ J/K} \quad (6.52)$$

The two pressure curves are represented in Figure 6.34. The curves are nearly superposed (blue curve and red curve), showing the equivalence of the two models.

A comparison of the curves for the temperature is also done (Equation 6.50 and Figure 6.33) and is represented in Figure 6.35. The values are here also similar (575.0 K for the analytic model and 576.6 for the TIM).

A further comparison of models is made for the compression of hydrogen. In this case, two TIMs are compared, namely a TIM using the Van der Waals state equation and a TIM using the Beattie–Bridgeman state equations. The TIM with Van

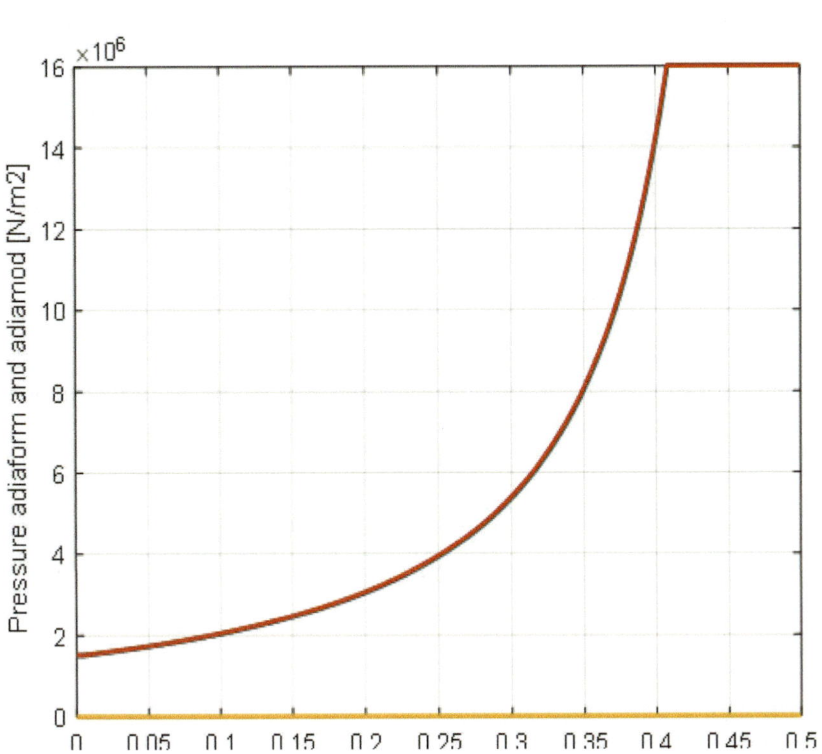

FIGURE 6.34 Pressures in adiabatic mode with the analytic formula (Equation 6.39) and with the TIM (air, **FIGURE 6.33**). (The blue and red curves are practically superposed.)

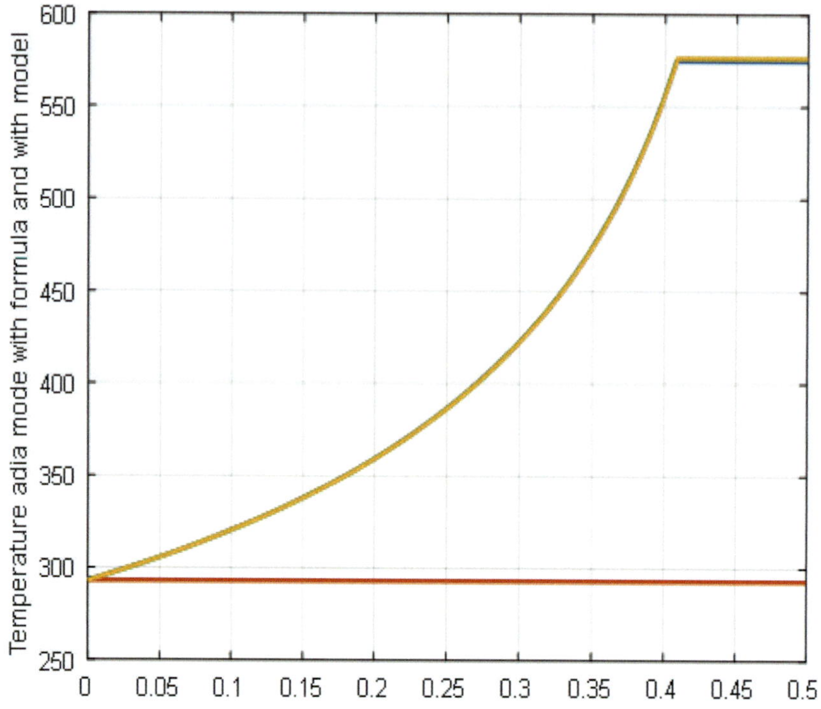

FIGURE 6.35 Temperatures in adiabatic mode with the analytic formula (blue curve) and with the TIM (air) (yellow curve). The red curve represents the isothermal case.

der Waals state equation is represented in Figure 6.36, and the TIM with Beattie–Bridgeman state equation is represented in Figure 6.37.

In these diagrams, the pressure during compression is calculated from the volume (or mass volume) with the help of the state equation. In these equations, the temperature is involved and needs an additional evaluation for it. From the pressure, the compression force is evidently calculated, and then, the mechanical compression power is calculated in a similar way as for the TIM of Figure 6.33.

The thermal capacity is in this case (hydrogen):

$$C_{th_{H2}} = m \cdot c_v = 0.664 \cdot 10^{-4}\,\text{kg} \cdot 10.16\,\text{J/(g} \cdot \text{K)} = \\ 0.664 \cdot 10^{-1}\,g \cdot 10.16\,\text{J/(g} \cdot \text{K)} = 0.674\,\text{J/K} \tag{6.53}$$

The thermal capacity of the hydrogen mass (in the initial volume and at initial pressure) is similar to the thermal capacity of air (identical volume and pressure). This is due to a much higher mass of air (0.954 g) than hydrogen (0.06 g) when the massic heat values are in inverse proportions (0.718 J/gK for air versus 10.1 J/gK for H_2).

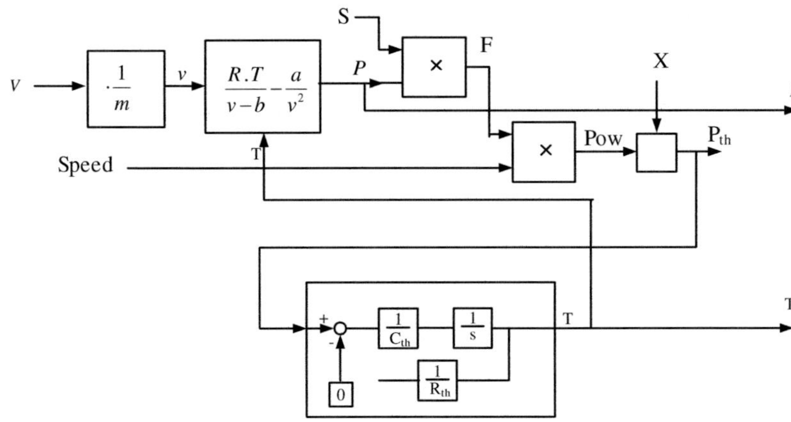

FIGURE 6.36 TIM with Van der Waals state equation.

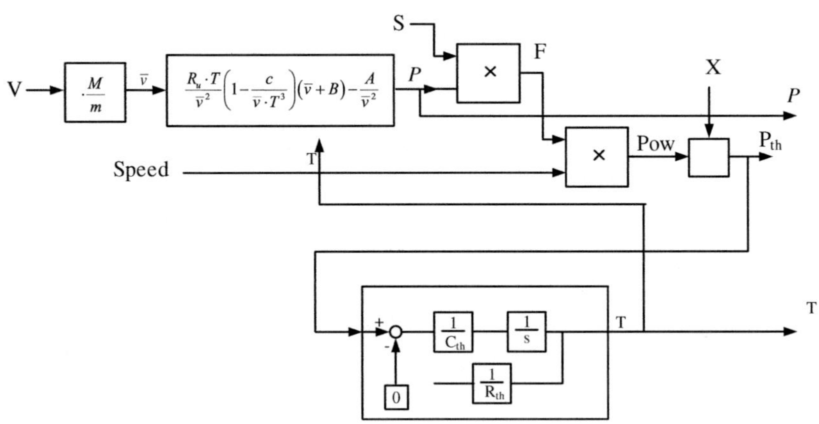

FIGURE 6.37 TIM with Beattie–Bridgeman state equation.

The pressure curves of the two TIMs are represented in Figure 6.38 and are totally superposed. When these curves are compared with the evolution of the pressure calculated with the analytic model (Figure 6.32), one can see that the pressure value of 160 bar is reached in these cases earlier (0.4 s instead of 0.41 s).

Regarding the evolution of the temperature, the same remarks are valid as for the pressure (same values and earlier reach of the limit). The temperature curves calculated with the Van de Waals model and with the Beattie–Bridgeman models are represented in Figure 6.39.

The delay of reaching the pressure limit between the analytic model and the two TIMs (VDW and B-B state equations) can be seen on the curves in Figure 6.40.

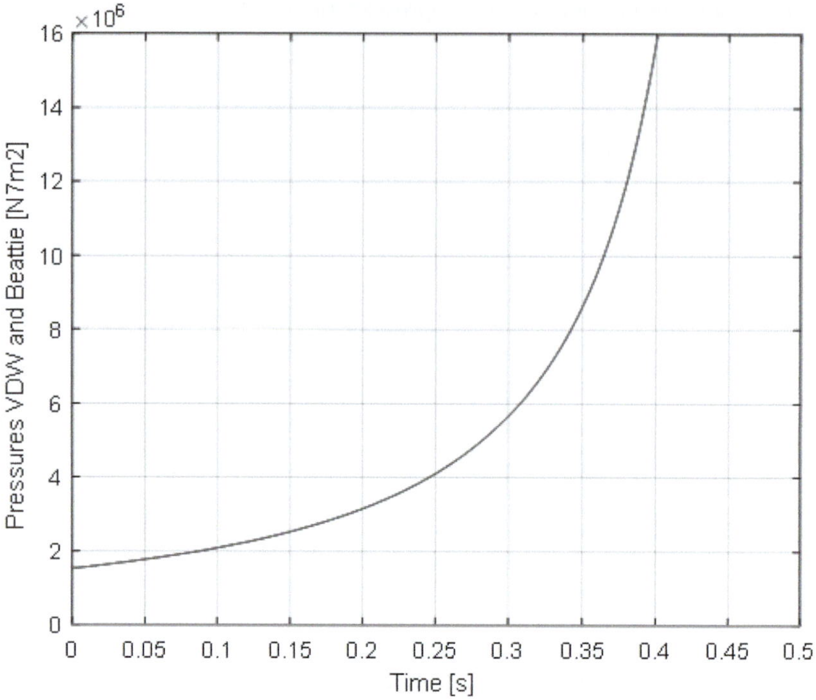

FIGURE 6.38 Pressures H_2 adiabatic VDW and Beattie (totally superposed curves).

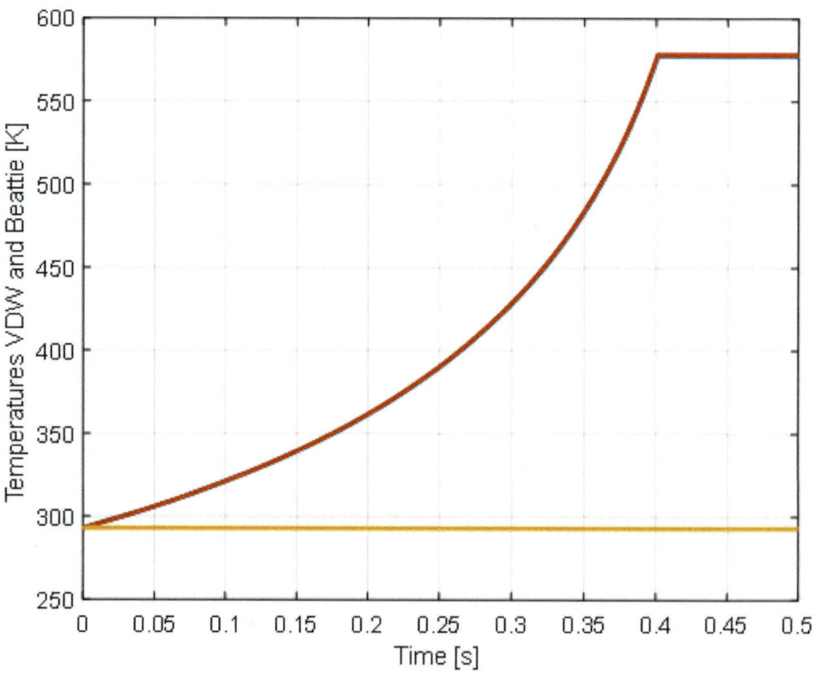

FIGURE 6.39 Curves for the temperature of hydrogen, Van der Waals (blue curve) and Beattie–Bridgeman (red curve) TIMs. The yellow curve corresponds to the isothermal case.

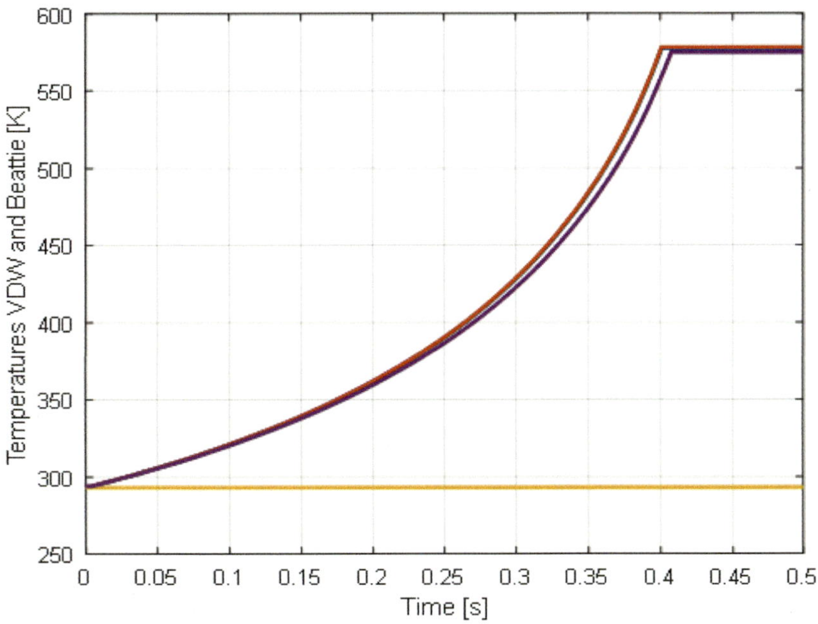

FIGURE 6.40 Temperatures H_2 adia VDW and Beattie (red, the two curves are superposed) and comparison with the analytic model (blue curve). The yellow curve corresponds to the isothermal case.

6.2.7.3 Expansion

For the simulation of the expansion phenomenon, the same cylinder as for compression is used. The system used for the expansion is represented in Figure 6.41. The intake of the high-pressure gas is controlled with the intake valve V_{in} and the exhaust by the valve V_{out}. The control of the valves is represented in the lower diagram of Figure 6.41. The expansion from an initial pressure of 160 bar is simulated. The volumetric expansion is defined with a ratio of 1 to 3. For that, the gas is injected in the cylinder during the first third of the stroke from a reservoir with constant pressure. The position x of the piston evolves from zero up to the position le. Then, the inlet valve is closed during the remaining two thirds of the same stroke. The position of the piston is increased up to the full length of the stroke. As a result, the volume of the gas increases. During these three thirds of the positive stroke the exhaust valve is closed. Then during the return stroke the expanded gas is exported through the opened exhaust valve V_{out} to the exhaust reservoir. In order to prevent the opening of the exhaust valve from an additional dynamic transient, the pressure of the exhaust reservoir is supposed to correspond to the pressure level after expansion.

The maximum volume of the cylinder is identical to the previous value according to Equation 6.37

$$V_{cyl} = 53.5 \cdot 10^{-6} m^3$$

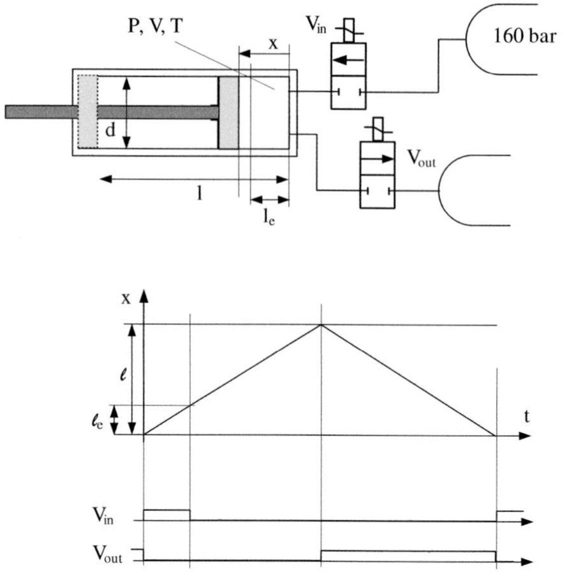

FIGURE 6.41 Structure of the expansion cylinder with sequential diagram of the control of the valves.

The volume of the gas taken in at high pressure (first third of the stroke) becomes

$$V_e = 53.5 \cdot 10^{-6} m^3 / 3 = 17.83 \cdot 10^{-6} m^3.$$

The expansion velocity has in this case the same value as for the compression but with an opposite sign, namely –0.18 m/s. The duration of the intake is 0.5 s/3 or 0.166 s. The evolution of the gas volume inside of the cylinder is represented in Figure 6.42. Between 0 and 0.166 s, the constant value of the volume does not illustrate the real volume of the cylinder but means that the cylinder is filled with gas under constant pressure up to the indicated value of the volume. The resultant expansion ratio is represented in Figure 6.43, where the value of 1 is characteristic of the filling duration (0–0.166 s) where the pressure is constant and is imposed by the upstream reservoir.

The evolution of the pressure and of the temperature during expansion can be calculated with the classic relations given in Equations 6.39 and 6.50. In those relations only the variation of the volume is considered and the nature or the mass of the gas does not appear in the expressions. But for the use of a more accurate model such as the Van der Waals or the Beattie Bridgeman model the mass volume v or the molar volume $\overline{v}$ must be known. The relation of these variables to the mass of gas is given through:

$$v = \frac{V}{m} \text{ and } \overline{v} = M \cdot v = M \cdot V / m \qquad (6.54)$$

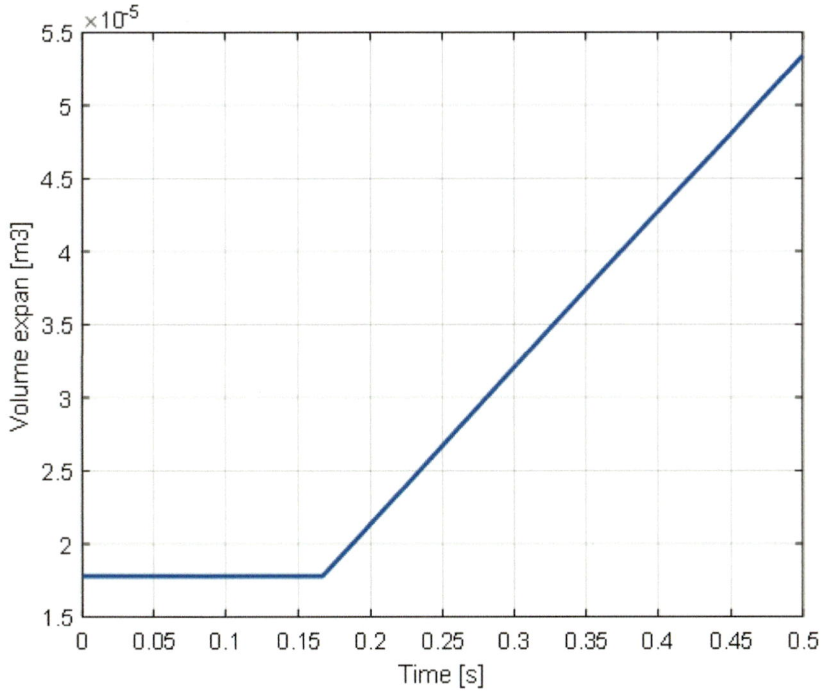

FIGURE 6.42 Volume of expansion.

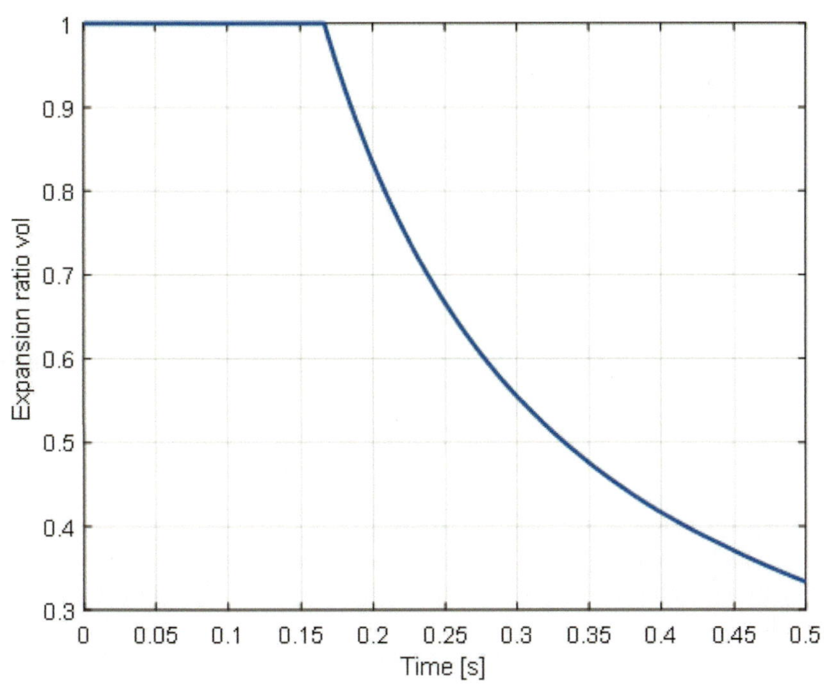

FIGURE 6.43 Volume ratio of the expansion.

The volume V and the pressure P are given, but the mass m must be evaluated. A first approach uses the ideal gas law and gives in this case:

$$m = \frac{P \cdot V}{R \cdot T} = \frac{160 \cdot 10^3 N/m^2 \cdot 53.5 \cdot 10^{-6} m^3 \cdot \frac{1}{3}}{4.124 kJ/kg \cdot K^{-1} \cdot 293K} = 2.361 \cdot 10^{-4} kg \quad (6.55)$$

Considering the use of the Van der Waals model, and the initial conditions P and V before the expansion, the temperature of the gas can be calculated with Equation 6.56 and should correspond to the atmospheric temperature.

$$T = \frac{\left(P + \frac{a}{v^2}\right)}{R}(v - b) \quad (6.56)$$

From Equation 6.7 and with the values of Table 6.1, the factors a and b become

$$a = \frac{27R^2 \cdot T_{cr}^2}{64P_{cr}} = \frac{27 \cdot (4.124 \cdot kJ/kgK)^2 \cdot (33.3K)^2}{64 \cdot 1.3MPa} = 6.120 kPa \cdot m^6/kg^2 \quad (6.57)$$

$$b = \frac{R \cdot T_{cr}}{8P_{cr}} = \frac{4.124 kJ/kgK \cdot 33.3K}{8 \cdot 1.3MPa} = 0.0132 m^3/kg \quad (6.58)$$

Still in the situation of constant pressure before the expansion, the mass volume is

$$v = \frac{V}{m} = \frac{53.5 \cdot 10^{-6}\,\mathrm{m}^3/3}{2.361 \cdot 10^{-4} kg} = 0.0755 m^3/kg \quad (6.59)$$

The temperature calculated with the Van der Waals model based on this value of the mass volume v is:

$$T = \frac{\left(P + \frac{a}{v^2}\right)}{R}(v - b) = \frac{160 \cdot 10^5 N/m^2 + (6.120 kPa \cdot m^6/kg^2)/(0.0755 m^3/\mathrm{kg})^2}{4.124 kJ/kgK} \cdot (0.0755 m^3/kg - 0.0132 m^3/kg) = 257.5\,\mathrm{K} \quad (6.60)$$

This value does not correspond to the temperature of the gas taken in and shows the difference between the ideal gas model and the Van der Waals model, especially for high pressures. Figure 6.20 shows the difference between the two models. During filling with constant pressure (0–0.166 s), the difference of the temperature is significant (257.5K instead of 293K) because of the chosen value of the gas mass. A correct modeling of the temperature should include the real value of the gas mass. Consequently, the value of the mass volume should be calculated using Equation 6.20

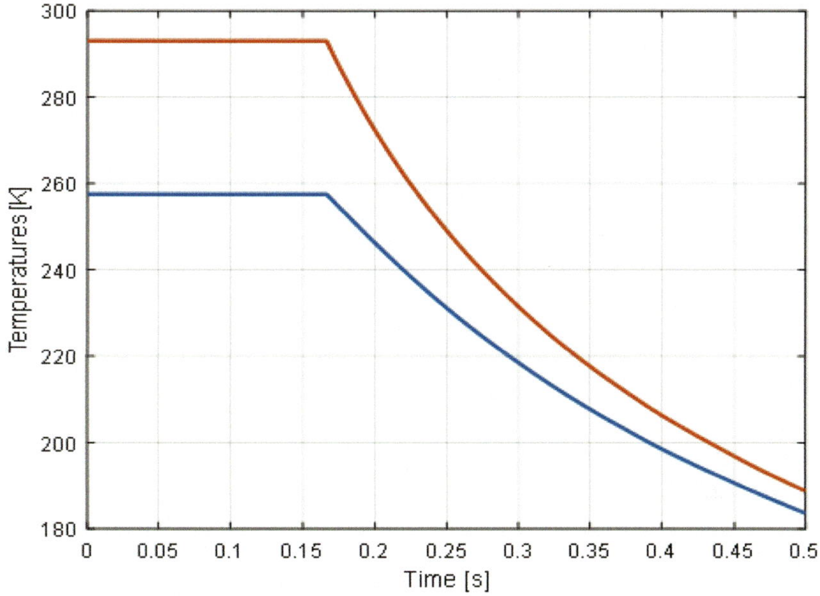

FIGURE 6.44 Evolution of the temperature calculated with the ideal gas law (red) and with the uncorrected Van der Waals model (blue).

with imposed pressure and temperature. The form of this relation, however, does not allow a resolution versus v and needs another method for the evaluation of the right value of the gas mass in the cylinder.

In the present study, an iterative method based on the simulation of Equation 6.20 is used and leads to a corrected value of the mass of $2.1 \cdot 10^{-4}\,kg$ of hydrogen instead of the $2.361 \cdot 10^{-4}\,kg$ calculated previously.

Figure 6.45 shows three different evaluations of the temperature with three different models, namely the ideal gas law (yellow), a Van der Waals model (with corrected value of the gas mass) where the evolution of the pressure is calculated with the ideal gas law (blue) and a Van der Waals model completed with integrated thermal model (TIM) (red curve).

The structural diagram of the TIM used here for the expansion is identical to the model represented in Figure 6.36 and previously used for compression.

The Van der Waals TIM of Figure 6.12 does not only represent the evolution of the temperature but gives also the evolution of the pressure (Figure 6.46).

The state equations used in the simulation of compression of air and hydrogen have been implemented in simulations where the mechanical variables as the volumetric ratio is imposed. The different models bring similar results, especially in the isothermal mode and when the pressure is not going above the values of 50 to 100 bar. For the simulation of the variation of the temperature, an easy-to-implement thermal model (TIM) is used where the energy corresponding to transferred mechanical work to the gas is considered. The thermal model is inserted "in-the-loop" with the

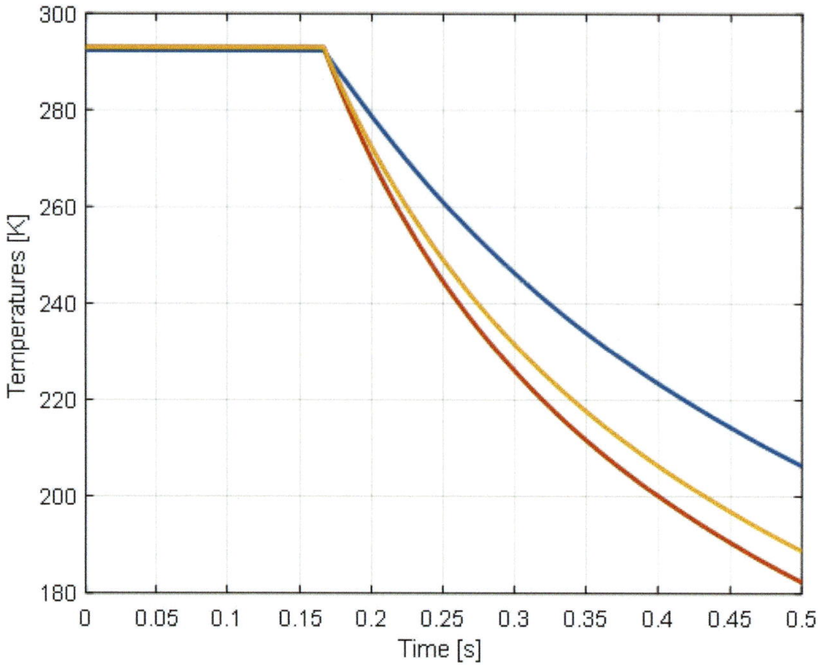

FIGURE 6.45 Evolution of the temperature with the ideal gas law (yellow), Van der Waals with idealized pressure (blue) and Van der Waals with TIM (red).

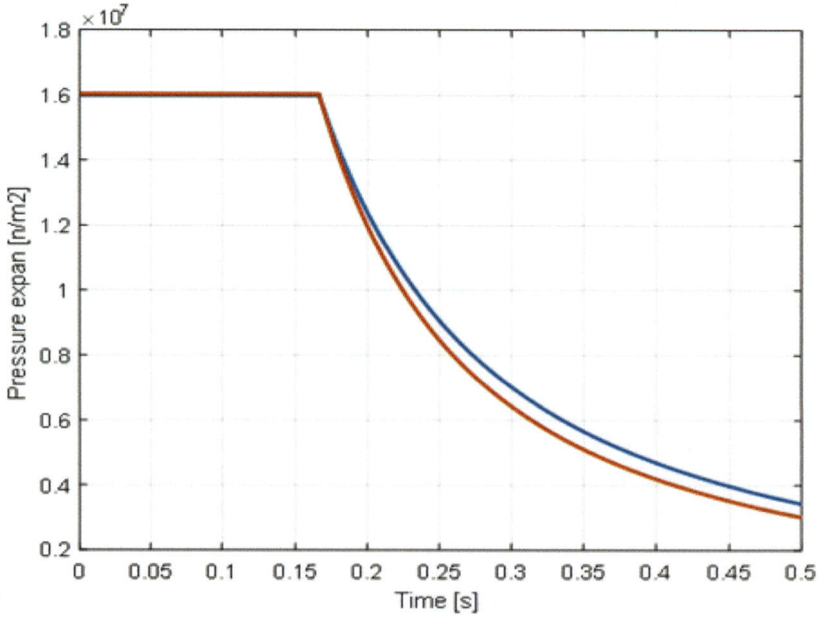

FIGURE 6.46 Pressure of expansion in adiabatic mode with VDW TIM (blue) and ideal gas law (red).

functional diagrams representing the state equations. Regarding the results of simulations, only small differences are observed, in the order of a few percent. In the present study only the adiabatic mode is simulated, but any polytropic phenomena can be represented with the TIM as long as the thermal resistance between compressed gas and the surroundings would be known.

For the simulation of the expansion mode the same models are used. One larger difference is observed in relation to the considered mass of gas, which is initially in the volume to be expanded. The estimation of that mass depending on the initial pressure and temperature and from the state equations has been realized by successive iterations by variation of the factor representing the mass volume v. For the simulation of the temperature, a comparison is made between the ideal gas law model and the Van der Waals model in open loop and where the pressure is calculated with the analytic relation. A further comparison is made with the Van der Waals model completed by the TIM. The results show differences on the order of 10 to 15%.

For the representation of the variation of the pressure in expansion mode, the comparison is made between the ideal gas law and the completed TIM. Only small differences are observed in the range of 3 to 5%. Detailed information on the differences between the state equations used for the present study can be found in Reference [20].

6.3 FROM POOR TO INCREASED ENERGETIC PERFORMANCE OF COMPRESSED AIR ENERGY CONVERSION

6.3.1 Energy Loss Due to the Use of a Pressure Reduction Valve

As indicated in Section 6.2.4, the energy density of CAES is increased using a higher storage pressure. This, however, leads to the use of complex and heavy compression and expansion equipment. For the compression device, multistage compressors can be used, and the storage reservoir must be designed in consequence. For the inverse conversion (the expansion), cylinders or expansion stages are often used where their operating pressure is reduced. Between the storage reservoir and the expander, a pressure reduction valve (PRV; Figure 6.47) must be used. This component, however, has consequences for global energy efficiency.

In Figure 6.47, the converting element at the output is a conventional pneumatic actuator. Such devices usually operate at constant pressure, which also has an effect on the conversion efficiency, as will be discussed in the next section.

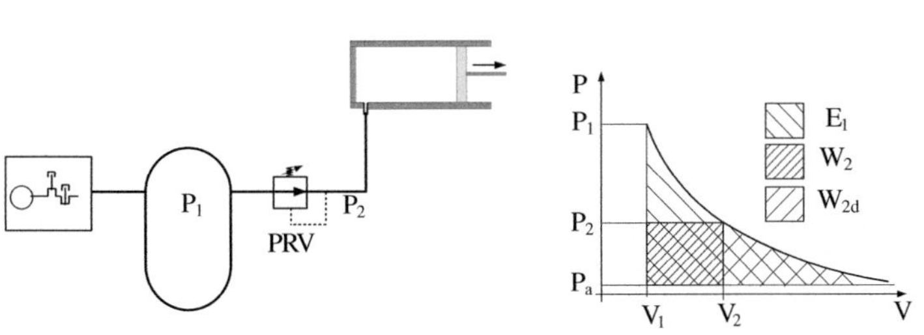

FIGURE 6.47 Use of a pressure reduction valve between reservoir and expander.

Let us first consider the effect of the use of a PRV valve on the usable energy in the discharge process.

The *P-V* diagram in Figure 6.47 shows the different quantities of energy to be recovered from a reservoir at a pressure P_1 in a volume V_1. The maximum amount of energy is noted with E_1 and corresponds to the expansion energy from P_1 down to the atmospheric pressure P_a under isothermal conditions. This value can be calculated as

$$E_1 = P_1 V_1 \left(\ln \frac{P_1}{P_a} - 1 + \frac{P_a}{P_1} \right) \tag{6.61}$$

W_2 is the mechanical work produced with the cylinder while the pressure is kept constant through the PRV valve. It is calculated through

$$W_2 = (P_2 - P_a) \cdot (V_2 - V_1) \tag{6.62}$$

For the calculation of the effect of the PRV valve alone, it must be considered that the energy conversion at constant pressure in the cylinder is completed by an expansion process that uses the rest of the energy amount available. The last surface W_{2d} of the diagram in Figure 6.47 is then considered in the efficiency calculation.

This last surface corresponds again to an isothermal expansion from P_2 down to P_a and is equal to

$$W_{2d} = P_2 V_2 \left(\ln \frac{P_2}{P_a} - 1 + \frac{P_a}{P_2} \right) \tag{6.63}$$

Finally, the efficiency of the discharge process while using a PRV valve can be estimated through

$$\eta = \frac{W_2 + W_{2d}}{E_1} \tag{6.64}$$

Figure 6.48 shows the influence of the use of a pressure reduction valve on the discharge process from a reservoir at P_1 = 300 bar. The equivalent efficiency factor is given through Equation 6.64.

6.3.2 Poor Energetic Performance of Classical Pneumatic Actuators

In a classical pneumatic actuator, the mechanical work is obtained from the displacement of the piston under constant pressure. At the end of the stroke, the pressure in the fully deployed cylinder is released to the atmosphere by opening the exhaust valve, allowing the free return of the piston. This corresponds to renounce to the pneumatic energy content inside the cylinder. The pneumatic energy content of the deployed cylinder can be illustrated by the W_{2d} surface in Figure 6.47. In other

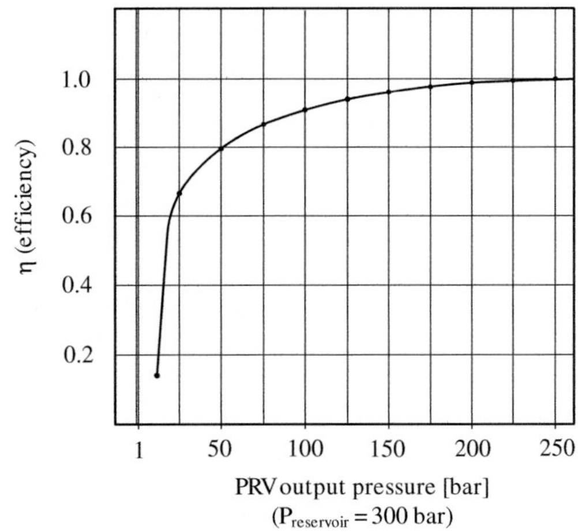

FIGURE 6.48 Equivalent efficiency due to the PRV valve.

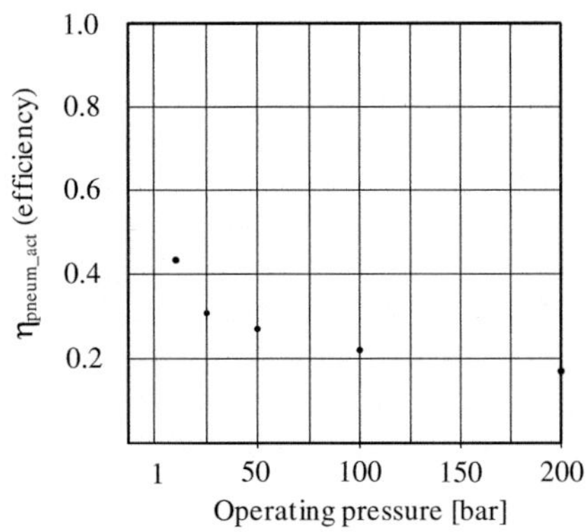

FIGURE 6.49 Energy loss factor of a classical pneumatic actuator.

words, it corresponds to the expansion energy of a V_2 volume from a P_2 pressure. A corresponding efficiency or energy loss factor can be defined as

$$\eta_{pneum\_act} = \frac{W_2}{W_2 + W_{2d}} \tag{6.65}$$

The energy loss factor of the classical pneumatic actuator is represented in Figure 6.49. This diagram shows that such actuators can only be designed for pressure levels under 50 bar. The energetic performance is still very poor.

6.3.3 Addition of an Expansion Chamber (Motor Development International (MDI) Motor)

In order to realize a pneumatic motor with a higher efficiency, an original concept has been proposed in the context of the development of the compressed air car [8].

In such a concept, an additional expansion cylinder is coupled to the cylinder producing displacement work, and the energy recovery from the pneumatic level to the mechanical output can be increased, respectively a large part of the lost energy (W_{2d}, Figure 6.47) can be sequentially recovered. The principle of the Motor Development International (MDI) motor is represented in Figure 6.50. The four steps are noted:

A: Filling of the main cylinder with air at pressure p_{in}
B: Expansion due to an air transfer from V_1 to $V_1 + V_2$
C: Exhaust
D: Return to the initial state. The complex movement of the pistons is achieved through a dedicated mechanism (special crankshaft)

In the MDI motor, the recovery of expansion energy increases the efficiency significantly, but the expansion process is still achieved in non-isothermal conditions.

6.3.4 Principles of the Liquid Piston with Integrated Heat Exchanger

In order to benefit from both advantages, the high energy density of the storage system, together with the properties of isothermal compression and expansion, a new system has been proposed where the volumetric change is produced via a so-called liquid piston [1]. Simultaneously, the properties of isothermal transformations are obtained from an integrated heat exchange device inside of the cylinder. Such a system is represented in Figure 6.51, where two compression chambers are represented (1), (2), which are operated in a reciprocating mode. The pressure is generated by

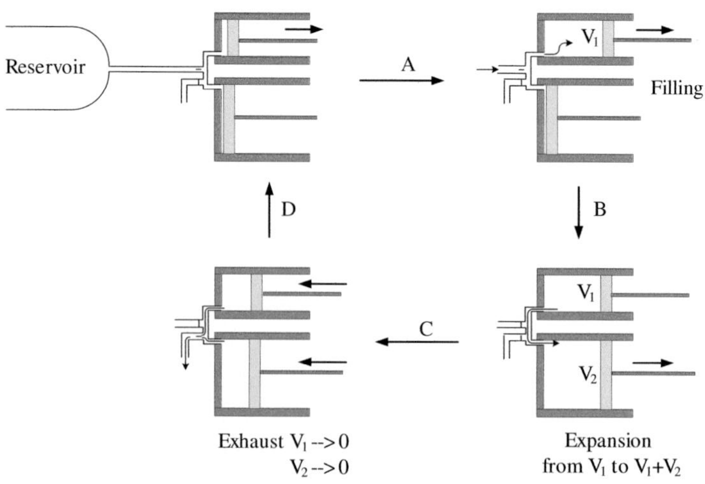

FIGURE 6.50 Principle of the MDI motor with "active chamber".

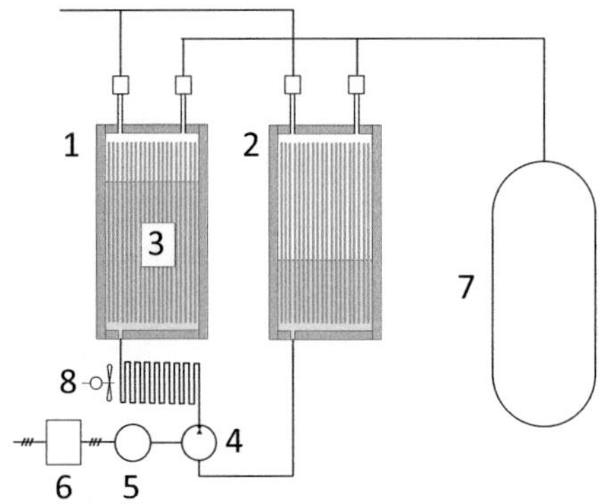

FIGURE 6.51 The principle of the liquid piston with integrated heat exchanger.

the rise of the liquid level (3) inside of the compression chamber, while the liquid is moved from one cylinder to the other. A hydraulic pump (4) generates the compression effort and can also be operated in the reversal mode (motor) during expansion. The hydraulic pump/motor is directly coupled to an electric motor/generator (5) from which power can be exchanged through one dedicated power electronic circuit (6). The energy is stored inside the high-pressure reservoir (7).

The heat exchanger is composed of a high number of laminations inside of the cylinder. These laminations absorb the compression heat of the air during compression (elevation of the level of the compressing liquid). They must have a high convection factor to absorb heat from the compressed air (above the liquid level). These laminations have their temperature kept nearly constant through the rising liquid level inside of the compression chamber. After one complete compression process, nearly all the compression heat will be transferred to the liquid.

During the next sequence, namely the second stroke of compression using the second cylinder, the liquid is transferred from the first one to the second, moved by the hydraulic pump. In the transfer path from cylinder one to cylinder two, a dedicated heat exchanger (8) exports the compression heat to the surrounding.

In Figure 6.11, the energetic efficiency of a hydraulic pump is shown. Related to the pressure variations in the storage reservoir, a specific control strategy for a maximum efficiency point tracking (MEPT) is proposed [1, 14]. The variation of the hydraulic average power is made with an ON-OFF control, and an auxiliary storage is added to the system for smoothing the pulsations.

During the ON state, the hydraulic pump can be run at its maximum efficiency (where the power is defined according the pump characteristics (see Figure 6.11). The average power is imposed through the duty cycle of the ON-OFF process.

6.3.5 The Concept of the Finned Piston

A second approach of realizing a compression/expansion device able to achieve the process in nearly isothermal conditions is given by the concept of the finned piston [2, 9]. The basic idea of the finned piston is to increase the convection heat transfer area using multiple annular differential pistons. The principle of the differential piston is explained through Figure 6.52. In such a system, a compression chamber of an annular cylinder is built instead of a circular cylinder, where the same compression surface is acting as for the normal cylinder but where the convection surface is strongly increased.

In Figure 6.53, several circular and annular front surfaces of pistons are represented. For all cases, the force produced by the pressure on the piston surfaces is identical. From left to right in the first line (conventional piston), the pressure varies from 10 bar to 250 bar, and in order to keep the force constant, the piston surface A_{press} is decreased from 23.7 cm^2 to 0.94 cm^2. The corresponding convection surface changes from 142 cm^2 to 20 cm^2.

The second row of Figure 6.53 represents the differential piston. The piston effective pressure surfaces A_{press} are identical to the surfaces of the first row, calculated

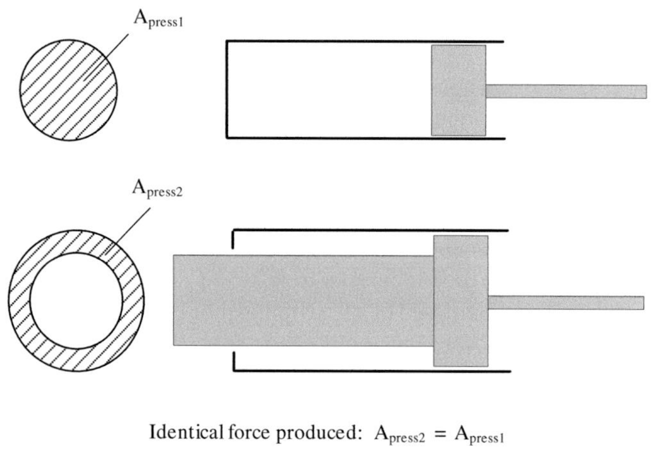

FIGURE 6.52 Normal and differential pistons.

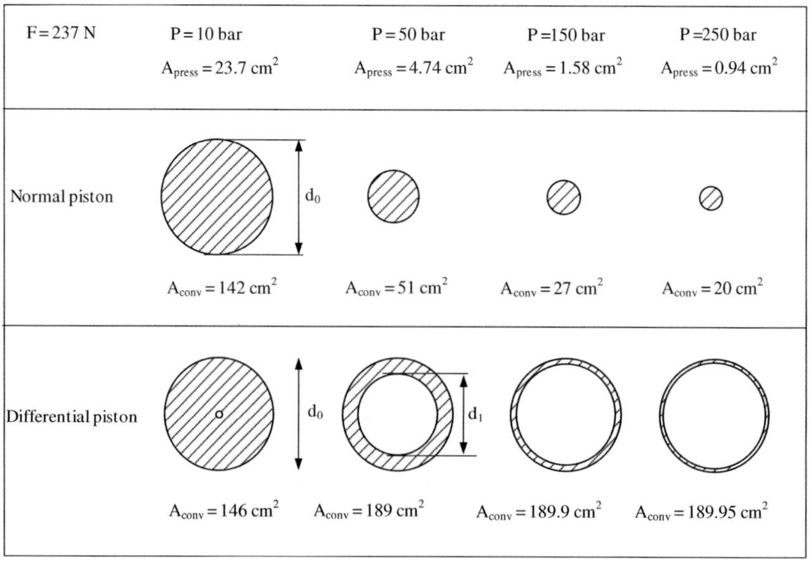

FIGURE 6.53 Normal and differential piston geometries.

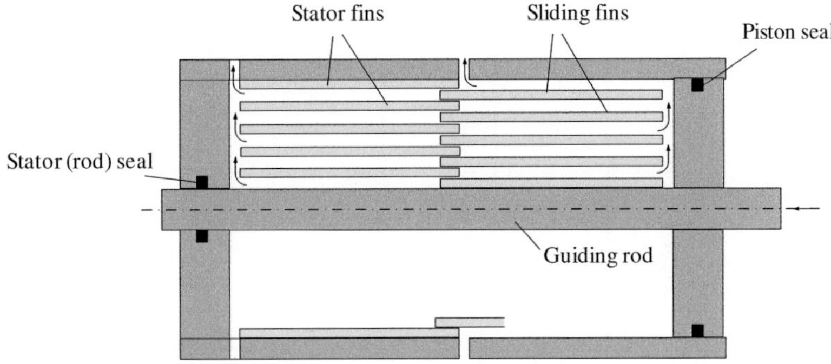

FIGURE 6.54 Multilayer differential annular piston/cylinder.

for a force of 237 N for the different pressures. In this case, the exterior diameter d_0 is kept constant, and the interior one d_1 is varied in function of the pressure level. For the differential piston, the convection surface A_{conv} is also given. For all pressure levels, the convection surface is much higher than for the classical piston.

Finally, the multilayer differential annular piston/cylinder is shown in Figure 6.54, where a set of mobile and stationary assembly of concentric fins are moving one in the other for the volumetric compression and expansion, with the benefit of a highly increased convection performance [9].

6.3.5.1 Performances of the Finned Piston System

In Figure 6.55, the pressure in the chambers of multilayer finned piston are compared with the pressure in the chamber of a classical one.

Figure 6.56 gives the curve of the associated temperatures. The effect of a higher convection surface is evident.

Figure 6.57 gives the rate of heat transfer, and finally, the work spent for the compression is given in Figure 6.58. The benefit of the finned piston is in this case around 30%.

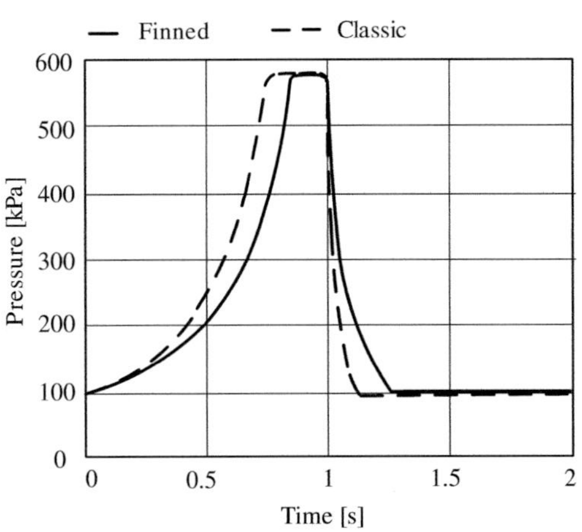

FIGURE 6.55 Pressures versus time.

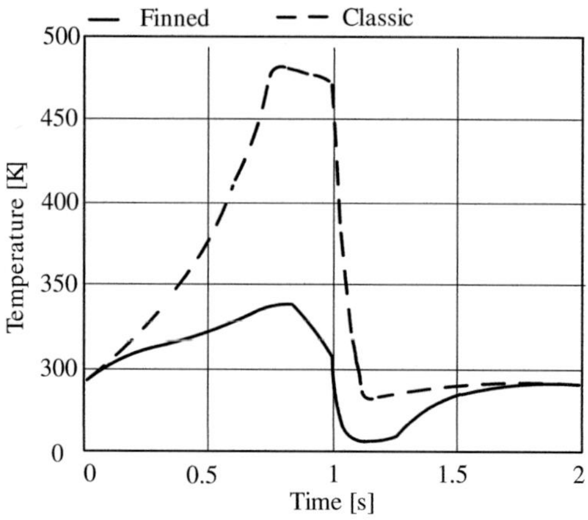

FIGURE 6.56 Temperatures versus time.

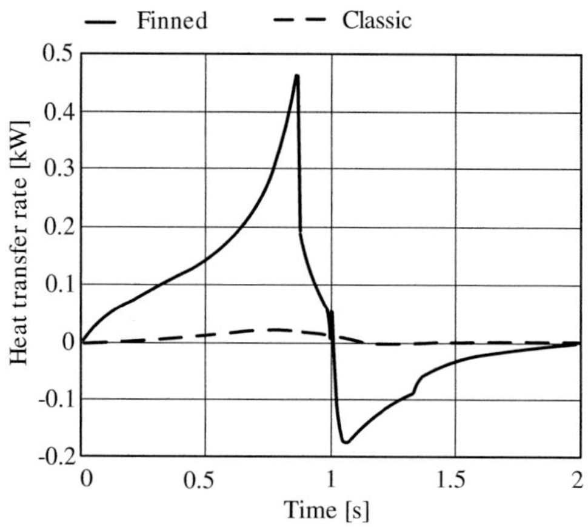

FIGURE 6.57 Heat transfer rate.

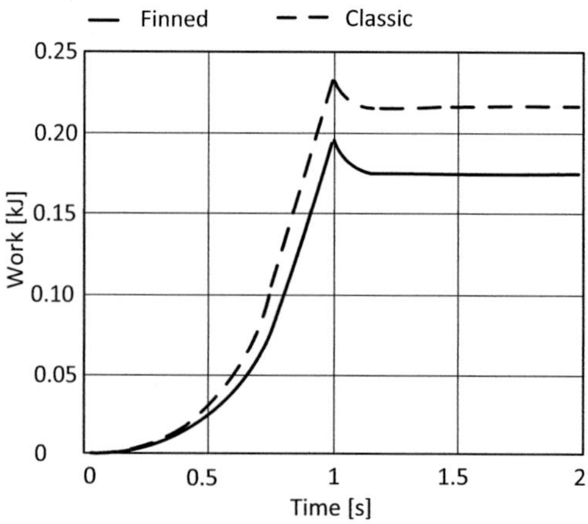

FIGURE 6.58 Mechanical work.

6.4 MEDIUM SCALE COMPRESSED AIR WITH "CLOSED" GAS CYCLE

A medium scale CAES with closed gas cycle is described in Reference [21]. In such a system, a given volume of water is pumped into a set of storage cylinders. Such cylinders can be identical to those used in gas pipelines. Figure 6.59 gives a simplified diagram of the system, and Figure 6.60 shows how it could be implemented in reality.

For the realization of the proposal in Figure 6.60, a natural lake or river can serve as the low pressure reservoir. The storage tubes correspond to the high-pressure volume in which the water is injected using the hydraulic pump(s). In regenerative mode, hydraulic motor(s) can be used if the machine used for the pump is not a reversible one. The hydraulic machines must of course be designed for the maximum pressure of the storage tubes. In the industrial world, increasingly more water hydraulic machines are developed for high pressure [22].

For the system described in Figure 6.59, the amount of water pumped into the reservoir can be optimized. A small amount of water can be pumped at a high level

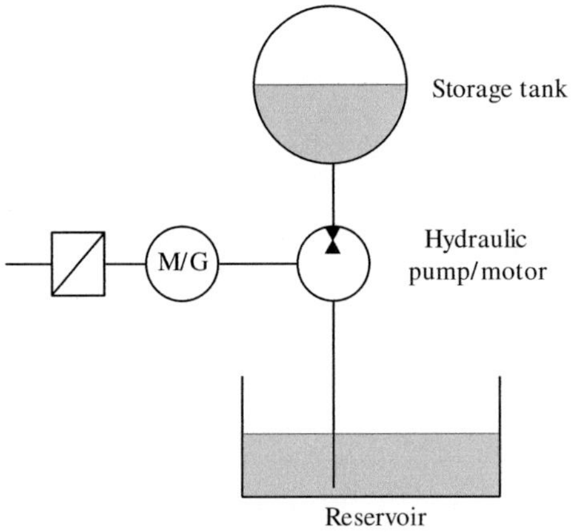

FIGURE 6.59 Simplified diagram of the closed cycle system.

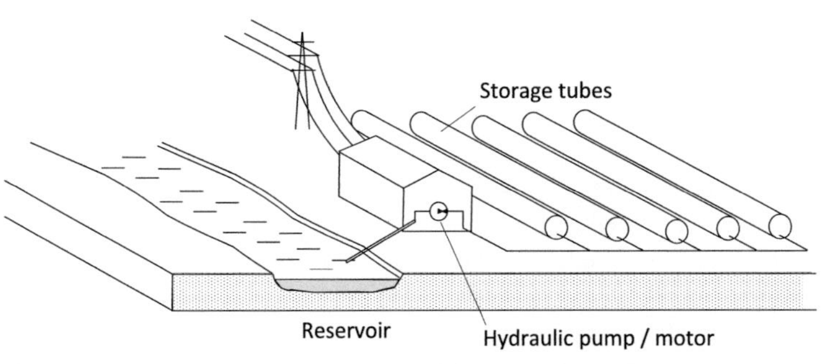

FIGURE 6.60 Medium-scale storage system with closed gas cycle [21].

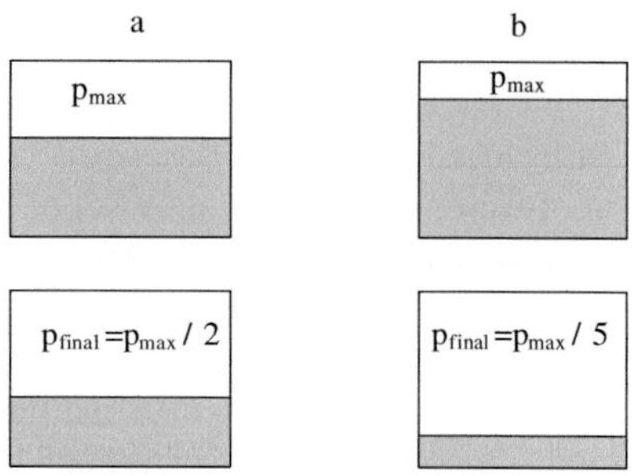

FIGURE 6.61 Optimizing the amount of water. (a) Ratio of the pressures = P_{max}/P_{final} = 2 and (b) ratio of the pressures = P_{max}/P_{final} = 5.

of the mean pressure, or a higher amount of water can be pumped into the reservoir at lower mean pressure (Figure 6.61).

For the calculation of the maximum storage energy, a simple maximization function can be defined:

$$W_{max} = P_{max} \cdot (V - V_0) \frac{\ln(\alpha)}{\alpha} \tag{6.66}$$

where α is the compression ratio, V is the storage volume and p_{max} is the maximum working pressure

The maximum is obtained for $\alpha = e = 2.718$

$$W = \frac{1}{e} P_{max} \cdot (V - V_0) = 0.368 P_{max} (V - V_0)$$

In kWh:

$$W = \frac{P_{max} \cdot (V - V_0)}{100} \text{ with } p_{max} \text{ in bar} \tag{6.67}$$

Practically, the result corresponds to

$$1 \text{ kWh per m}^3 \text{ and hectobar}$$

Table 6.4 indicates the order of magnitude that can be realized depending on the maximum working pressure level.

From Table 6.4, and the definitions given in Chapter 7, a comparison can be made between gravitational (pumped hydro storage) and pressurized (hydropneumatic) storage:

For the classical pumped hydro storage with a head (fall) of 250 m, a storage quantity of 1 MWh needs a volume of 2000 m^3 of water pumped.

For a hydropneumatic storage as described in this section (closed gas cycle), at a maximum pressure of 200 bar, the same quantity of energy (1 MWh) would need a storage volume of 500 m^3 (pipes volume in Figure 6.60) in which 200 m^3 of water should be pumped.

TABLE 6.4
Stored Energy in a Hydropneumatic System

Maximum Working Pressure p_{max}	Available Energy per m³	Stored Energy in 100 m³
50 bars	0.5 kWh	50 kWh
120 bars	1.2 kWh	120 kWh
200 bars	2 kWh	200 kWh
300 bars	3 kWh	300 kWh

For the realization of such a reservoir, an example with 48" pipes (122 cm diameter) can be designed.

The needed volume means a length of 430 m linear, or 35 × 12 m pipes.

Placed in seven rows of five pipes, the needed area would correspond to a 10 m × 60 m surface.

6.5 UNDERWATER CAES

Energy storage based on compressed air in a cavern has been proposed for long time, utilizing a water column to maintain the pressure constant. The same principle of constant pressure but utilizing an immersed variable volume for the storage of compressed air has been proposed more recently with the concept of immersed energy bags [23, 24].

The intent of the submerged installation is to use the surrounding hydrostatic pressure for compensation of the pressure of the stored air, with the advantages of highly reducing the requirements of the vessel's structure. The function of the energy bag is reduced to providing a membrane boundary between air and water. In the system represented in Figure 6.62, the compression/expansion station is on shore and the pressurized air is fed down to the energy bag through a pipe.

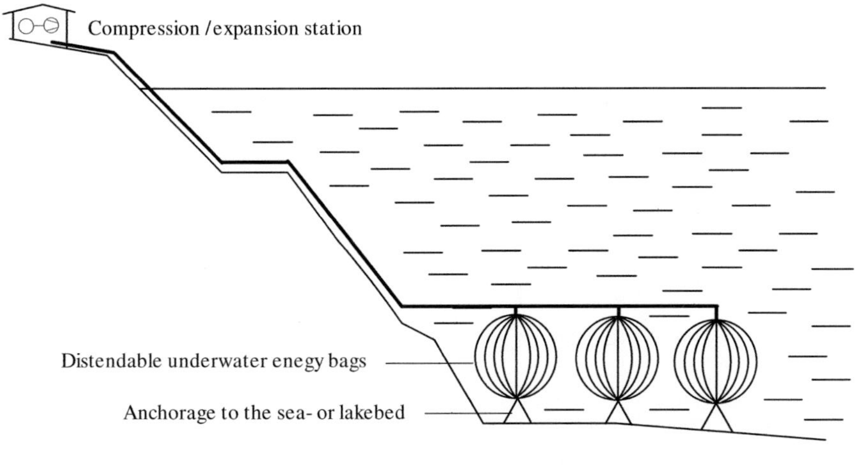

FIGURE 6.62 Underwater CAES system using energy bags.

Due to the Archimedes principle, the immerged volume undergoes a buoy effect that needs a strong anchoring to a sufficient heavy ballast.

The energy bags can be realized as scalable and flexible reservoirs, as described in Reference [24].

6.5.1 Model of the Storage Infrastructure

The energy bag or other expandable reservoirs can be modeled according the scheme of Figure 6.63. The compression/expansion machine is a high-performance machine achieving compression/expansion under isothermal conditions. In Figure 6.63, the compression and expansion machines are represented separately, but a reversible volumetric machine can also be used. Depending on the depth of the energy bag, multiple-stage compression/expansion equipment must be used.

The storage reservoir can be modeled as an expandable reservoir undergoing constant pressure.

The pressure related to the immersion depth of the energy bags depends on the atmospheric one and on the gravimetric pressure of the water.

The pressure around the bag is given by

$$P_{abs} = P_{atm} + \rho \cdot g \cdot h \tag{6.68}$$

ρ is the density of the water (1.025 for sea water)
g is the standard gravity
h is the depth of the bag

The energy amount that is stored in the underwater bag corresponds to the sum of the work realized by the volume variation under constant pressure W_{displ} and of the energy content of the corresponding volume being expanded under isothermal (or polytropic) conditions of the compression/expansion machine E as defined in Equation 6.1.

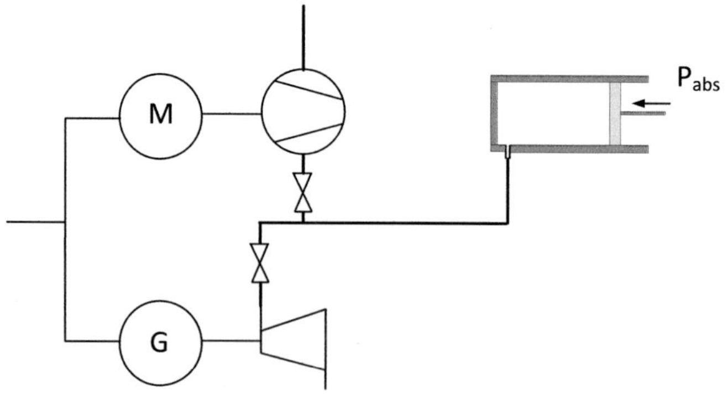

FIGURE 6.63 Model of the UWCAES.

$$E_{tot} = W_{displ} + E_{\exp an} = P_{abs} \cdot V_1 + P_{abs} \cdot V_1 \left[\ln\left(\frac{P_{abs}}{P_a} \right) - 1 + \frac{P_a}{P_{abs}} \right] \tag{6.69}$$

From Equation 6.69, it is easy to calculate the energy density e_{uw} of the UWCAES. The value of e_{uw} depending on the depth is represented in Figure 6.64.

The UWCAES can be compared with the medium scale system described in Section 6.4. regarding the occupied volume. The same amount of energy (1 MWh) is used for the comparison (Table 6.5).

Figure 6.65 shows a prototype energy bag realized by Red Line Aerospace for the University of Nottingham [24].

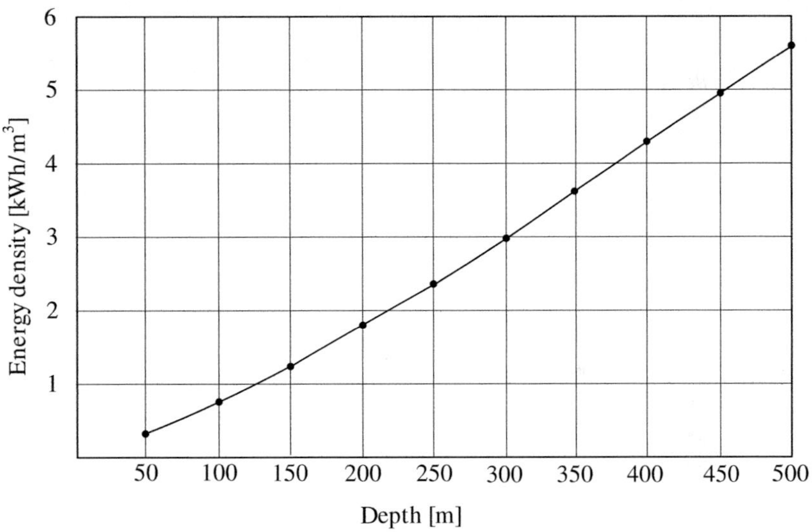

FIGURE 6.64 Energy density of the UWCAES.

TABLE 6.5
Volume of the Storage Vessel

System	*h* = 50 m	*h* = 100 m	*h* = 200 m	h = 500
Medium-scale hydro-pneumatic (200 bar)	500 m³			
UWCAES	3125 m³	1428 m³	555 m³	234 m³

FIGURE 6.65 Thin Red Line Aerospace energy bag designed and fabricated for University of Nottingham UWCAES.

6.6 COMPRESSED AIR FOR MOBILITY

6.6.1 Compressed Air Car

As an alternative to the electric car, a compressed air car has been proposed [8].

Such a car corresponds to a zero emissions car, and the compressed air storage and propulsion system presents very low aging phenomena in comparison with battery based EVs. Figure 6.66 shows the Airpod from MDI, a small personal transport means dedicated to short range applications.

For this vehicle, many questions on the possible range and content of the reservoir have been addressed [25]. However, the principle of compressed air and the possibility of refilling the vehicle's reservoir in short time from an intermediary storage tank is an interesting property.

A fast refill station for the compressed air car was discussed in [26]. It is based on the use of an intermediary reservoir filled by an isothermal compression facility. The fast transfer of the high pressure air from the stationary reservoir to the car's reservoir can be realized in just a few minutes. This fast transfer, however, has some drawback like the heating of the air in the car's vessel, a well-known phenomenon in the area of rapidly filling high-pressure diving equipment. An interesting solution to this problem is to add a return pipe to the line between the car and the station allowing exchange, after the equalization of the pressure, of the heated air of the car with cold air from the station's reservoir. The benefit is that the car can leave the station with a higher energy content (higher air density). The energy loss due to the cooling of the air in the car is avoided.

FIGURE 6.66 The compressed air car (MDI).

Exercise 5: Compressed Air Energy Storage

Basics of Compression and Expansion

Figure 6.67 shows the P-V diagram of the compression and expansion round trip cycle. Visibly, in compression phase since the gas gets hot, we need to cool down the gas using coolers. In contrast, during expansion, since the gas gets cool, we need to implement a heating system.

In Figure 6.67a to d, the compression process is represented and in Figure 6.67e to h, the corresponding expansion is also illustrated. Four different cases are illustrated, first the ideal isothermal process (curves (a) and (e)), and second the one-stage polytropic process (curves (b) and (f)), where the shaded surfaces represent the spent respectively recovered energy. The compression and expansion surfaces correspond to the done or to be extracted work from the system for a state change from point 1 to 2 or in the inverse sequence. The third case (curves (c) and (g)) represents a three-stage compression expansion machine with interstage heat exchangers.

Finally, the fourth case represented in Figure 6.67d and h corresponds to a compression/expansion machine with increased thermodynamic performance due to the use of directly integrated exchangers (DIEs).

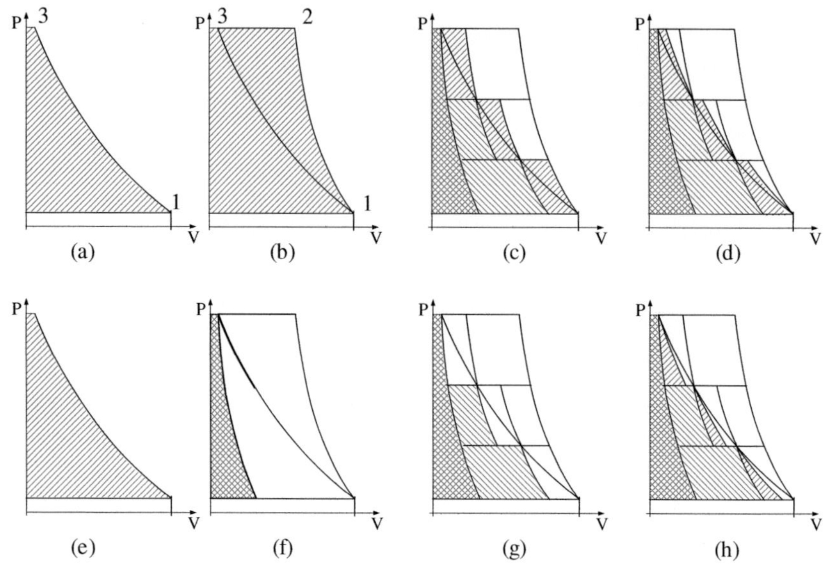

FIGURE 6.67 P-V diagram of compression and expansion.

Isothermal Expansion

The isothermal expansion of compressed air can be defined as an expansion process during which the temperature of air will stay constant and delivers the maximum extractable work, for example, using a piston in a cylinder, as described in Sylvain Lemofouet's PhD thesis "Investigation and Optimisation of Hybrid Electricity Storage Systems Based on Compressed Air and Supercapacitors" [1]. Sylvain Lemofouet describes in the introduction of his thesis the calculation of the energy amount provided by a pressurized reservoir of a given volume and at a given pressure level which corresponds to the mechanical work that can be extracted during an expansion at constant temperature conditions. The calculation proposed in this document is slightly different from the usual formulas, because it takes into account the atmospheric pressure acting on the rear side of the piston.

The goal of this exercise is to demonstrate Equation 6.1 of Section 6.1.1.

Exercise 5.1 Calculation of the Energy Content of a Pressurized Reservoir

Description of the System

The considered system comprises one air reservoir of a volume V_1, connected to a cylinder/piston system, as represented in Figure 6.68. The pressure level at t_0 is equal to P_1. The system is idealized as a piston/cylinder of infinite length. The section of the piston is defined as A_p.

For such a system, the following hypotheses are made:

Isothermal expansion: In both the reservoir and the piston system, there is a decrease of the pressure and would be a corresponding decrease of the

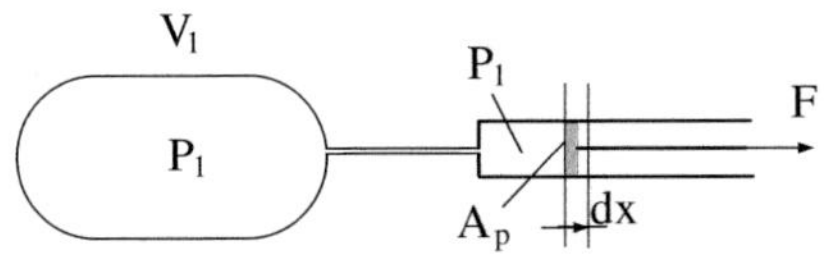

FIGURE 6.68 Idealized vessel and infinite cylinder.

temperature in relation to the ideal gas law. It is assumed that temperature remains constant during the entire expansion process, since it is a sufficiently slow expansion (quasi-equilibrium), allowing constant compensation of the temperature decrease by heat transfer from the surrounding.

Infinite cylinder: Particularly for high-pressure cylinders, the volume variation of a cylinder of real dimensions only allows a limited pressure decrease for one stroke. In practice, the principle of successive expansions and fillings is used, which includes additional dynamic phenomena of pressure exchange and gas circulation into and out of the cylinder of finite volume. In order to evaluate the ideal potential of the pressurized reservoir, it is considered that the expansion is made in a single cylinder of given section and of infinite length, allowing expansion of the air in a high (or large) pressure ratio.

Friction: The losses due to mechanical friction are neglected, as well as the air friction (viscous dissipation) inside the different air ducts.

Mechanical Work by Expansion

Goal: Calculation of the mechanical work to be produced with the piston/cylinder system when the pressure is decreasing from P_1 to the atmospheric pressure P_{at}.

The calculation should use the following equation:

$$W_{piston} = \int_0^{Xat} F(x)\cdot dx = \int_0^{Xat} (P_{res} - P_{at})\cdot A_p \cdot dx = \int_0^{Vat} (P_{res} - P_{at})\cdot dv \qquad (6.70)$$

Exercise 5.2 The Basic Thermodynamic Equation Explained through C_v and C_p

The specific heat is defined as the energy required to raise the temperature of a unit mass of a substance by one degree. In general, this energy depends on how the process is executed. In thermodynamics, we are interested in two kinds of specific heats:

C_v : Specific heat at constant volume

C_p : Specific heat at constant pressure

Physically, the specific heat at constant volume C_v can be viewed as the energy required to raise the temperature of the unit mass of a substance by one degree as the volume is kept constant (Q_1). The energy required to do the same as the pressure is kept constant (Q_2) is the specific heat at constant pressure C_p.

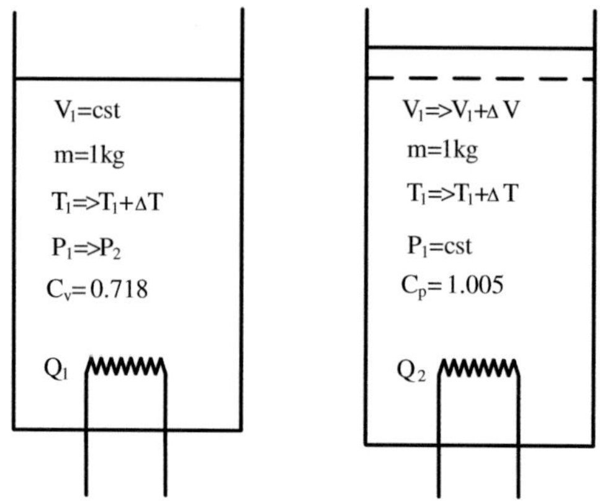

FIGURE 6.69 Heating the air under constant volume and constant pressure.

Figure 6.69 illustrates the two situations where the mass of air is heated up by transferring a heat amount. The amount of heat can be transferred to the system by the heating resistor R_1.

The goal of the exercise is to demonstrate that the physical properties of the air are the same in both cases but that it must be considered that the constant pressure condition related to C_p can only be achieved by changing the air volume when the heat quantity is added to the system.

TASKS

1. Calculate the volume of one mass of air equal to one kg at ambient temperature (293K) and ambient pressure ($P_1 = 10^5\,\text{N/m}^2 = 10^5\,\text{Pa}$).
2. Calculate the pressure level P_2 that is reached at constant volume after injection of a given heat quantity to the system that produces an increase of temperature of 1K.
3. Calculate the heat quantity Q_1 that produces the temperature raise of 1 K

$$C_v = 0.718\ \text{kJ}/(\text{K*kg}) \tag{6.71}$$

4. Calculate the heat quantity Q_2 that produces a temperature rise of 1K of the same mass of air under the condition of constant pressure

$$C_p = 1.005\ \text{kJ}/(\text{K*kg}) \tag{6.72}$$

5. For the same case as in 4, calculate the volume increase ΔV that is necessary to maintain the pressure in the reservoir when Q_2 is injected.
6. The final calculation should verify that the heat quantity Q_2 is equal to the sum of Q_1 and the mechanical work related to the volume increase under the pressure P_1.

Exercise 5.3:

Part a) Maximizing the Expansion Work

When the changes in kinetic and potential energies are negligible, the expansion work is given by

$$W_{out} = \int_1^2 v \cdot dp \tag{6.73}$$

Increasing the work output of an expander requires that the gas be heated as it is expanded.

To have a better understanding of the effect of heating during the expansion process, we compare the work output for three kinds of processes: an isentropic process (involves no heating), a polytropic process (involves some heating) and an isothermal process (involves maximum heating).

Assuming all three processes are executed between the same pressure levels (p_1 and p_2) in an internally reversible manner and the gas behaves as an ideal gas $(PV = RT)$ with constant specific heats,

1. Determine the expansion work for each process by performing the integration in Equation 6.73.

   Polytropic $(PV^n = condtant)$:

   Isentropic $(PV^k = condtant)$

   Isothermal $(PV = condtant)$:

2. Plot the three processes on a P-V diagram for the same inlet state and exit pressure. On a P-V diagram, the area to the left of the process curve is the integral of vdp. Thus it is a measure of the steady flow expansion work. Which process delivers the maximum work? Which one delivers the minimum work? Which one is between these two?

Part b) Multistage Expansion with Inter-Heating

It is clear from the arguments in part a) that heating a gas as it is expanded is desirable since this increases the work output from the expander. However, often it is not possible to have adequate heating through the casing of the expander, and it becomes necessary to use other techniques to achieve effective heating. One such technique is multistage expansion with inter-heating, where the gas is expanded in stages and heated between each stage by passing it through a heat exchanger called an inter-heater. Ideally, the heating process takes place at constant pressure, and the gas is heated to the initial temperature T_1 at each inter-heater. Multistage expansion with inter-heating is especially attractive when a gas is to be expanded from very high pressures.

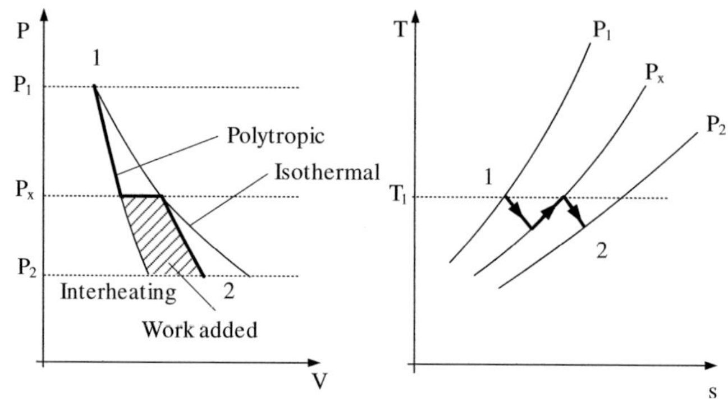

FIGURE 6.70 P-V and T-s diagrams for a two-stage expansion process.

The effect of inter-heating on expander work is graphically illustrated on P-V and T-s diagrams in Figure 6.70 for a two-stage expander. The gas is expanded in the first stage from P_1 to an intermediate pressure P_x, heated at constant pressure to the initial temperature T_1 and expanded in the second stage to the final pressure P_2. The expansion processes, in general, can be modeled as polytropic ($PV^n = condtant$), where the value of n varies between k and 1. The shaded area on the P-V diagram represents the work added as a result of two-stage expansion with the inter-heater. The process paths for single-stage isothermal and polytropic processes are also shown for comparison.

1. Determine the intermediate pressure P_x under which the work output is maximized. (Hint: set the differentiation of work with respect to P_x equal to zero.)
2. What is the relation between compression work at each stage?

SOLUTIONS TO EXERCISE 5

Exercise 5.1:

The expansion volume considers the presence of an initial volume V_1 (volume of the reservoir) in which the pressure is initially P_1. The displacement of the piston leads to an increase of the volume towards a new value V_{atm}, to be reached when the internal pressure reaches the atmospheric pressure P_{atm}. This volume can be easily calculated for an isothermal process:

$$V_{atm} = \frac{V_1 \cdot P_1}{P_{atm}} \tag{6.74}$$

The mechanical work produced by the pistons force can be calculated through

$$W_{piston} = \int_{V_1}^{Vatm} (P_{res} - P_{atm}) \cdot dv = \int_{V_1}^{Vatm} ((\frac{P_1 \cdot V_1}{v}) - P_{atm}) \cdot dv \tag{6.75}$$

$$W_{piston} = \int_{V_1}^{Vatm} \left(\frac{P_1 \cdot V_1}{v} \right) \cdot dv - \int_{V_1}^{Vatm} P_{atm} \cdot dv =$$

$$= \int_{V_1}^{Vatm} \left(\frac{P_1 \cdot V_1}{v} \right) \cdot dv - P_{atm} \cdot V_{atm} + P_{atm} \cdot V_1 = \tag{6.76}$$

$$= P_1 \cdot V_1 \cdot \ln \frac{V_{atm}}{V_1} - P_1 \cdot V_1 + \frac{P_{atm}}{P_1} P_1 \cdot V_1$$

$$W_{piston} = P_1 \cdot V_1 \cdot (\ln \frac{P_1}{P_{atm}} - 1 + \frac{P_{atm}}{P_1}) \tag{6.77}$$

Let us call this expression Lemofouet's formula.

In opposition to this expression, the common formulas give different results, mainly if the ratio of the pressures is low.

In Reference [27], the relation gives the "necessary work for isothermal compression"

$$W_i = \int_{p_1}^{p_2} mRT_1 \frac{dp}{p} = mRT_1 \cdot \ln \frac{p_2}{p_1}, \text{ with } mRT_1 = P_1 V_1 \tag{6.78}$$

This relation can be used for expansion and corresponds to the first term of Equation 6.77. It does not consider the effect of the external pressure on the rear side of the piston. In fact, this relation gives the maximum of work that can be extracted from the reservoir by expansion from a pressure P_2 to a pressure P_1 under isothermal conditions, corresponding to a situation where the external pressure is zero.

Comparison of the Results

One reservoir of 200 L at 300 bar contents according the Equation 6.79:

$$\mathrm{W}_i = 0.2 * 300 * 10^5 \ln(300) = 34{,}2 \text{ MJ} \tag{6.79}$$

According to Lemofouet's formula (6.77), it is possible (at atmospheric pressure) to extract

$$\mathrm{W}_{\text{piston}} = 0.2 * 300 * 10^5 \left(\ln(300) - 1 + 1/300 \right) = 28{,}24 \text{ MJ}$$

For the same volume of 200 L, the storage at 8 bar gives:

According to Equation 6.77: $\mathrm{W}_{\text{piston}} = 192$ kJ

According Equation 6.78: $\mathrm{W}_i = 332$ kJ

Exercise 5.2:

1. The volume of the air at atmospheric pressure is

$$PV = mRT \Rightarrow V = \frac{mRT}{P} = \frac{1kg \cdot 0.287 \frac{kJ}{kg \cdot K} \cdot 293K}{10^5 \frac{N}{m^2}} = 0.84m^3 \quad (6.80)$$

2. The pressure P_2 after the injection of the heat for a ΔT of 1K:

$$\frac{P_1V_1}{T_1} = \frac{P_2V_2}{T_2}; V_1 = V_2 \Rightarrow P_2 = \frac{T_2}{T_1}P_1 = \frac{294K}{293K} \cdot 10^5\,Pa = 1.0034 \cdot 10^5\,Pa \quad (6.81)$$

3. This quantity of heat (Q_1) is

$$Q_1 = mC_v\Delta T = 1kg \cdot 0.718 \frac{kJ}{kg \cdot K} \cdot 1K = 0.718kJ \quad (6.82)$$

4. The heat quantity (Q_2) that produces the same ΔT but with constant pressure

$$Q_2 = mC_p\Delta T = 1kg \cdot 1.005 \frac{kJ}{kg \cdot K} \cdot 1K = 1.005kJ \quad (6.83)$$

5. To keep the pressure constant in the previous case (4), the volume must be increased:

$$\frac{P_1V_1}{T_1} = \frac{P_2V_2}{T_2}; P_1 = P_2 \Rightarrow V_2 = \frac{T_2}{T_1}V_1 = \frac{294K}{293K} \cdot 0.84m^3 = 0.842866m^3 \quad (6.84)$$

6. The mechanical work related to the change of volume

$$\Delta V = 0.00287m^3$$

is

$$W = P\Delta V = 10^5 \frac{N}{m^2} \cdot 0.00287m^3 = 0.287kJ \quad (6.85)$$

and the difference between Q_2 and Q_1

$$Q_2 - Q_1 = 1.005kJ - 0.718kJ = 0.287kJ$$

$$\Rightarrow W = Q_2 - Q_1$$

Q_2 is greater than Q_1 because at constant pressure, the system is allowed to expand, and for this expansion, work must be supplied to the system.

EXERCISE 5.3:

Part a): Maximizing the Expansion Work

Polytropic (PV^n constant):

In a polytropic process, pressure and volume are related by $PV^n = C$, where n and C are constants.

The pressure for a polytropic process can be expressed as:

$$P = CV^{-n} \tag{6.86}$$

$$W_b = -\int_1^2 P dV = -\int_1^2 CV^{-n} dV = -C\frac{V_2^{-n+1} - V_1^{-n+1}}{-n+1} = \frac{P_2V_2 - P_1V_1}{1-n} \tag{6.87}$$

since $C = P_1V_1^n = P_2V_2^n$. For an ideal gas $(PV = mRT)$, this equation can also be written as

$$W_{\exp,out} = \frac{mR(T_2 - T_1)}{n-1} = \frac{mRT_1}{n-1}\left[\left(\frac{P_2}{P_1}\right)^{(n-1)/n} - 1\right] \tag{6.88}$$

where m is the mass in [kg]

T_1 is the temperature of the gas before the expansion, and R is the ideal gas constant

Isentropic PV^k = constant

For the case where $n = k = c_p / c_v$

$$W_{\exp,out} = \frac{mR(T_2 - T_1)}{k-1} = \frac{mRT_1}{k-1}\left[\left(\frac{P_2}{P_1}\right)^{(k-1)/k} - 1\right] \tag{6.89}$$

Isothermal PV = constant

For the special case of n = 1 the boundary work becomes

$$W_b = \int_1^2 P dV = \int_1^2 CV^{-1} dV = PV \ln\frac{V_2}{V_1} \tag{6.90}$$

and as a specific work $w_{\exp,out}$

$$w_{\exp,out} = RT \ln\frac{P_2}{P_1} \tag{6.91}$$

The curves related to the three cases are represented in Figure 6.71.

b) Multistage Expansion with Inter-Heating

The size of the shaded area (the added work output, Figure 6.70) varies with the value of the intermediate pressure P_x, and it is of practical interest to determine the

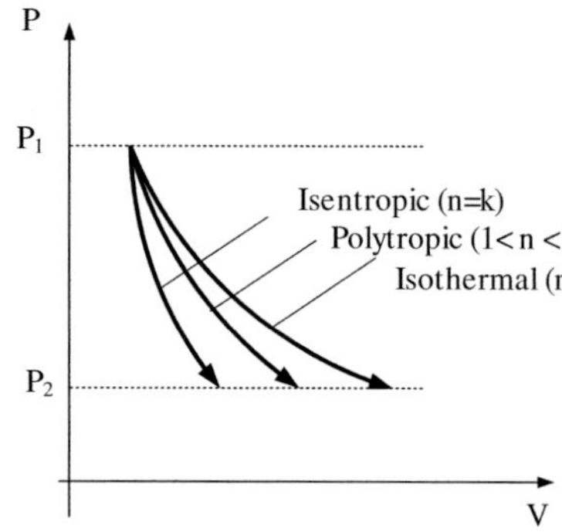

FIGURE 6.71 P-V diagrams of isentropic, polytropic and isothermal expansion processes between the same pressure limits.

conditions under which this area is maximized. The total work output for a two-stage expander is the sum of the work outputs for each stage of expansion, as determined from

$$W_{\exp,out} = W_{\exp I,out} + W_{\exp II,out} = \frac{mRT_1}{n-1}\left[\left(\frac{P_2}{P_x}\right)^{(n-1)/n} - 1\right] + \frac{mRT_1}{n-1}\left[\left(\frac{P_x}{P_2}\right)^{(n-1)/n} - 1\right] \quad (6.92)$$

The only variable in this equation is P_x. The value of P_x that maximizes the total work is determined by differentiating this expression with respect to P_x and setting the resulting expression equal to zero. It yields

$$P_x = (P_1 / P_2)^{1/2} \quad \text{or} \quad \frac{P_x}{P_2} = \frac{P_1}{P_x} \quad (6.93)$$

That is, to maximize expansion work during two-stage expansion, the pressure ratio across each stage of the expander must be the same. When this condition is satisfied, the expansion work at each stage becomes identical, that is,

$$W_{\exp I,out} = W_{\exp II,out} \quad (6.94)$$

REFERENCES

[1] Lemofouet, S., Investigation and optimisation of hybrid electricity storage systems based on compressed air and supercapacitors, Thesis EPFL, Lausanne, Switzerland, 2006.

[2] Heidari, M., Contribution to the technique of compressed air energy storage: The concept of finned piston, Thesis EPFL No. 6738, Lausanne, Switzerland, 2015, http://infoscience.epfl.ch/record/211042/files/EPFL_TH6738.pdf. Accessed on September 22, 2017.

[3] Radeska, T., The Mekarski system—compressed-air propulsion system for trams, *The Vintage News*, October 17, 2016 [Online], www.thevintagenews.com/2016/10/17/the-mekarski-system-compressed-air-propulsionsystem-for-trams/ . Accessed on July 28, 2017.

[4] Lehmann, J., Air storage gas turbine power plants, a major distribution for energy storage, *International Conference on Energy Storage*, Brighton, UK, April 1981, pp. 327–336.

[5] The Mc Intosh CAES Power Plant, www.powersouth.com/files/CAES%20Brochure%20[FINAL].pdf. Accessed on September 22, 2017.

[6] Bradshaw, D. T., Pumped hydroelectric storage (PHS) and compressed air energy storage (CAES), *Power Engineering Society Summer Meeting*, Seattle, WA, July 16–20, 2000, vol. 3, pp. 1551–1573.

[7] Kermit, A., CAES: The underground portion, *IEEE Transactions on Power Apparatus and Systems*, PAS-104(4), 809–812, 1985, https://doi.org/10.1109/TPAS.1985.319078.

[8] Negre, G., Negre, C., Engine with an active mono-energy and/or bi-energy chamber with compressed air and/or additional energy and thermodynamic cycle thereof, PCT filed: PCT/FR2004/002929, Carros (CEDEX), US Patent, US 7469527 B2, MDI Motor Development International S. A., Luxemburg, January 17, 2007/December 30, 2008.

[9] Heidari, M., Rufer, A., Fluid flow analysis of a new finned piston reciprocating compressor using pneumatic analogy, *International Journal of Materials, Mechanics and Manufacturing*, 2(4), 297–301, 2014.

[10] Giannoni, M., Un systeme original de stockage d'energie, *La Revue POLYTECHNIQUE*, 1812(4), 2016 [Online], http://polymedia.ch/RP/Articles/view/858. Accessed on September 22, 2017.

[11] Haislip, B., Energy-storage startup LightSail plots long-term game plan, *The Wall Street Journal*, February 21, 2016 [Online], www.lightsail.com/press/lightsailenergy-plots-long-term-game-plan. Accessed on May 9, 2017.

[12] Garvey, S. et al., *Compressed Air Energy Storage has Bags of Potential*, University of Nottingham, Nottingham, UK, 2012, www.theengineer.co.uk/issues/25april-2011/compressed-air-energy-storage-has-bags-of-potential/. Accessed on September 22, 2017.

[13] Favrat, D., *Thermodynamics and Energy Systems Analysis: From Energy to Exergy*, EPFL Press, Lausanne, Switzerland, 2010.

[14] Lemofouet, S., Rufer, A., A hybrid energy storage system based on compressed air and supercapacitors with maximum efficiency point tracking (MEPT), *IEEE Transactions on Industrial Electronics*, 53(4), 1105–1115, 2006.

[15] Kabinga, R. K. N., Tartibu, L. K., Okwu, M. O., Development and performance evaluation of a dynamic compressed air energy storage system, *Proceedings of the International Mechanical Engineering Congress and Exposition*, Portland, OR, USA, November 16–19, 2020.

[16] Cheng, Y., Chen, K., Chan, C., Bouscayrol, A., Cui, S., Global modeling and control strategy simulation, *IEEE Vehicular Technology Magazine*, 4, 73–79, 2009.

[17] Shearer, M., Gremaud, P., Kleiner, K., Periodic motion of a mass–spring system, *IMA Journal of Applied Mathematics*, 74, 807–826, 2009.

[18] REFPROP, *NIST Reference Fluid Thermodynamic and Transport Properties Database (REFPROP)*, www.nist.gov/srd/refprop.

[19] Rufer, A., Increasing the energy efficiency of gas boosters for hydrogen storage and for refueling stations, *Energies*, 16, 1763, 2023.

[20] Cengel, Y. A., Boles, M. A., *Thermodynamics: An Engineering Approach*, 6th edition, McGrow-Hill Companies, New York, NY, USA, 2008, ISBN 978-0-07-352921-9.

[21] Lefevre, P., Rufer, A., Le stockage hydropneumatique, in *Technologies du stockage d'énergie* (edited by Y. Brunet), Traite EGEM, serie Genie electrique, Lavoisier, Paris, France, November 2009.

[22] Hitchcox, A., Water hydraulics: Benefits and limitations, *Hydraulics & Pneumatics*, December 31, 2011 [Online], www.hydraulicspneumatics.com/200/TechZone/HydraulicFluids/Article/False/6452/TechZone-HydraulicFluids. Accessed on July 28, 2017.

[23] Garvey, S. et al., *Compressed Air Energy Storage has Bags of Potential*, University of Nottingham, 2012, www.theengineer.co.uk/issues/25-april-2011/compressed-air-energy-storage-has-bags-of-potential/.
[24] de Jong, M., Commercial grid scaling of energy bags for underwater compressed air energy storage, *Proceedings of the 2014 Offshore Energy & Storage Symposium, UWCAES Society*, Windsor, Ontario, Canada, July 10–11, 2014, www.tandfonline.com/doi/full/10.1080/00207233.2014.947726?scroll=top&needAccess=true. Accessed on September 22, 2017.
[25] Fairley, P., Deflating the air car, Green cars could run on compressed air instead of batteries. But don't rely on the new AirPod minicars to prove it, *IEEE Spectrum*, November 2009 [Online], http://spectrum.ieee.org/energy/environment/deflating-the-air-car. Accessed on July 29, 2017.
[26] Rufer, A., Lemofouet, S., Habisreutinger, M., Heidari, M., Leuba, A., Driving and filling personal vehicles—the questions of energy—and power—density (a fast filling station for the compressed air car), *World Engineer's Convention*, Geneva, Switzerland, September 4–9, 2011.
[27] Gieck, K., *Technical Formulae*, Karl Hanser Verlag GmbH & Co., Munich, Germany, April 2009.

7 Hydropower and Gravitational Storage Facilities

7.1 GENERAL VIEW

Hydropower has been used since ancient times as a renewable energy source for the operation of various mechanical devices such as flour mills, saw mills, trip hammers or many other applications. The energy reservoirs behind the different devices converting potential energy from water falls or kinetic from fast-running streams have been realized through building dams on rivers, in valleys or using natural basins. Filling these reservoirs has been based in most cases on the natural flow of water from creeks and rivers, collecting rainfall or water from snowmelt.

Purely mechanical hydropower systems have already included the principle of energy storage and for many of them also the concept of controllable power flow. Figures 7.1 and 7.2 illustrate the use of hydropower over the centuries.

After the realization of these early water turbines, the reverse transformation of power has appeared for pumping water into reservoirs with many kinds of norias, mainly used for irrigation.

Figure 7.3 shows an early noria used for feeding a waterduct from the river Orontes (Hama, Syria).

At the end of the 19th century, hydropower became an important source for the generation of electricity. Over the years, the technique of electrical hydropower has been intensively developed, from tens of kW (Figure 7.4) to tens of GW (Figure 7.5).

The significant development of hydro-electrical energy storage dates from the second half of the 20th century in the context of matching the variable power demand (daily cycles) with the installation of nuclear power plants, known for their mostly constant power production. Figure 7.6 shows a typical power profile of the weekly consumption, Figure 7.7 illustrates a pumped-storage power plant (PSP) built in Luxemburg in this context. Many other PSPs have been realized with the same motivations, as the examples of the Hongrin-Leman facility in Switzerland; the Raccoon Mountain Pumped-Storage Power Plant west of Chattanooga, Tennessee, US; or the Yagisawa Power Station in Japan [1–3]. A historical and geographical perspective on pumped storage power plants is given in Reference [4].

DOI: 10.1201/9781003736486-7

FIGURE 7.1 Use of hydropower over the centuries. (a) Roman turbine mill at Chemtou, Tunisia. The tangential water inflow of the millrace made the horizontal wheel in the shaft turn like a true turbine, the earliest known. (b) A Northern Song era (960–1127) water-powered mill for dehusking grain with a horizontal wheel.

FIGURE 7.2 A flour mill from the 19th/20th century (Le Moulin, Villars-sous-Yens, Switzerland).

FIGURE 7.3 Noria at Hama, Syria.

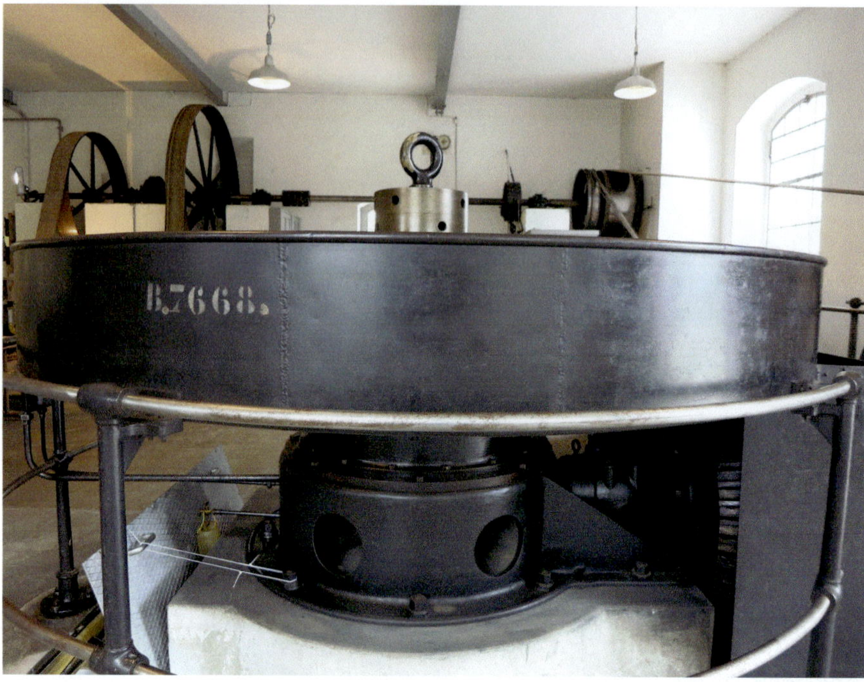

FIGURE 7.4 Small electrical generation.

FIGURE 7.5 The world's largest-capacity hydropower plant, Three Gorges, China, 22 GW. (By Christoph Filnkössl.)

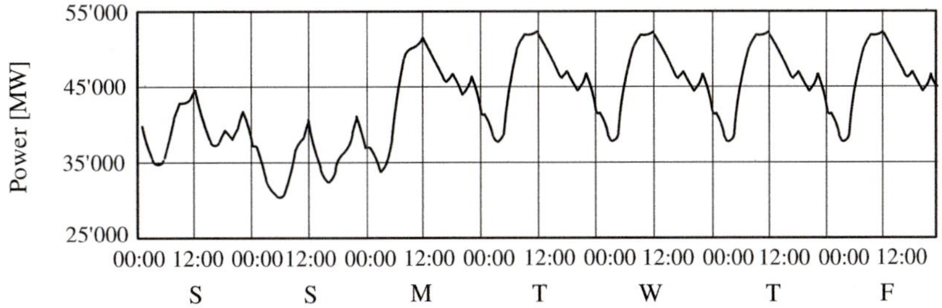

FIGURE 7.6 Daily variations of the power demand, profile over one week.

FIGURE 7.7 Pumped-storage plant (Vianden, Luxemburg).

7.1.1 General Properties

Hydraulic pumped storage facilities allow the storage of a tremendous amount of energy, the supply of the network over several days, months or compensating in some cases for the seasonal demand [5].

The ratings of hydropower were briefly presented in Section 3.1. A simpler representation is given in Figure 7.8. Pumped storage power plants have used conventional machinery for a long time, where the electric motor/generator has been directly connected to the grid. For such facilities, even when the power level in generation mode can be controlled through the governor, the variation of power in pumping mode is not possible due to the fixed rotational speed of the equipment.

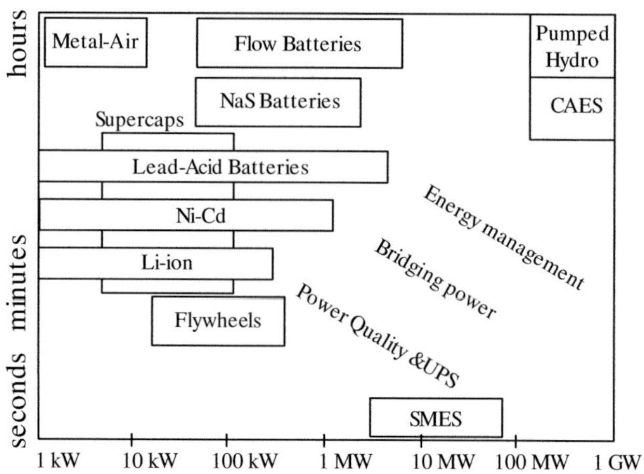

FIGURE 7.8 Ratings of storage facilities. (Adapted from [5].)

Controlling power in pumping mode is the main reason for modern pumped storage equipment to feature variable speed of the rotating machinery. Furthermore, variable-speed technology made it possible to handle larger variations of headwater and tailwater reservoir levels.

Pumped storage facilities benefit from good conditions regarding energetic efficiency and can be operated over several decades concerning the machinery, even if some periodic revisions are needed. Regarding the realized dams, the lifetime is in the range of centuries. The lifetime and efficiency parameters are presented in Figure 3.14.

7.2 PRINCIPLES OF OPERATION

7.2.1 Components of a Hydropower Plant

Figure 7.9 represents a classical pumped storage facility, where water flows from the lower reservoir (tailwater reservoir) to the upper one (headwater reservoir) in pumping mode and in the inverse direction for generating mode [6].

From the headwater reservoir, the water flows through a tunnel and a penstock to the hydraulic machine. This hydraulic machine, which is generally a reversible machine (pump turbine), is coupled to an electric generator/motor that is generally a synchronous machine.

For modern variable-speed systems, the synchronous machine can be replaced by an asynchronous one, a so-called doubly fed induction motor generator (DFIMG). For such a system, the needed power electronic converter is designed for only a fraction of the full power [7]. Another solution for variable-speed operation is to

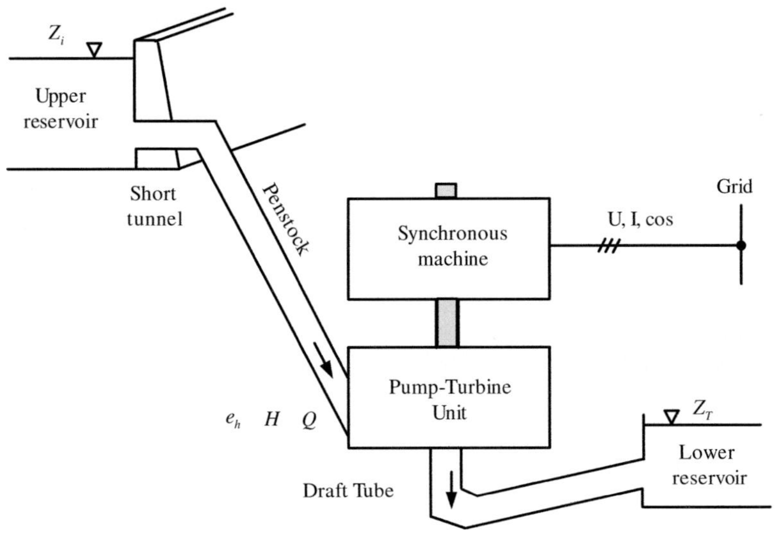

FIGURE 7.9 Schematics of a pumped-storage facility.

use a synchronous machine with a power-electronic converter designed for the full power [8].

7.2.2 Water Power, Definitions

Figure 7.9 provided the following definitions:

Head: $H = H_1 - H_2$ [m] where the head is derived from the headwater and tailwater reservoir levels Z_i and Z_T, respectively, as follows

In generating mode, $H = H_1 - H_2 = (Z_i - Z_T) - H_r^{Gen}$ with H_r^{Gen}, the head loss in the hydraulic circuit in turbine mode

In pumping mode ,
$H = H_1 - H_2 = (Z_i - Z_T) + H_r^{Pump}$ with H_r^{Pump}, the head loss in the hydraulic circuit in pumping mode

Throughout this document, the head losses are considered negligible.

Discharge: $Q[m^3/s]$

Specific hydraulic energy: $e_h = g \cdot H = g \cdot (H_1 - H_2)\,[(m/s^2) * m] = [J/kg]$

with water density $\rho\,[kg/m^3]$ and acceleration due to gr $g = 9.81\,[m/s^2]$.

7.3 ELEMENTARY RELATIONS OF THE HYDROPOWER GENERATION

Potential energy:

The potential energy of a mass of water accumulated through elevation is described ideally by the following relation:

$$E = m \cdot e_h = m \cdot g \cdot h \tag{7.1}$$

where *eh* is the specific hydraulic energy

Example

One mass of water of 1000 kg (1 m^3) at a height of 1000 m has a potential energy of

$$E = 1000kg \cdot 9.81m/s^2 \cdot 1000m = 9.81 \cdot 10^6\,Ws = 2.73kWh$$

Kinetic energy:

If a mass of water begins to set in motion (friction free), its accumulated potential energy is converted into kinetic energy according to:

$$E = \frac{m \cdot v^2}{2} \tag{7.2}$$

So the velocity reached by one mass of water through conversion from potential fall energy into kinetic energy is ideally calculated from:

$$m \cdot g \cdot h = \frac{m \cdot v^2}{2} \tag{7.3}$$

leading to (Figure 7.10):

$$v = \sqrt{2 \cdot g \cdot h} \tag{7.4}$$

Examples

The theoretical velocities obtained from falls equal to 10 m, 100 m and 1000 m are:

$$10m : v = \sqrt{2 \cdot 9.81m \cdot s^{-2} \cdot 10m} = 14.0m \cdot s^{-1}$$

$$100m : v = \sqrt{2 \cdot 9.81m \cdot s^{-2} \cdot 100m} = 44.3m \cdot s^{-1}$$

$$1000m : v = \sqrt{2 \cdot 9.81m \cdot s^{-2} \cdot 1000m} = 140.0m \cdot s^{-1}$$

Pressure:

Considering one reservoir as represented in Figure 7.10, the pressure at the bottom of the reservoir can be calculated according to:

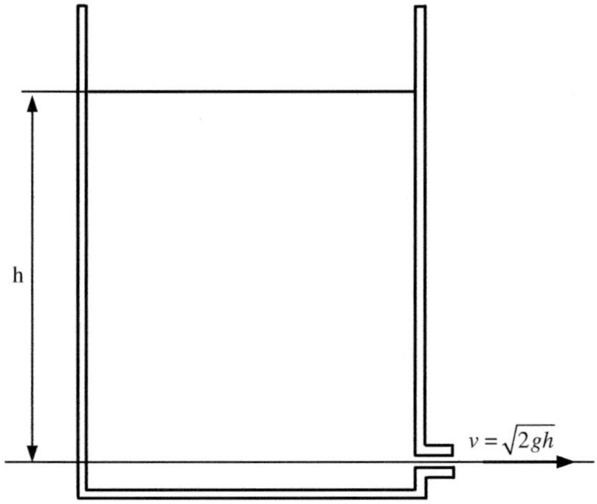

FIGURE 7.10 Conversion from potential to kinetic energy.

$$p = \rho \cdot g \cdot h \tag{7.5}$$

Example

The pressure at the bottom of a reservoir having a height of 100 m is equal to

$$\begin{aligned} p &= 1000kg \cdot m^{-3} \cdot 9.81m \cdot s^{-2} \cdot 100m \\ &= 9.81 \cdot 10^5 \, Pa \\ &= 9.81bar \end{aligned}$$

According the Bernoulli equation:

$$p + \frac{1}{2}\rho \cdot v^2 + \rho \cdot g \cdot h = Cst \tag{7.6}$$

it can be written for the outcoming fluid jet:

$$\rho \cdot g \cdot h = \frac{\rho \cdot v^2}{2}$$

leading to the same result:

$$v = \sqrt{2 \cdot g \cdot h}$$

The velocity of the water in the lower penstock section is calculated as:

$$velocity = \frac{Flowrate}{Section\_of\_the\_tube}$$

Example

Twelve turbines each absorb 3.75 m^3/s, diameter of the lower section 2.9 m (Nendaz/Grande-Dixence, head 1000 m)

$$v = \frac{3.75m^3/s \cdot 12 \cdot 4}{2.9^2 m^2 \cdot 3.14} = 6.8m/s$$

Power

A turbine is ideally converting the potential energy $m \cdot g \cdot h$ of the water into mechanical energy in a given time unit t. Its power (friction not considered) is defined as the per-second converted amount of energy

$$P = \frac{m}{t} \cdot g \cdot h = \rho \cdot Q \cdot e_h \quad (7.7)$$

Example

2 m^3/s from 500-m head

$$P = \frac{2000kg}{1s} \cdot 9.81m/s^2 \cdot 500m = 9.81 \cdot 10^6 \, Nm \cdot s^{-1} = 9.81 \cdot 10^6 \, W = 9.81MW$$

For a real turbine, the converted power is affected by the turbine efficiency

$$P = \eta \cdot \rho \cdot Q \cdot e_h \quad (7.8)$$

7.4 PUMP TURBINES

7.4.1 Classification of Turbines

The different types of hydraulic machines are represented in Figure 7.11. The classification is made using a specific parameter of hydraulic machines, the specific speed.

The specific speed is a dimensionless quantity used to characterize hydraulic turbo-machines [9]. According to the IEC standard, n_{QE} is given by

$$n_{QE} = \frac{n \cdot Q^{1/2}}{e_h^{3/4}} \quad (7.9)$$

where

n_{QE} is the specific speed (dimensionless)
n is the runner/impeller rotational speed (Hz)
Q is the flowrate (m^3/s) at the point of best efficiency
H is the head (m) per stage at the point of best efficiency

Alternatively, hydraulic engineers often use the unit-specific speed n_Q, defined as

$$n_Q = N \frac{Q^{1/2}}{H^{3/4}} \quad (7.9bis)$$

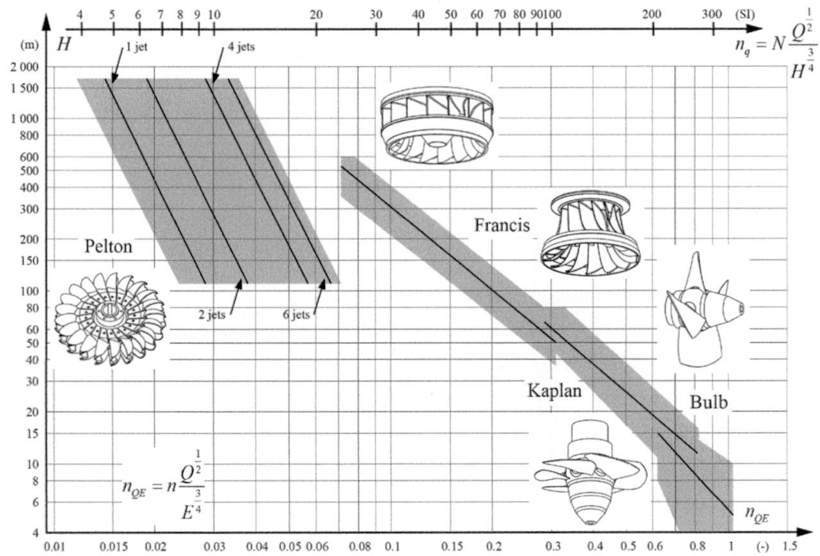

FIGURE 7.11 Head versus specific speed for hydraulic turbines and the corresponding shape of the runners.

where

n_{QE} is the specific speed (dimensionless)
N is the runner/impeller rotational speed (min^{-1})
Q is the flowrate (m^3/s) at the point of best efficiency
H is the head (m) per stage at the point of best efficiency

There are three main types of hydraulic machines and the corresponding shape of the runner: the Pelton turbine, the Francis turbine and the Kaplan turbine.

1. The Pelton turbine is an impulse turbine with tangential flow and is suited for high head applications.
2. The Francis turbine is a reaction machine with radial flow and is suited for medium head applications.
3. The Kapan turbine is another reaction machine but with axial flow. It is suited for low head applications.

The different shapes of the runners of each type are also represented in Figure 7.11.

7.4.2 Ternary Pump Turbine Units

Pump turbine units based on the use of Pelton machines don't benefit from the reversibility of this type of turbine. As a consequence, the train must be completed by a separated pump.

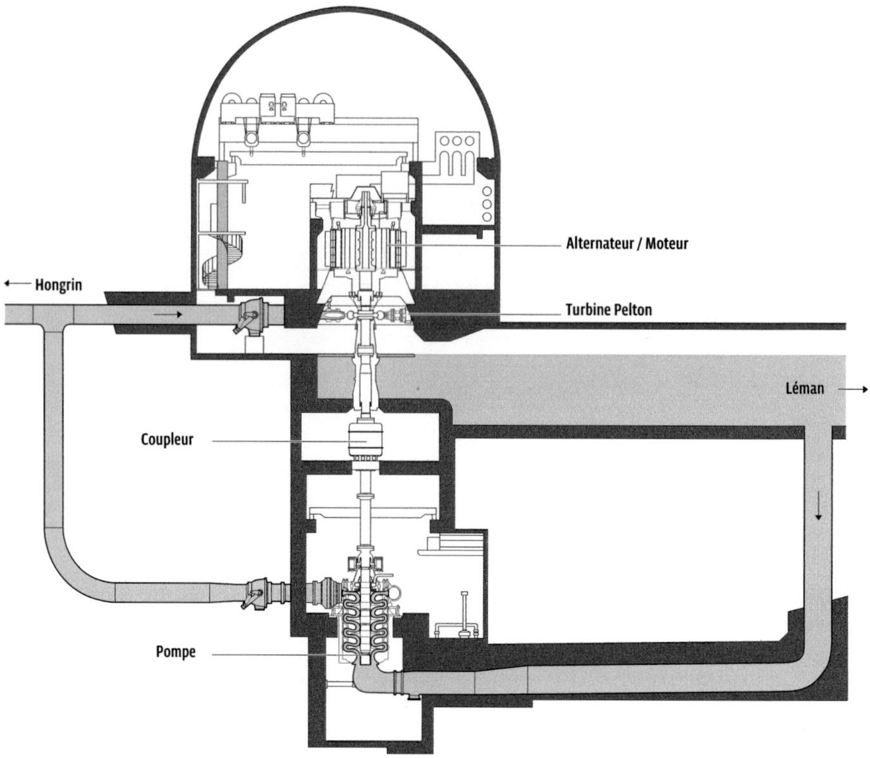

FIGURE 7.12 Example of the 120 MW pump turbine ternary arrangement of FMHL (Forces Motrices Hongrin Leman, Switzerland).

The drawback of this solution is evidently the higher length of the rotating train. However, the train can be started by the turbine without an additional start-up machine.

The rotating train has only one direction of rotation, and a changeover from turbine mode to pump mode is possible without changing the direction of rotation.

The system is characterized by a high-grade control for the Pelton turbine and the hydraulic bypass.

Figure 7.12 shows an example of a ternary unit as is realized in the FMHL$^+$ h plant in Veytaux Switzerland [10]. The FMHL plant uses an upper basin called Lac de l'Hongrain, and the lower basin is the lake of Geneva (Lac Léman). One of the characteristics of the plant is the submerged pump under the level of the lake.

7.4.3 Reversible Pump Turbine Units

Figure 7.13 illustrates an example of a reversible Francis machine, a 250-MW pump turbine of a 500-rpm facility.

FIGURE 7.13 Reversible pump-turbine unit of 250 MW/500 rpm. (Courtesy GE/Alstom, Grenoble, France.)

By reversible pump turbine sets, the direction of rotation must be reversed for the changeover from turbine mode to pump mode. This can be achieved using a so-called pony-motor for the acceleration of the train from standstill to the synchronous rotational speed. Other solutions exist where the start-up is realized in the pump mode through a static frequency converter feeding the synchronous machine with variable frequency.

As will be explained in Section 7.5, reversible pump-turbines are traditionally operated with constant speed. But they can also be operated with variable speed, as will be described in Section 7.6. Two solutions will be discussed, the so-called doubly fed asynchronous motor/generator and the synchronous motor-generator with full-conversion principle, where the static frequency converter is designed for the full power of the motor/generator. For both solutions, the reversal of the train is achieved through the main frequency converter.

7.5 FIXED-SPEED PUMPED STORAGE

In Figure 7.14, the structural diagram of a fixed-speed pumped storage plant is shown. The synchronous machine serves as well as a motor in the pump mode as a generator for the turbine mode.

The synchronous machine is connected to the grid by its armature and is coupled to the reversible hydraulic machine.

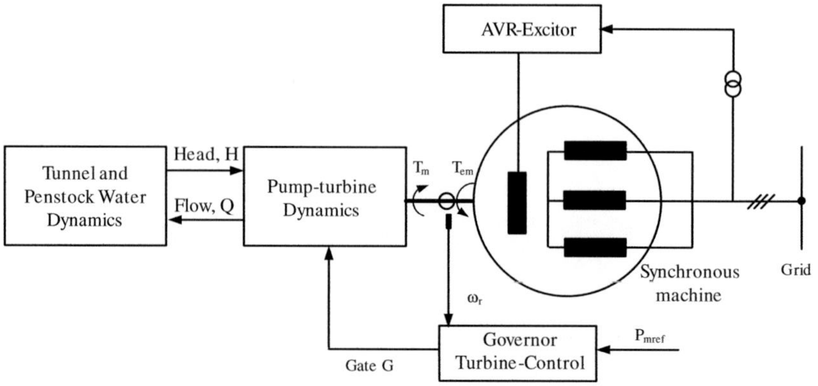

FIGURE 7.14 Fixed-speed pump-turbine system.

The active power is controlled through the governor and turbine control. The imposition of the voltage as well as the control of the reactive power is achieved through the machine's excitation system.

Because the pump discharge (water flowrate) is linked to the rotation speed of the machines, the level of the power in the pump mode cannot be varied when the motor runs at the synchronism of the grid frequency. This limitation is one of the reasons variable-speed pump storage facilities were constructed.

The domain of fixed-speed pump-storage, as in general the domain of hydropower generation, is a well-known subject. But in relation to new methods of exploitation linked with the integration of renewable sources, new approaches and studies are emerging nowadays.

In Reference [7], several models are detailed for the system transient and long-term dynamic analysis.

7.6 VARIABLE-SPEED PUMPED STORAGE

In order to be able to vary the power in pumping mode, but also for optimizing turbine mode, variable-speed pump storage plants are realized [2].

For variable-speed pumped storage facilities, two different solutions are used for the energy conversion from the mechanical shaft to the electric grid. The simplest method for the adaptation of the variable frequency of the synchronous motor/generator to the grid frequency is to use a static frequency converter carried out with power electronic modules. This frequency converter transforms the entire electric power from the stator side of the machine where the stator frequency is linked to the rotation speed of the machines to the electrical grid with its constant frequency.

A second system used for variable pump storage plants is the so-called doubly fed induction motor/generator. In such a system, the electric machine is of the asynchronous type where the wound rotor is interconnected to a frequency converter via slip rings and brushes. The machine's stator is directly connected to the grid.

The bidirectional active feeding of the machine's rotor makes it possible to cover significant speed excursions in the hyper- and hyposynchronous domain.

7.6.1 Variable-Speed Pump Storage with Full Conversion

Variable-speed pump storage facilities with full conversion generally use voltage source converters (VSCs) connected back to back at the level of an intermediary DC circuit. Such converters are also called pulse width modulation (PWM) converters due to their switched mode of operation [8].

At the side of the grid, the converter is operated at grid synchronism and overtakes the functions of active and reactive current control.

At the motor/generator side, the converter adapts the stator frequency and also the stator voltage level [8].

The flux in the generator is additionally controlled through the excitation winding (DC) fed from a third converter also connected to the grid.

More recent developments propose advanced converter technology using modular multilevel converter (MMC) architectures [11].

Figure 7.15 suggests one possible configuration for variable-speed operation of a pumped storage with control system for operation in pumping mode [12].

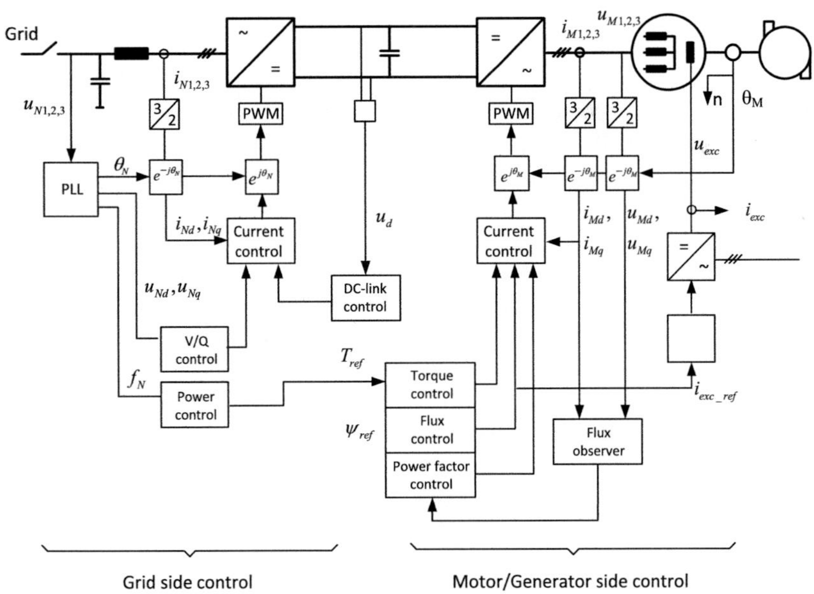

FIGURE 7.15 Suggested configuration for variable-speed operation of pumped storage with control system for operation in pumping mode. (Adapted from Suul, J. A., Uhlen, K., Undeland, T., Wind power integration in isolated grids enabled by variable speed pumped storage hydropower plant, *IEEE Conference on Sustainable Energy Technologies, ICSET 2008*, Singapore, 2008.)

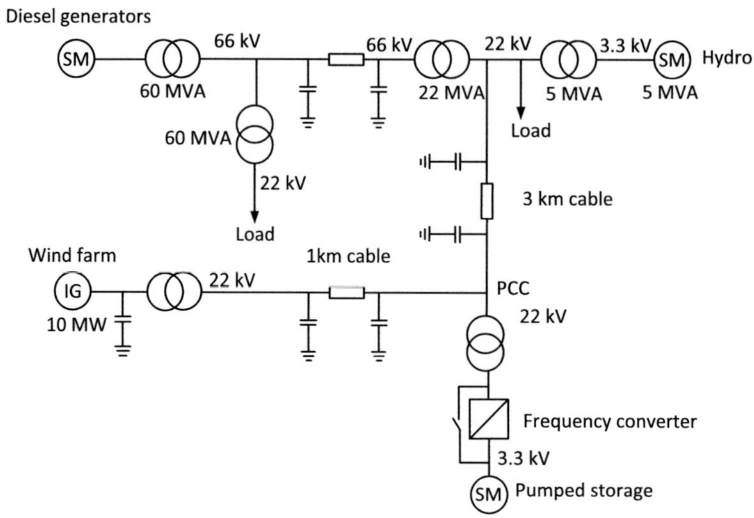

FIGURE 7.16 Simplified grid model representing the main part of the power system on the Faroe Islands. (Adapted from Suul, J. A., Uhlen, K., Undeland, T., Wind power integration in isolated grids enabled by variable speed pumped storage hydropower plant, *IEEE Conference on Sustainable Energy Technologies, ICSET 2008*, Singapore, 2008.)

In Figure 7.16, the context of application of a small pumped storage system is represented; see Reference [12].

Pumped storage plants with high power have also been carried out recently as in the Grimsel 2 facility of the KWO plant (Kraftwerke Oberhasli) in the Bernese Oberland of the Swiss Alps. The variable frequency operation of the 100 MW synchronous machine is possible through the use of a high-end power electronics converter. The very high-power VSC used for the synchronous machine is carried out through the coupling of submodules with transformers. Two line side transformers feed two line side converters. They are connected to the machine side converters and transformers via two separated DC links. The so-called three-level technology is based on a positive and a negative voltage intermediary DC circuit. The schematic of the complete installation of the Grimsel 2 facility is represented in Figure 7.17. The full conversion variable-speed operation range is defined from 690 to 765 rpm, corresponding to stator frequencies between 46 and 51 Hz.

Figure 7.18 shows more details on the KWO 100 MW facility, with the left side of the figure the variable speed rotating machinery and the right side of the figure the volume occupied in the cavern by the transformers and the frequency converter.

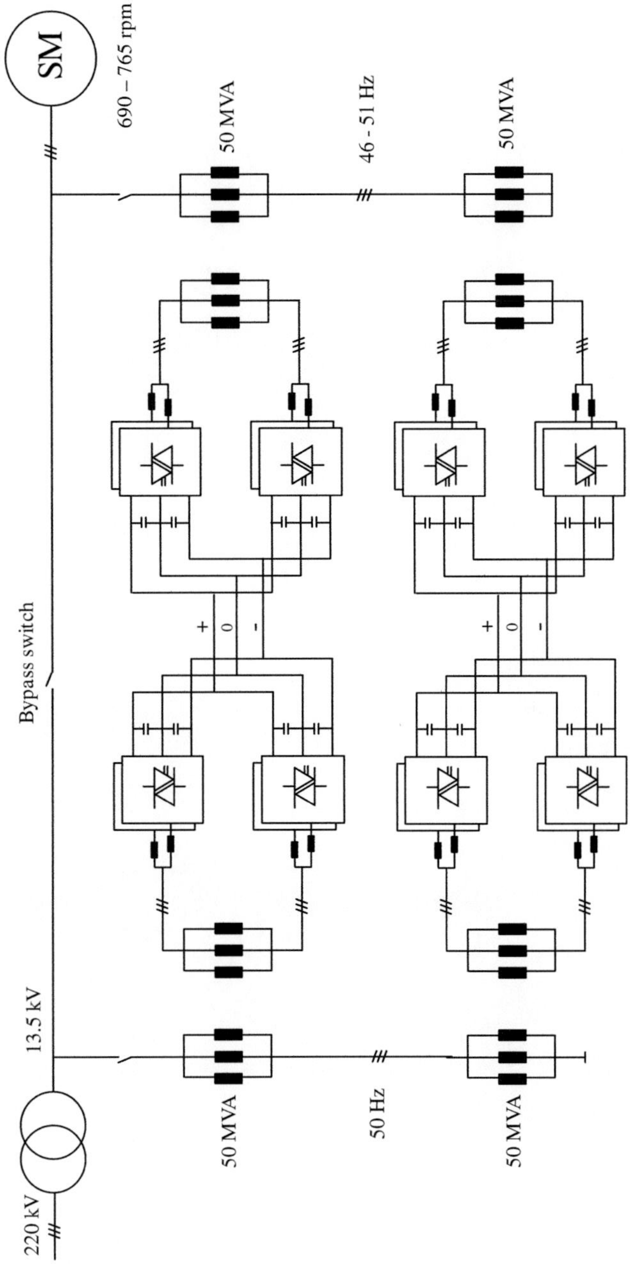

FIGURE 7.17 Schematic of the 100 MW pumped-storage facility of KWO/Grimsel 2 plant.

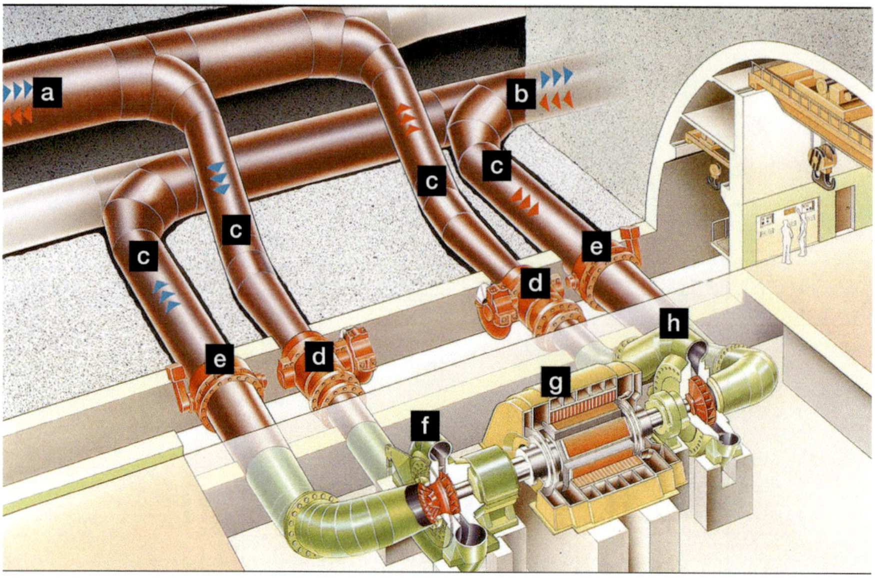

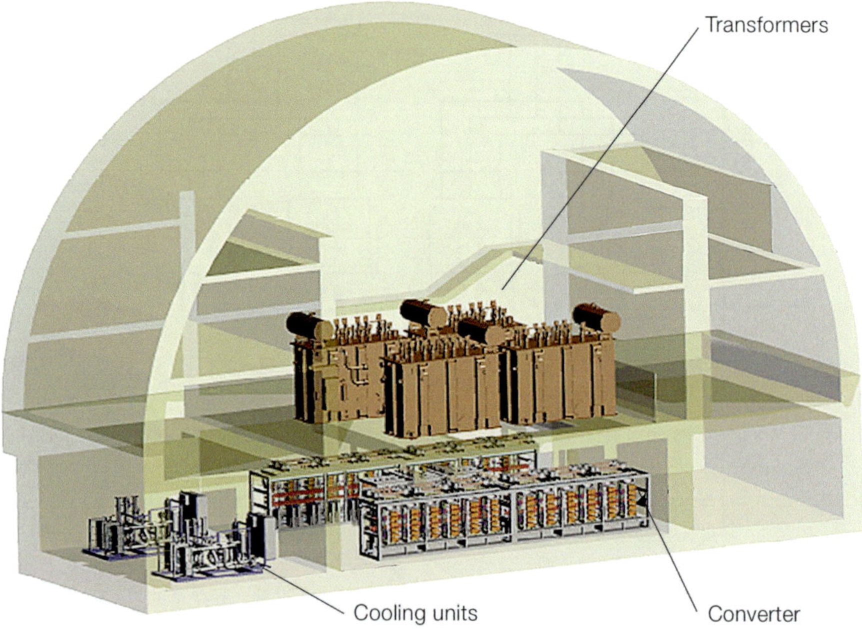

FIGURE 7.18 Rotating machinery and frequency converter with transformers of the 100 MW facility of KWO (courtesy ABB Review [8]). (a) Pressure shaft to/from Lake Oberaar, (b) pressure shaft to Lake Grimsel, (c) connection pipes, (d) spherical valve, (e) throttle, (f) turbine, (g) generator/motor, (h) pump. (From Schlunegger, H., ABB Review, 2(14), 42, 2014.)

7.6.2 Pumped Storage with Variable-Speed DFIMG

In Figure 7.19, the block diagram of a DFIMG is given. The stator of the machine is directly connected to the grid, and the rotor of the machine is connected to the converter through slip-rings. The rotor-side converter is designed for a fraction of the total power, namely slip power. As a result, the energy efficiency of the total system is sensibly higher than the efficiency of a full conversion system, because the conduction and switching losses of the converter affect only one fraction of the transformed power [13, 14].

While traditional realizations have used thyristor-based cyclo-converters for a long time, VSCs are today well established [15]. The flows of power in a DFIMG are represented for generator operation in Figure 7.20.

- P_s is the active power flow from the stator to the grid.
- P_r is the active power flow through the rotor and the converter chain.
- P_{mech} is the mechanical power on the shaft
- P_{EEO} is the total active power from the DFIMG to the grid.
- n is the rotation speed
- n_s is the synchronous speed

In Figure 7.20, generating mode is represented. The rotation speed is negative as well as the mechanical power. The expression for the rotor power is given by

$$P_r = P_s\left(\frac{n}{n_s} - 1\right) \tag{7.10}$$

In supersynchronous operation, the negative rotor power is added to the negative stator power, resulting in an increase of the transmitted amount of (negative) grid power.

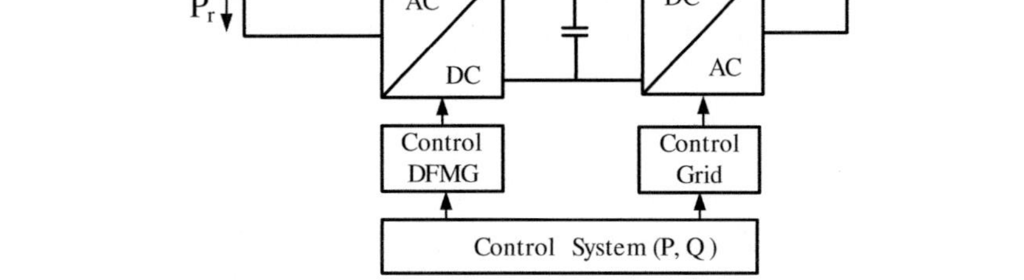

FIGURE 7.19 Block diagram of a DFIMG.

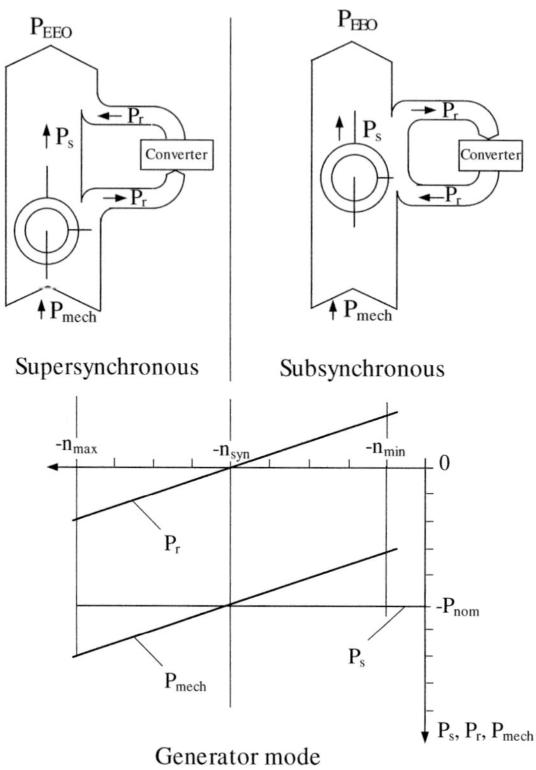

FIGURE 7.20 Power flows within a DFIMG generating mode.

In subsynchronous operation, the rotor power is positive and consequently subtracted from the stator power, resulting in a decreased amount of (negative) grid power.

For pumping mode (Figure 7.21), the mechanical power is defined as positive, and the rotation speed is positive. The increased amount of (positive) mechanical power results from the sum of stator power and rotor power in supersynchronous operation. For the subsynchronous domain, the decreased mechanical power results from the addition of (positive) stator power and (negative) rotor power.

The rotor power can be expressed by

$$P_r = P_s \cdot slip \tag{7.11}$$

7.7 EXAMPLE OF A MODERN PUMPED STORAGE FACILITY

A recent example of a large pumped storage facility is represented through Figures 7.22 and 7.23.

The new facility is based on an existing hydro-plant transformed into a pump storage installation (Nant-de-Drance, South-West Switzerland). The 900 MW plant

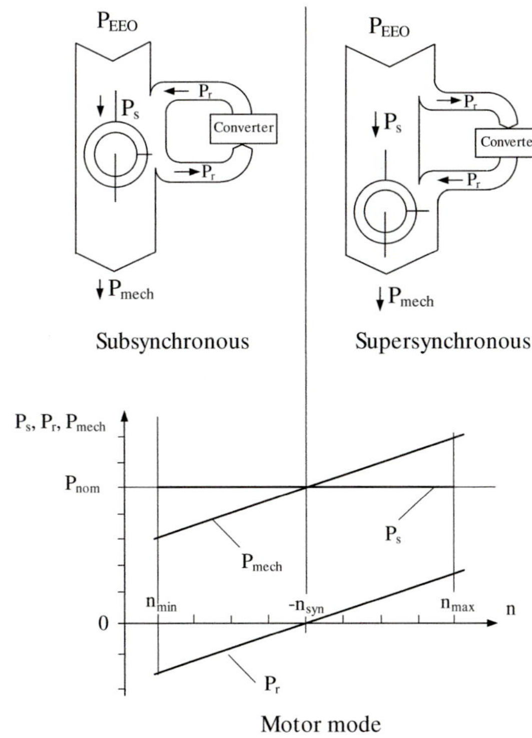

FIGURE 7.21 Power flows within a DFIMG pumping mode.

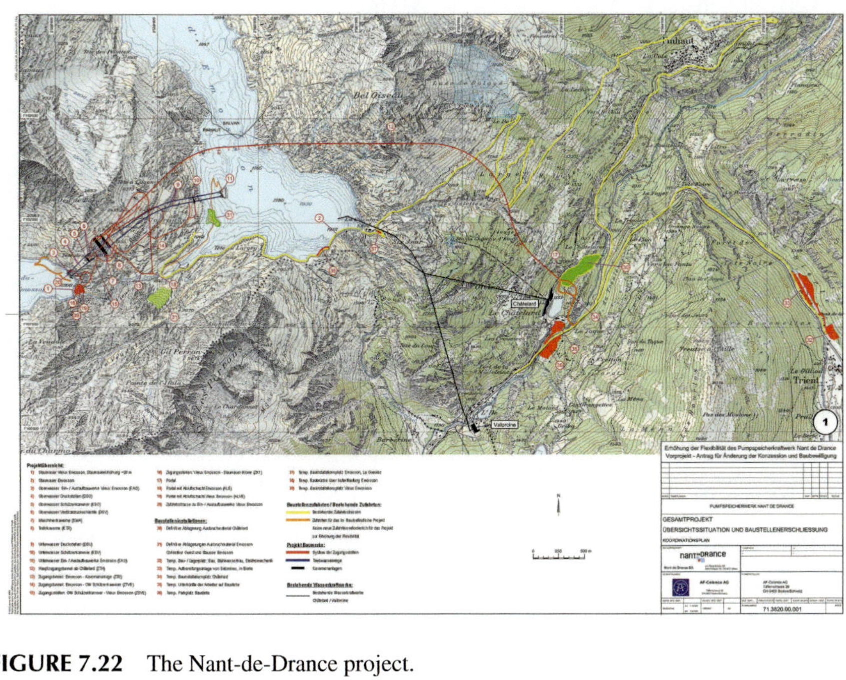

FIGURE 7.22 The Nant-de-Drance project.

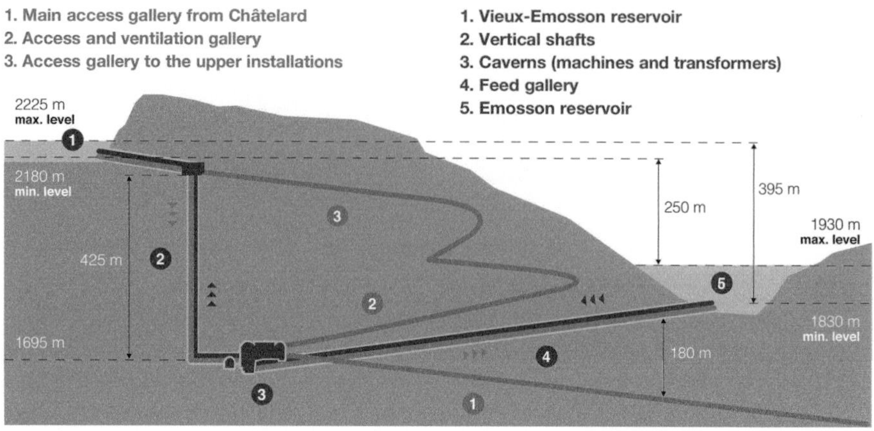

FIGURE 7.23 The Nant-de-Drance project (cut).

is composed of six units, each 150 MW. The machines used are of the reversible Francis pump-turbine type.

Main data of the Nant-de-Drance plant

- 2.5 billions of kWh/year	
- 900 MW	
- Six machines with each	175 MVA
- Efficiency	80%
- Vertical penstock	425m
- Cavern	194 m long, 32 m wide, 52 m high
- 17 km of underground galleries	
- 1.7 million m^3 of excavated rock	
- Investment	1.9 billion CHF

Exercise 6: Nant-de-Drance Pumped Storage Power Plant

The Nant-de-Drance pumped storage power plant is presented as an example of modern achievement.

Figure 7.23 shows the global structure of the plant with upper lake, penstock and lower lake with dimensions (altitudes, differences of height).

The rated power is equal to 900 MW.

Table 7.1 gives the main parameters of the upper lake.

Problem

On the basis of the indicated parameters, and with the help of the elementary formulas given in Chapter 7, calculate the duration the plant is able to operate at rated power when the level of the upper lake varies from its highest (2225 m) to its lowest value (2180 m).

TABLE 7.1
Upper Lake (Vieux-Emosson)

Upper Reservoir	
Year of construction	1955
20 m dam height increase	2014
Final height	76.5 m
Length of crown	205 m
Final capacity	25 Mio m^3
Surface	550,000 m^2
Length	1.5 km
Catchment area	4.6 km^2
Maximum level	2225 m
Minimum level	2180 m
Lower Reservoir	
Altitude of the lower basin	1930 m

During this phase of operation, it is supposed that the level of the lower lake is not changing (its surface is equal to 3,270,000 m^2).

SOLUTION TO EXERCISE 6

The energy capacity of the Nant-de-Drance pumped storage power plant is calculated through the volume of water that is moved from the upper lake to the lower lake.

The surface of the upper lake is given as

$$S_{\text{upp\_lake}} = 550{,}000 \text{ m}^2$$

and the moved volume of water is obtained by multiplying this value by the difference between the upper and lower level of this basin. This simplified calculation considers the walls of the reservoir vertical.

$$\begin{aligned} V_{upp\_lake} &= S_{upp\_lake} \cdot (Lev_{\max} - Lev_{\min}) = 55 \cdot 10^4 m^2 \cdot (2225m - 2180m) = \\ &= 24'750'000 m^3 \end{aligned} \tag{7.12}$$

The moved mass is then

$$m_{upp\_lake} = V_{upp\_lake} \cdot 1000 kg / m^3 = 24.75 \cdot 10^9 kg \tag{7.13}$$

The energy capacity will be calculated according a simplified model where the head losses are neglected and the head is considered constant and equal to its average value:

$$H = \frac{Lev_{\max} + Lev_{\min}}{2} - Lev_{low\_lake} = \frac{2225m + 2180m}{2} - 1930m = 272.5m \quad (7.14)$$

The energy capacity is

$$E = m_{upp\_lake} \cdot g \cdot H = 24.75 \cdot 10^9 kg \cdot 9.81 \frac{m}{s^2} \cdot 272.5m = 66.16 \cdot 10^6 MWs \quad (7.15)$$

The autonomy of the plant running at rated power (900 MW)

$$t = \frac{66.16 \cdot 10^6 MWs}{900MW} = 73'513s \Rightarrow 20.42h \quad (7.16)$$

In reality, the calculation should consider

- The losses of in all different components of the chain
- The head variation
- The real profile of the lake

7.8 DRY GRAVITATIONAL STORAGE SYSTEMS

Beneath the classical and well-known gravitational hydraulic pump storage systems, dry gravitational storage facilities have recently been studied and proposed [16–18]. In such systems, the gravitational force and the height-related potential energy are the carriers of the amount of energy to be stored. Elevating a given mass from one altitude to a higher one corresponds to the energy accumulation process, while lowering it back to the initial level is the inverse phenomenon of the energy discharge. The motivation of the development of dry gravimetric systems is to run a similar process as in hydraulic pumped storage and to install such systems in places with flat topography. Materials with a higher specific weight are used in order to elevate the energy density or to reduce the footprint of such installations. The gravitational force has been used for decades, even centuries, as the motor of so-called Morbier wall clocks, where 5-kg weights moved over around 2 meters height have provided

TABLE 7.2
Comparison Of Storage Systems

GES: Gravitational
PHES: Pumped Hydro
EC: Electrochemical batteries

System	Energy Density	Efficiency	Lifetime or Cycle
GES	0.2–3 Wh/l	85%	50 years
PHES	0.13–0.5 Wh/l	65–85%	40–60 years
EC	30–300 Wh/l	80–90%	1000–3000 cycles

only 100 Joules, leading to an autonomy of 24 to 36 hours [19]. In recent proposals, masses of hundreds of tons are considered, moving inside available vertical shafts of abandoned mines. Also, ambitious crane-based towers of stacked concrete blocks are in development with the goal to store up to 35 MWh of energy [20, 21]. Performance comparison and system description models of gravitational storage systems for industrial and domestic applications are given in [22–24].

In order to roughly situate the possible energy density of the gravitational storage systems, a comparison of parameters is given in Table 7.2 [25].

7.8.1 Principles

According to the basic rules of mechanics and kinematics, the increase of the stored energy amount is defined by the instantaneous power transferred to the moving mass, which corresponds to the product of the gravity force multiplied by the vertical translation speed. The vertical translation speed is further the result of an acceleration process, the acceleration force being the difference of the gravitation force and of the traction force applied by the electromechanical driving system. Thus, the variation of the loading/unloading power of the energy storage device is directly related to the variation of the speed, which is governed by Newton's law. The inertia of the system and the applied force define the possible dynamics of the controllable power. Similar limitations corresponding to well-known properties can be found within the behavior of hydro-power installations where the inertia of the upstream water column can lead to a so-called water hammer and needs an equilibrium chamber.

7.8.2 Structure of the Gravitational Energy Storage System

The main component of the gravitational storage system is a moving mass m, guided in a vertical shaft and set in movement by a cable system operated by a winder, itself driven by an electric motor/generator. The motor/generator is coupled to the electrical grid via a frequency converter allowing a variable speed of the winder and moving mass. A structural diagram of the gravitational energy storage system is represented in Figure 7.24.

The mass m undergoes two main forces, first the gravitational force F_v equals

$$F_v = m \cdot g \tag{7.17}$$

where

m is the mass in kg

g is the acceleration of the earth

F_v is expressed in N

Second, the traction force F_t, which is exerted from the winder and motor/generator. During the accumulation process, the electric machine is working as a motor; during the energy recovery, the machine works as a generator.

The speed of the mass is the result of the integration of the resulting force exerted on the mass, divided by the mass (7.18).

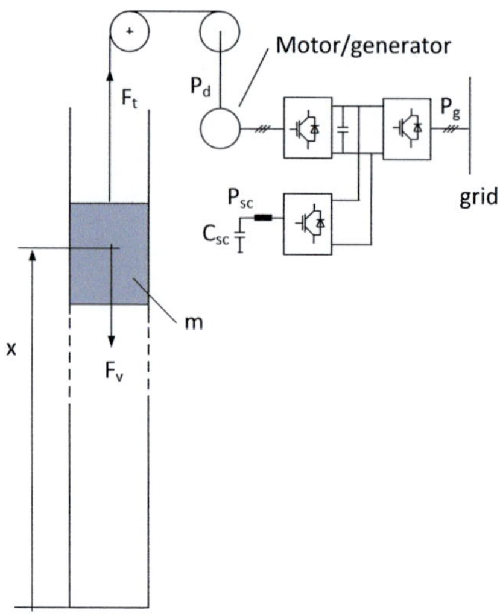

FIGURE 7.24 Structural diagram of the dry gravitational energy storage system.

$$V = \frac{1}{m}\int_0^t (F_t - F_v)dt \tag{7.18}$$

With respect to the standard conventions on the units associated to the variables and constants, the speed V is expressed in m/s.

Then, the elevation of the mass corresponds to the integration of the speed (7.19)

$$X = \int_0^t V \cdot dt \qquad X \text{ is expressedin m} \tag{7.19}$$

Finally, the stored potential energy *Epot* is calculated

$$E_{pot} = X \cdot m \cdot g \qquad E_{pot} \text{ is expressed in Joules or Watt-second,} \tag{7.20}$$

During the vertical movement of the mass, another energy quantity must be considered, the kinetic energy of the moving mass. This quantity is equal to

$$E_{kin} = \frac{1}{2} \cdot m \cdot V^2 \tag{7.21}$$

This additional energy amount has to be provided by or recovered to the external source through the traction force F_t. The total power can be decomposed into two terms according to Equation 7.5. The first term corresponds to the product of a constant force F_{vk} which compensates the gravitation force, and the second term corresponds to the acceleration force. For the power, they are multiplied by the velocity.

$$P_{tot} = F_t \cdot V = (F_{vk} + F_{acc}) \cdot V = P_{sto} + P_{acc} \tag{7.22}$$

In steady state, the mass moves at constant speed and

$$F_t = F_{vk} = F_v$$

Then, the integration with time of the power gives

$$\begin{aligned} \int_0^t P_{tot}(t) \cdot dt &= \int_0^t (F_{vk} + F_{acc}) \cdot V(t) \cdot dt = \int_0^t (F_{vk} \cdot V(t) \cdot dt + \int_0^t (F_{acc} \cdot V(t) \cdot dt \\ &= m \cdot g \cdot \int_0^t V(t) \cdot dt + \int_0^t (m \frac{dV(t)}{dt} \cdot V(t)) \cdot dt \\ &= m \cdot g \cdot X + m \frac{V^2(t)}{2} \end{aligned} \tag{7.23}$$

With regard to Equations 7.3 and 7.4, the total energy transferred to the system becomes

$$E_{tot} = m \cdot g \cdot X + \frac{1}{2} m \cdot V^2 = E_{pot} + E_{kin} \tag{7.24}$$

For a stationary state of the storage process where the storage power P_{sto} is constant, the traction force F_t compensates the gravity force F_v, and the mass evolves at constant speed.

After a discharge process operated at a negative speed of the mass, when the power is intended to be set to zero, the kinetic energy E_{kin} must be recovered. This means that the recovering power from the storage device cannot simply be reduced to zero from a given (negative) set value but must be first increased in its absolute value during the interval following the desired power reduction. A solution to this typical difficulty based on the use of an auxiliary storage facility has been proposed in [26].

7.8.3 Example of a Large Gravitational Energy Storage System

After realizing different crane-based open-sky demonstration projects, the company Energy Vault has started the construction of an advanced building-based system in China. The Rudong project has a power capability of 25 MW and an energy capacity of 100 MWh [27]. A view of the plant during construction is given in Figure 7.25.

FIGURE 7.25 Gravity energy storage plant in Rudong (China) during construction.

REFERENCES

[1] Schleicher, M., Modern practice in Germany and the European continent with regard to supervisory control systems as applied to large interconnected supply areas, *Electrical Engineers, Journal of the Institution of Year*, 75(456), 710–721, 1934, https://doi.org/10.1049/jiee-1.1934.0181.

[2] Tanaka, H., *An 82 MW Variable Speed Pumped-Storage System*, Water Power & Dam Construction, November 1991.

[3] Beyer, T., *Goldisthal Pumped-Storage Plant: More than Power Production, HRW-Hydro Review Worldwide*, www.hydroworld.com/articles/print/volume-15/issue-1/articles/goldisthal-pumped-storage-plant-more-than-power-production.html. Accessed on August 3, 2017.

[4] Guittet, M., Capezzali, M., Gaudart, L., Romerio, F., Vuille, F., Avellan, F., Study of the drivers and asset management of pumped-storage power plants historical and geographical perspective, *Energy*, 111, 560–579, 2016, Elsevier.

[5] Soerensen, B., *Renewable Energy, Its Physics, Engineering, Environmental Impacts, Economics and Planning*, 2nd edition, Academic Press, London, UK, 2000.

[6] Liang, J., Harley, R.-G., Pumped storage hydro-plant models for system transient and long-term dynamic studies, *IEEE Power and Energy Society General Meeting*, 2010, pp. 1–8.

[7] Hodder, A., Simond, J.-J., Schwery, A., Double-fed asynchronous motor-generator equipped with a 3-level VSI cascade, *Industry Applications Conference. 39th IAS Annual Meeting*, Seattle, WA, 2004, vol. 4, pp. 2762–2769.

[8] Schlunegger, H., Pumping efficiency—a 100 MW converter for the Grimsel 2 pumped storage plant, *ABB Review*, 2(14), 42–47, 2014.

[9] IEC 60193, International Standards CEI-IEC 60193, 1999.

[10] Nicolet, C. et al., Transient analysis of FMHL+ pumped-storage power plant and new surge tank design, *Congress SHF: "Enhancing Hydropower plants"*, Grenoble, April 9–10, 2014.

[11] Steimer, P. et al., Converter-Fed synchronous machine for pumped hydro storage plants, *IEEE Energy Conversion Congress and Expo, ECCE*, Pittsburgh, PA, September 14–18, 2014, pp. 4561–4567.

[12] Suul, J. A., Uhlen, K., Undeland, T., Wind power integration in isolated grids enabled by variable speed pumped storage hydropower plant, *IEEE Conference on Sustainable Energy Technologies, ICSET*, Singapore, 2008.

[13] Azbe, V., Mihalic, R., Dynamic behavior of a Doubly-Fed motor-generator during and after the fault, *IEEE Powertech 2011*, Trondheim, Norway, 2011.

[14] Pannatier, Y., Kawkabani, B., Nicolet, C., Simond, J. J., Schwery, A., Investigation of control strategies for variable –speed pump-turbine units by using a simplified model of the converter, *IEEE Transactions on Industrial Electronics*, 57, 3039–3049, 2010.

[15] Hodder, A., Schwery, A., Double-Fed asnchronous motor-generator equipped with a 3-level VSI cascade, *IEEE IAS Annual Meeting*, Seattle, WA, 2004.

[16] Botha, C. D., Kamper, M. J., Capability study of dry gravity energy storage, *Journal of Energy Storage*, 23, 159–174, June 2019.

[17] Morstyn, T., Chilcott, M., McCulloch, M. D., Gravity energy storage with suspended weights for abandoned mine shafts, *Applied Energy*, April 2019, https://doi.org/10.1016/j.apenergy.2019.01.226.

[18] Ares—the Power of Gravity [Online], www.aresnorthamerica.com/. Accessed on 2018.

[19] Van Veldhoven, L., *Mayet Morbier Comtoise. Birth and Life of a Legendary Clock* (edited by the author), 2014.

[20] Rathi, A., Stacking concrete blocks is a surprisingly efficient way to store energy, *Quartz*, August 18, 2018, https://qz.com/1355672/stacking-concrete-blocks-is-a-surprisingly-efficient-way-to-store-energy/.

[21] Moody, O., Giant battery uses gravity to convert stored energy into electricity, Berlin, *The Times*, Monday February 17, 2020, 12.01 GMT [Online], www.thetimes.co.uk/article/giant-battery-uses-gravity-to-convert-stored-energy-into-electricity-hq3r89b9r.

[22] Ruoso, A. C., Caetano, N. R., Oliveira Rocha, L. A., Storage gravitational energy for small scale industrial and residential applications, *MDPI Inventions*, 4, 64, 2019, https://doi.org/10.3390/inventions4040064.

[23] https://heindl-energy.com/technical-concept/basic-concept/.

[24] Aneke, M., Wang, M., Energy storage technologies and real life applications—a state of the art review, *Applied Energy*, 179, 350–377, 2016.

[25] Nikolaidis, P., Poullikkas, A., Cost metrics of electrical energy storage technologies in potential power system operations, *Sustainable Energy Technologies and Assessments*, 25, 43–59, 2018, https://doi.org/10.1016/j.seta.2017.12.001.

[26] Rufer, A., Design and control of a KE (kinetic energy) compensated gravitational energy storage system, *EPE ECCE Europe, European Power Electronics Conference*, Lyon, September 2020, https://doi.org/10.23919/EPE20ECCEEurope43536.2020.9215714.

[27] www.energyvault.com/projects/cn-rudong. Accessed on March 1, 2025.

8 Flywheel Systems

8.1 INTRODUCTION

A flywheel system is generally composed of an electric machine coupled to a rotating mass. With the help of a power electronic converter, the driving torque of the variable-speed machine (positive for charging and negative for discharging) can be imposed precisely and allows controlling the exchanged power level. Flywheels have been realized for a long time based on normal steel and reduced speed, but modern equipment benefits from advanced materials like carbon composites and fast-running permanent magnet motors that can rotate at several hundred thousand of revolutions per minute. Partially evacuated encapsulations reduce the aerodynamic losses of the flywheel and motor. A schematic representation of a flywheel system is given in Figure 8.1.

8.2 BASIC RELATIONS

The energy stored in a rotating mass as kinetic energy also called rotational energy is given by the basic relation (Equation 8.1)

$$E_k = \frac{1}{2} J \Omega^2 \tag{8.1}$$

Ω is the angular velocity

J is the moment of inertia of the mass around the center of rotation.

8.2.1 Rotor Dynamics

The angular velocity is the state variable of the dynamic mechanical system, and its evolution can be described through the rotational movement differential equation as

$$J \frac{d\Omega}{dt} = M_e - M_r \tag{8.2}$$

M_e is the acceleration torque imposed by the coupled electric motor/generator, and M_r represents the load torque.

In a flywheel energy storage system, the load torque is generally equal to zero, but in the further developments, M_r will be used for parasitic braking torque due to aerodynamic friction.

In order to simplify the description of the behavior of the system, normalized variables are introduced (per-unit system).

DOI: 10.1201/9781003736486-8

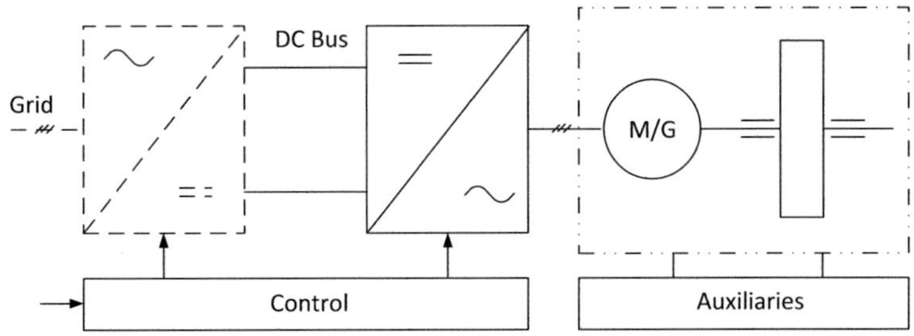

FIGURE 8.1 Flywheel system.

The definitions of $\Omega = n \cdot \Omega_n$, $M_e = m_e \cdot M_n$, $M_r = m_r \cdot M_n$ are used with the values of the rated velocity Ω_n and the rated torque M_n. The normalized variables (per-unit variables) are defined for the rotational velocity n, the torque m_e and the braking torque m_r.

The movement differential equation becomes

$$T_m \frac{dn}{dt} = m_e - m_r \tag{8.3}$$

with the mechanical time constant T_m as

$$T_m = \frac{J\Omega_n}{M_n} \tag{8.4}$$

The rated torque is obtained from the rated power and rated angular velocity

$$M_n = \frac{P_n}{\Omega_n} \tag{8.5}$$

The angular velocity is defined from the mechanical frequency f_{mech} (Hz) or (rev/s)

$$\Omega_n = 2\pi \cdot f_{mech} \tag{8.6}$$

8.2.2 Moment of Inertia

The moment of inertia of the rotating mass can be calculated with the help of the relations given in Table 8.1 for the different shapes of the spinning object described in Figure 8.2. For the calculation of the system properties, SI units are used:

- Kilogram for the mass (kg)
- Meters for the radius (m)
- Rad/sec for the angular velocity (rad/s)

The energy amount is expressed in Joules (J).

TABLE 8.1
Different Types of Shapes

Type of Spinning Object	Moment of Inertia
Solid cylinder	$J = \frac{1}{2}mr^2$
Thin-walled empty cylinder	$J = mr^2$
Thick-walled empty cylinder	$J = \frac{1}{2}m(r_{ext}^2 + r_{int}^2)$

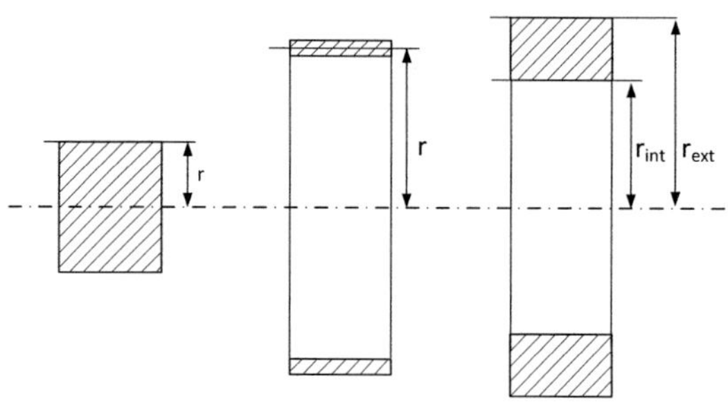

FIGURE 8.2 Different types of shapes.

8.2.3 Specific Energy of a Flywheel

The maximum specific energy of a flywheel rotor is mainly dependent on the rotor's geometry and on the properties of the used material. For an isotropic rotor, the expression of the specific energy can be expressed by Equation 8.7

$$e_{fw} = \frac{E}{m} = K\left(\frac{\sigma}{\rho}\right) \tag{8.7}$$

where

E : is the kinetic energy of the rotor (J)
M: is the mass of the rotor (kg)
K : is the geometric form factor (dimensionless)
σ : is the tensile strength of the material (Pa)
ρ : is the density of the used material (kg/m^3)

The form factor K depends on the geometry of the flywheel and is also called the flywheel shape factor [1]. Essentially, the value of K comes from the expression of the moment of inertia. Values of K for different flywheel shapes are given in Table 8.2.

The tensile strength of the material dictates the upper limit of angular velocity

$$\sigma_t = \rho r^2 \omega^2 \tag{8.8}$$

TABLE 8.2
Flywheel Shape Factors

Flywheel shape	K
Constant stress disk	0.931
Flat unpierced disk	0.606
Thin rim	0.5
Rod or circular brush	0.333
Flat pierced disk $\left(r_{ext}/r_{int} = 1.1\right)$	0.305

The expression can be simplified using specific tensile strength σ_t/ρ and tangent velocity v_t.

$$\frac{\sigma_t}{\rho} = v_t^{\,2} \tag{8.9}$$

8.2.4 Aerodynamic Drag of a Flywheel

A flywheel with high value of specific energy is characterized through the high tangential velocity of the rotor. The friction losses due to the interaction with the surrounding air even if the rotor is running inside a partial vacuum chamber should not be underevaluated.

The drag force causing the aerodynamic losses can be calculated in a general case through Reference [2]

$$F_D = \frac{C_D \cdot \rho_g \cdot V^2 A}{2} \tag{8.10}$$

ρ_g is the density of the air
V is the velocity of the undisturbed fluid (m/s)
A is the projected area of objects as spheres, disks or plates with axes perpendicular to the flow (m^2)
C_D is a constant depending essentially on the Reynolds number

The constant C_D is calculated for the case of a flat plate placed parallel with the flow.

$$C_D = 1.33/\mathrm{Re}^{0.5} \tag{8.11}$$

The drag force expressed through Equation 8.10 uses the area of the plate parallel with the flow. For the Reynolds number, the fluid velocity is considered, as well as the dimension of the fluid streamline D. In addition, the kinematic viscosity ν of the fluid is also used:

$$\mathrm{Re} = VD/\nu$$

Different approaches are given in the literature; for a rotating disk in a free surrounding, the following relation for the braking torque is given [3–6]:

$$M_r = C_m \cdot \rho_g \cdot \omega^2 r^5 (1 + \frac{h}{r}) \tag{8.12}$$

C_m is calculated for different cases:

For a rotating disk:

$\mathrm{Re} < 3 \cdot 10^5$	$C_m = 3.87 \cdot \mathrm{Re}^{-\frac{1}{2}}$
$\mathrm{Re} > 3 \cdot 10^5$	$C_m = 0.146 \cdot \mathrm{Re}^{-\frac{1}{5}}$

For a cylinder:

$\mathrm{Re} < 170$	$C_m = 8 \cdot \mathrm{Re}^{-1}$
$\mathrm{Re} > 170$	$C_m = 0.146 \cdot \mathrm{Re}^{-\frac{1}{5}}$

Further, the air density around the rotor is taken as directly proportional to the pressure according the ideal gas law

$$\rho_g = \frac{P}{RT} \tag{8.13}$$

The aerodynamic drag is therefore reduced by running the flywheel system inside a partially evacuated enclosure [4, 6].

In addition, the real shape of the rotor must be considered, as will be shown in the example of Section 8.3.

8.3 EXAMPLE DESIGN OF A FLYWHEEL

The properties and the associated energetic characteristics of a flywheel will be presented in this section, according to a pre-designed flywheel geometry. The shape and dimensions of the rotor are given through Figure 8.3. The flywheel is driven by a 20-kW electrical motor/generator. Its nominal rotational speed is 3000 rev/min.

The nominal values are listed in Table 8.3.

The rotating part of the system comprises the outer ring (inertia J_1), the inner disk (inertia J_2) and the rotor of the motor/generator (inertia J_3). The inertia of the axle is neglected.

J_1:

The mass m_1 is given through

$$m_1 = \rho_{steel} \cdot V = \rho_{steel} \cdot \frac{\pi}{4}(d_{ext}^2 - d_{\mathrm{int}}^2) \cdot h \tag{8.14}$$

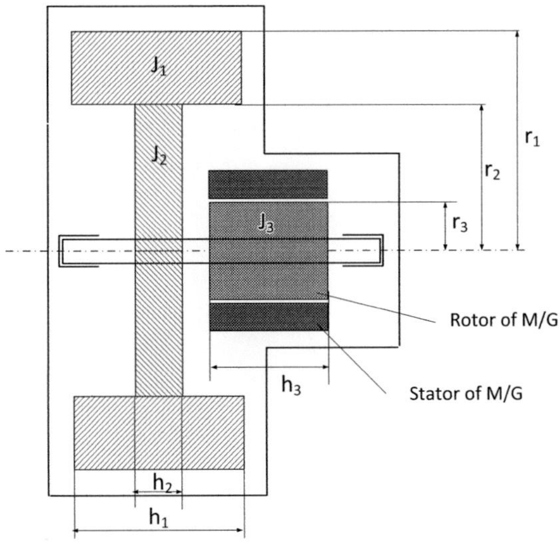

FIGURE 8.3 Flywheel and motor/generator.

TABLE 8.3
Technical Data of the Flywheel System

Rated speed	N	3000 rev/min
Rated power	Pn	20 kW
Flywheel		
r_1		450 mm
r_2		300 mm
r_3		100 mm
h_1		400 mm
h_2		100 mm
h_3		250 mm

with $d_{ext} = 2r_1$ and $d_{int} = 2r_2$

$$m_1 = 7.9 \cdot 10^3 \frac{kg}{m^3} \cdot \frac{\pi}{4}(0.9^2 - 0.6^2)m^2 \cdot 0.4m = 1116kg \tag{8.15}$$

and the moment of inertia J_1 becomes

$$J_1 = \frac{1}{2} m(r_{ext}{}^2 + r_{int}{}^2) = \frac{1}{2} \cdot 1116kg \cdot (0.202 + 0.09)m^2 = 163kgm^2 \tag{8.16}$$

J_2:

The mass m_2 is given through

$$m_2 = \rho_{steel} \cdot V = \rho_{steel} \cdot \frac{\pi}{4} \cdot d_{ext}^2 \cdot h = 7.9 \cdot 10^3 \frac{kg}{m^3} \frac{\pi}{4} 0.6^2 m^2 \cdot 0.1m = 223kg \quad (8.17)$$

and the moment of inertia J_2 becomes

$$J_2 = \frac{1}{2} m \cdot r^2 = \frac{1}{2} \cdot 223kg \cdot (0.3)^2 m^2 = 10kgm^2 \quad (8.18)$$

J_3:

The mass m_3 is given through

$$m_3 = \rho_{steel} \cdot V = \rho_{steel} \cdot \frac{\pi}{4} \cdot d_{ext}^2 \cdot h = 7.9 \cdot 10^3 \frac{kg}{m^3} \frac{\pi}{4} 0.2^2 m^2 \cdot 0.25m = 62kg \quad (8.19)$$

and the moment of inertia J_3 becomes

$$J_3 = \frac{1}{2} m \cdot r^2 = \frac{1}{2} \cdot 62kg \cdot (0.1)^2 m^2 = 0.31kgm^2 \quad (8.20)$$

Total moment of inertia:

$$J_{tot} = J_1 + J_2 + J_3 = 163kgm^2 + 10kgm^2 + 0.31kgm^2 = 173.3kgm^2 \quad (8.21)$$

8.3.1 Mechanical Time Constant

According to Equations 8.3 and 8.4, the mechanical time constant can be calculated

$$T_m = \frac{J\Omega_n}{M_n} = \frac{173.3kgm^2 \cdot 314s^{-1}}{63.7N \cdot m} = 854 \frac{kgm^2 \cdot s^2 \cdot s^{-1}}{kg \cdot m \cdot m} = 854s \triangleq 14.23 \min \quad (8.22)$$

where the rated torque is

$$M_n = \frac{P_n}{\Omega_n} = \frac{20 \cdot 10^3 W}{314s^{-1}} = 63.7 \frac{J \cdot s^{-1}}{s^{-1}} = 63.7Nm \quad (8.23)$$

8.3.2 Stored Energy

According to Equation 8.1, the stored energy is

$$\begin{aligned} E_k &= \frac{1}{2} J\Omega^2 = \frac{1}{2} \cdot 173.3kgm^2 \cdot 314^2 \cdot s^{-2} \\ &= 8.543 \cdot 10^6 \frac{N \cdot s^2}{m} m^2 \cdot s^{-2} = 8.543MJ \end{aligned} \quad (8.24)$$

8.3.3 Profile of the Discharge

The discharge of the flywheel can be done with great flexibility, depending on the power demand of the application.

Two specific cases are illustrated in Figure 8.4. The first discharge profile corresponds to a discharge with a constant torque. This does not really correspond to a real case of application but allows it to represent very easily the variation of the rotational velocity. The discharge time of the flywheel under a solicitation with its nominal torque corresponds to a discharge within the mechanical time constant T_m defined through Equation 8.4.

A more realistic discharge profile is that of a discharge with constant power. In such a case, the reduction of the rotational speed must be compensated through an increase of the deceleration torque.

In Figure 8.4, the variation of the rotational speed is defined between the nominal value and half of it. A deeper discharge is theoretically possible but needs a special design of the motor/generator due to the elevation of torque at reduced speed.

The profile of the electromagnetic torque during discharging the flywheel with constant power is represented in Figure 8.5.

Other control strategies can be used, such as the so-called field-weakening principle. This principle is easy to implement for synchronous machines with a separate excitation winding. Field-weakening strategies for permanent magnet synchronous machines (PMSMs) exist but often lead to a reduced efficiency of the electrical machine.

Table 8.4 represents the numerical values of the discharge with constant power (nominal value of 20 kW).

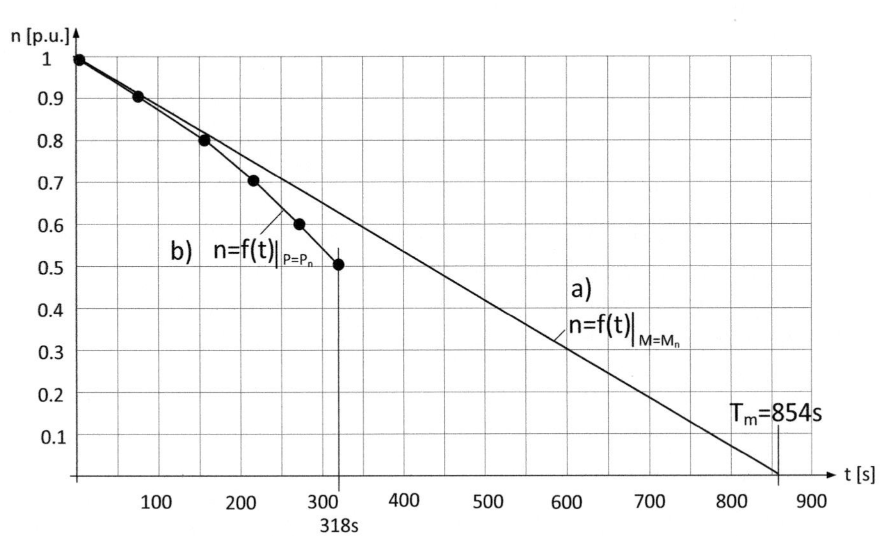

FIGURE 8.4 Rotational speed of the flywheel.

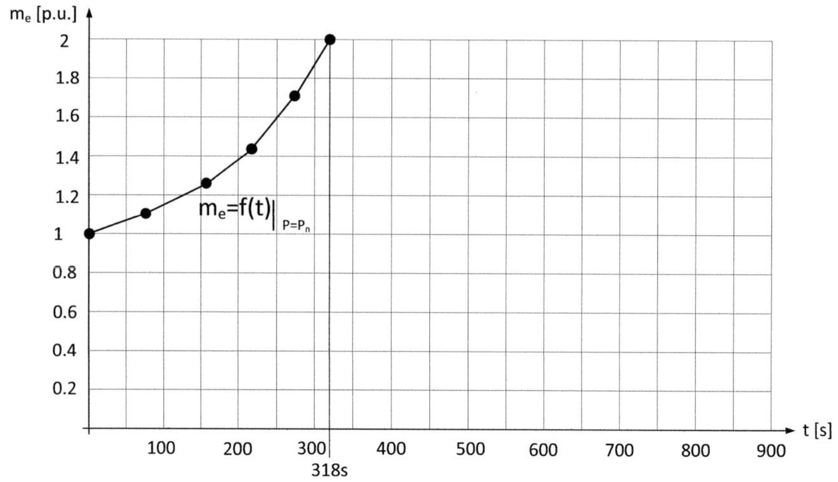

FIGURE 8.5 Electromagnetic torque by discharge with constant power.

TABLE 8.4
Discharge with Constant Power

Normalized Speed Reached [p.u.]	Extracted Energy [MJ]	Discharge Time [s]	Normalized Torque [p.u.]
1	0	0	1
0.9	1.62	81	1.11
0.8	3.05	153	1.25
0.7	4.35	217	1.42
0.6	5.46	273	1.66
0.5	6.37	318	2

8.3.4 Aerodynamic Drag of the Flywheel

Due to the high peripheric velocity of the flywheel, aerodynamic drag forces must be estimated. The aerodynamic breaking torque is calculated with the two main contributions, the friction torque of the outer band of the flywheel and the torque produced by the friction of the two sides of the outer ring. The friction torque of the inner disk and of the electric machine are neglected (Figure 8.6).

$$M_r = F_{tg} \cdot r_e + 2 \cdot F_{lat} \cdot \frac{(r_e + r_i)}{2} \tag{8.25}$$

For the first term corresponding to the influence of the tangential surface:

$$M_{tg} = F_{tg} \cdot r_e = C_D \cdot \rho \cdot V_{tg}^2 \cdot S_{tg} \cdot r_e = C_D \cdot \rho \cdot \Omega^2 \cdot r_e^2 \cdot 2\pi r_e h \cdot r_e \tag{8.26}$$

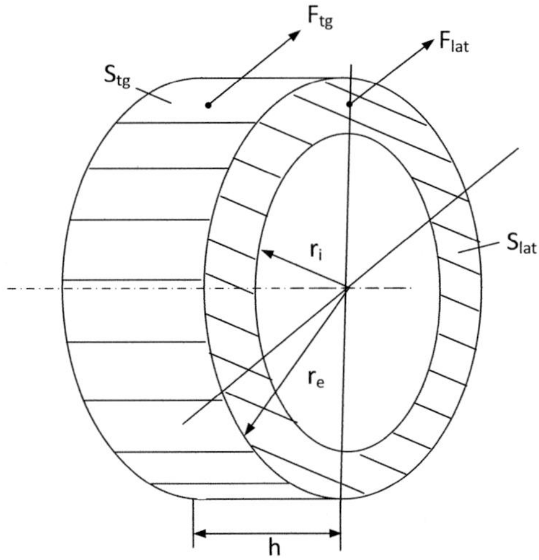

FIGURE 8.6 Drag forces, tangential, lateral.

with $\rho = 0.1 \cdot 1.225 \frac{kg}{m^3}$ (partial vacuum) and C_D being the drag factor

$$M_{tg} = C_D \cdot 0.1225 \cdot \frac{kg}{m^3} \cdot \frac{314^2}{s^2} \cdot 2\pi \cdot 0.45^4 \cdot m^4 \cdot 0.4 \cdot m = \\ = C_D \cdot 0.1225 \cdot \frac{N \cdot s^2}{m} \cdot \frac{1}{m^3} \cdot \frac{314^2}{s^2} \cdot 2\pi \cdot 0.45^4 \cdot m^4 \cdot 0.4 \cdot m = C_D \cdot 1244 Nm \tag{8.26}$$

The second term in Equation 8.25 corresponds to the influence of the two lateral surfaces

$$M_{lat} = 2F_{lat} \cdot (\frac{r_e + r_i}{2}) = 2 \cdot C_D \cdot \rho \cdot V_{lat}^2 \cdot S_{lat} \cdot r_{lat} = \\ = 2 \cdot C_D \cdot \rho \cdot \Omega^2 \cdot \left(\frac{r_e + r_i}{2}\right)^2 \cdot \pi (r_e^2 - r_i^2) \cdot \left(\frac{r_e + r_i}{2}\right) = \\ = C_D \cdot 2 \cdot 0.1225 \cdot \frac{N \cdot s^2}{m \cdot m^3} \cdot \frac{314^2}{s^2} \cdot 0.375^2 \cdot m^2 \cdot 0.35 \cdot m^2 \cdot 0.375 \cdot m = \\ = C_D \cdot 445.8 N \cdot m \tag{8.27}$$

The drag factor C_D is calculated in dependency on the Reynolds factor Re with Equation 8.28

$$C_D = 1.33 \ \mathrm{Re}^{-0.5} \tag{8.28}$$

and the Reynolds factor according to Equation 8.29

$$\mathrm{Re} = V \cdot D / v \tag{8.29}$$

where

V is the fluid velocity
D is the dimension of the stream line
v is the kinematic viscosity of the air

$$v = 1.56 \cdot 10^{-5} m^2 / s$$

From Equation 8.29

$$\mathrm{Re}_{tg} = V_{tg} \cdot D / v = \Omega \cdot r_e \cdot D / v = 314 \cdot \frac{1}{s} \cdot 0.45 \cdot m \cdot 0.05 \cdot m \cdot \frac{1 \cdot s}{1.56 \cdot 10^{-5} \cdot m^2} =$$

$$= 141 \cdot \frac{m}{s} \cdot 0.032 \cdot 10^5 \cdot \frac{s}{m} = 4.52 \cdot 10^5$$

$$C_{Dtg} = 1.33 \cdot \mathrm{Re}_{tg}^{-0.5} = \frac{1.33}{(4.52 \cdot 10^5)^{0.5}} = 0.00197$$

$$\mathrm{Re}_{lat} = V_{lat} \cdot D / v = \Omega \cdot \frac{(r_e + r_i)}{2} \cdot D / v = 314 \cdot \frac{1}{s} \cdot 0.375 \cdot m \cdot \frac{0.05 \cdot m \cdot s}{1.56 \cdot 10^{-5} \cdot m^2} =$$

$$= 117 \frac{m}{s} \cdot 0.032 \cdot 10^5 \cdot \frac{s}{m} = 3.75 \cdot 10^5$$

$$C_{Dlat} = 1.33 \cdot \mathrm{Re}_{lat}^{-0.5} = \frac{1.33}{(3.75 \cdot 10^5)^{0.5}} = 0.00217$$

Finally, for the drag torque, we get

$$\begin{aligned} M_r &= M_{tg} + M_{lat} = 0.00197 \cdot 1244 \cdot Nm + 0.00217 \cdot 445.8 \cdot Nm = \\ &= 2.45 \cdot Nm + 0.97 \cdot Nm = 3.42 \cdot Nm \end{aligned} \tag{8.30}$$

8.3.5 Efficiency

The energetic efficiency is calculated at one specific point of operation, namely at rated speed and for an extraction of power at the rated power level. The extraction is supposed to be short in comparison with the discharge time, so the speed variation is neglected:

$$\eta_{20kW} = \frac{M_n - M_r}{M_n} = \frac{63.7 \cdot Nm - 3.42 \cdot Nm}{63.7 \cdot Nm} = 0.946$$

In order to draw an modified Ragone representation (MRR) as defined in Section 3.6, the efficiency is recalculated for the same point at rated speed but for reduced power.

$$\eta_{10kW} = \frac{M_n - M_r}{M_n} = \frac{31.8 \cdot Nm - 3.42 \cdot Nm}{31.8 \cdot Nm} = 0.89$$

$$\eta_{4kW} = \frac{M_n - M_r}{M_n} = \frac{12.7 \cdot Nm - 3.42 \cdot Nm}{12.7 \cdot Nm} = 0.73$$

$$\eta_{2kW} = \frac{M_n - M_r}{M_n} = \frac{6.37 \cdot Nm - 3.42 \cdot Nm}{6.37 \cdot Nm} = 0.46$$

Figure 8.7 illustrates the typical behavior of a flywheel as already described in Section 3.5. The energy efficiency at part load (reduced power) is lower as for the rated value.

For a more accurate model, the variation of the rotational velocity should be taken into account while discharging, and the aerodynamic friction torque should be recalculated for all speed values. This is an additional information that the energy efficiency of a flywheel depends on its state of charge (SoC), because the aerodynamic friction torque strongly depends on the rotational velocity (Equations 8.27 and 8.27).

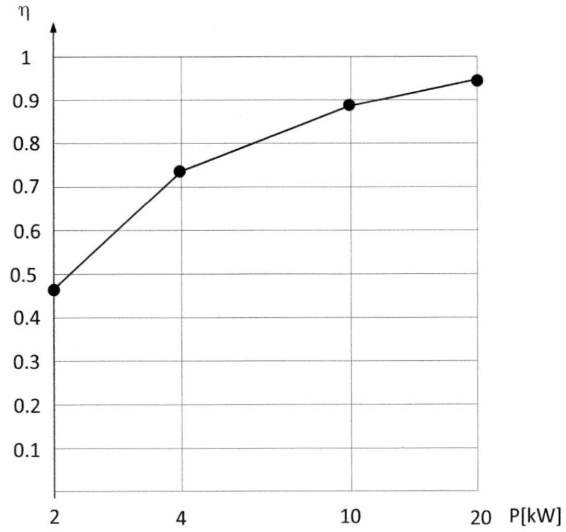

FIGURE 8.7 MRR (modified Ragone representation) of a flywheel (8.5 MJ/20 kW).

REFERENCES

[1] Ter Gazarian, A. G., *Energy Storage for Power Systems*, IEE The Institution of Electrical Engineers, Published by Peter Peregrinus Ltd, on behalf of the Institution of Electrical Engineers, London, UK, 1994.

[2] *Fundamentals of Engineering Supplied Reference Handbook*, 7th edition, National Council of Examiners for Engineering and Surveying (NCEES), Clemson, SC, 2005, http://ef.engr.utk.edu/ef402-2006-01/links/fe_reference_handbook.pdf. Accessed on August 4, 2017.

[3] Schlichting, H., *Grenzschicht Theorie*, Braun, Karlsruhe, Germany, 1994.

[4] von Burg, P. et al., *Kinetischer Kurzzeit Energie Speicher, NEFF Projekt 525, ENET, OFEN*, Swiss Federal Office for Energy, Berne, Switzerland, 1993.

[5] Ludlum, K., Optimizing flywheel design for use as a kinetic energy recovery system for a bicycle, Thesis, Copper Union University, New York, 2013.

[6] Genta, G., *Flywheel Energy Storage: Theory and Practice of Advanced Flywheel Systems*, Butterworth & Co., Cambridge, England, 1985.

9 Energy Storage Based on Hydrogen

9.1 INTRODUCTION

Hydrogen can be used for energy storage as an energy carrier due to its high energy density of 33 kWh/kg. This high value makes hydrogen a possible candidate for mid- and long-term storage, such as the so-called seasonal storage in the context of renewable energy sources [1].

Compared to other fuels, hydrogen contains three times more energy than diesel fuel, and 2.5 times more than natural gas. Hydrogen can be produced from electricity using electrolyzers, and the reverse transformation from hydrogen to electricity can be realized using a fuel cell.

However, hydrogen is difficult to store due to its very low weight density. At atmospheric pressure and ambient temperature, 1 kg of hydrogen needs a storage volume of 11 m^3. As a consequence, hydrogen is usually compressed at a high pressure level of 350 to 700 bar. This high pressure addresses the question of the volume and weight needed for the reservoir, as well as the needed energy for the compression.

Figure 9.1 shows the different energy densities of important energy vectors such as hydrogen, natural gas, propane, diesel fuel and ethanol.

In the first chapter of this book, energy densities of different storage means were presented, where typically a powerful Li-ion battery appears with "only" 0.15 kWh/kg.

As has been mentioned before, hydrogen must be conditioned in order to express a valid energy density corresponding to a real application, in the sense that the storage volume can be reduced.

Three main techniques can be used for the volume reduction:

1. Storage of hydrogen as compressed gas (350–700 bar)
2. Storage of hydrogen in its liquid phase (–253°C)
3. Storage of hydrogen in a solid form

As will be described later, the solid form of hydrogen corresponds to a metallic hydride, where different metals can be used such as Mg, Al or other metallic alloys.

Table 9.1 indicates the amount of hydrogen mass that can be stored in these three forms.

A more complete picture of the different possible forms of hydrogen is given though Figure 9.2 [2]. The lower part of the figure shows the volumetric density of hydrogen (kgH_2m^{-3}) and its gravimetric density (mass%) depending on the vessel pressure. Steel and composite materials are shown. The upper part shows different forms of adsorbed (chemisorbed) hydrogen.

DOI: 10.1201/9781003736486-9

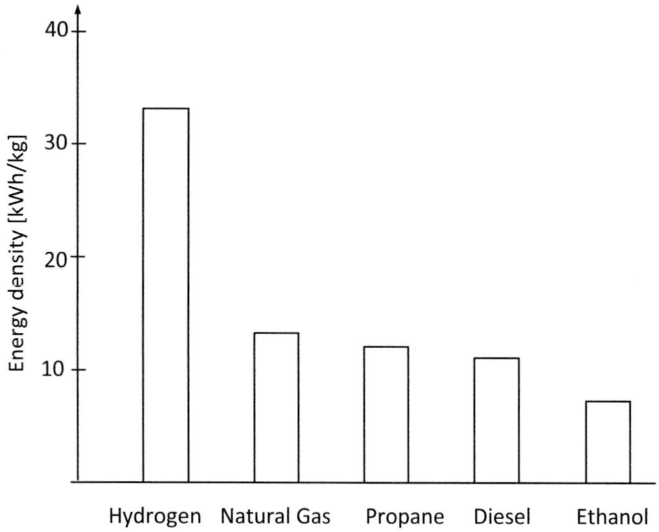

FIGURE 9.1 Energy densities.

TABLE 9.1
Various Forms of Hydrogen

Compressed hydrogen	700 bar	42 kg/m^3
Liquid hydrogen	–253°C	70 kg/m^3
Metal hydride MgH_2	Atmospheric pressure, ambient temperature	106 kg/m^3

The energy needed to compress hydrogen to the high pressure level corresponds to around 10 to 15% of its energy content depending on the quality of the compressor.

For the liquid form of hydrogen, one should note the high amount of energy needed to cool it down to –253°C. Additionally, the tank must be well isolated to prevent boil off.

In reference to Figure 9.1, liquid hydrogen has a lower energy density by volume than hydrocarbon fuels. There is actually about 64% more hydrogen in a liter of gasoline (116 grams hydrogen) than in a liter of pure liquid hydrogen (71 grams hydrogen). In addition, the carbon in the gasoline contributes to the energy of combustion.

9.2 POWER-TO-POWER STORAGE SYSTEM (ESS) BASED ON HYDROGEN

In Section 1.2.3, an electrical energy storage system (ESS) was defined. In such a system, the input and output of the storage system is interfaced to an electrical

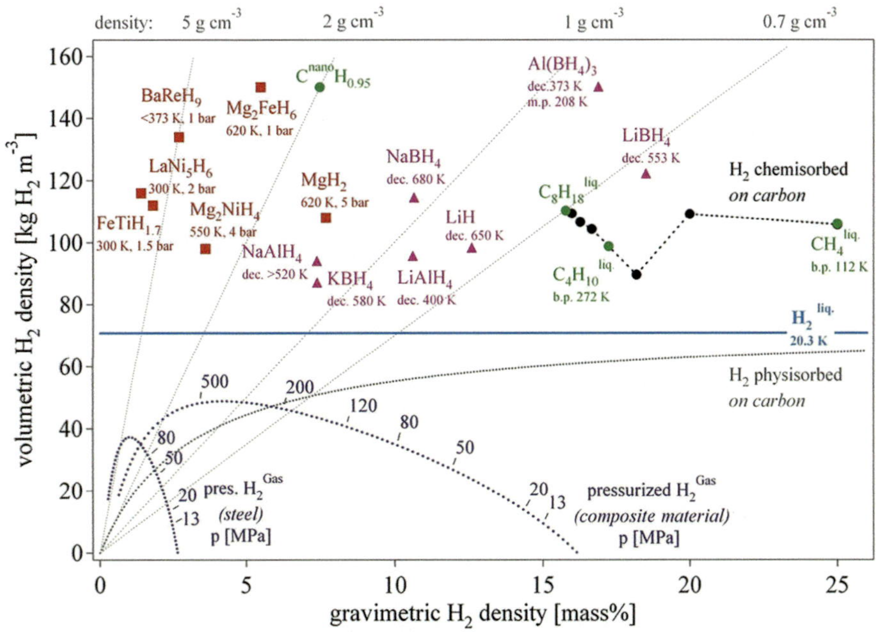

FIGURE 9.2 Volumetric and gravimetric hydrogen density of different selected hydrides. Mg_2FeH_6 shows the highest known volumetric hydrogen density of 150 kgm^{-3}, which is more than double that of liquid hydrogen. Pressurized gas storage is shown for steel and for a hypothetical composite material [3].

distribution system or grid. The storage form of energy, as well as the intermediate conversions, are mentioned.

An ESS or power-to-power storage system based on hydrogen can be realized using a water electrolyzer and a fuel cell for the intermediate conversions. In addition to these intermediate conversions hydrogen and oxygen, conditioning systems are needed in order to elevate the volumetric energy density. Figure 9.3 shows the elementary structure of an ESS based on hydrogen.

In Figure 9.3, the basic components of the system are the electrolyzer, the hydrogen storage and the fuel cell. Between the electrolyzer and the storage reservoir, a hydrogen conditioning device is represented (C_{H2s}), as well as between the storage reservoir and the fuel cell (C_{H2r}). The electrolyzer also produces oxygen that can be stored. In Figure 9.3, the oxygen path is represented with a dotted line due to the fact that in a high number of modern fuel cells, oxidation is achieved using the ambient air. For this purpose, an air compressor is represented (K). For the case of using stored oxygen, the conditioning blocks are represented (C_{O2s}, C_{O2r}).

The conditioning devices for the storage process (C_{H2s}, C_{O2s}) are generally compressors or liquefiers, while the recovery conditioning devices (C_{H2r}, C_{O2r}) are simple relief valves. In the case of solid storage of hydrogen in the form of metal hydrides, the conditioning processes are more complex.

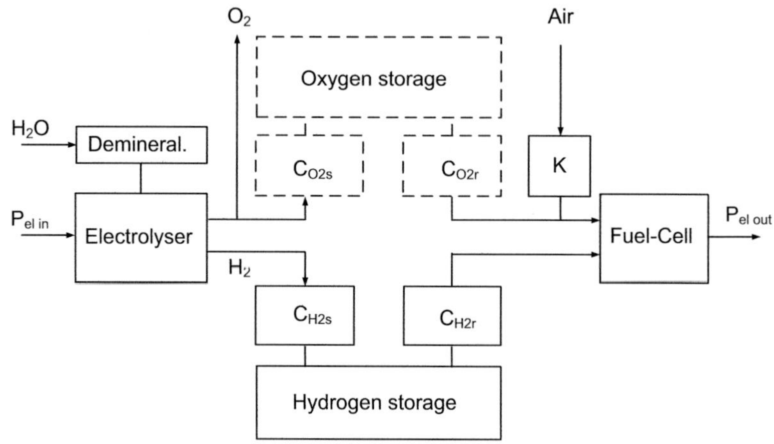

FIGURE 9.3 Structure of a storage system based on hydrogen.

9.3 ELECTROLYSIS OF WATER

Today, around half of the hydrogen produced around the world is obtained from reforming natural gas, which represents the most economic source. The other half of hydrogen is produced from coal or oil. Electrolysis of water is a more expensive process and today represents only a small percent of the total produced hydrogen. However, it allows conversion of electrical power into a chemical carrier characterized by a high specific energy. In recent days, there was a growing interest in this conversion, such as the example of recently built power-to-gas plants [4]. Electrolysis of water is a well-known process in the industry for the production of hydrogen and has been developed for the production of hydrogen of high purity.

An electrolyzer is a device for the chemical decomposition of water by circulation of an electric current. It comprises two electrodes, the anode and the cathode, separated by an electrolyte. The electrodes are connected to a DC current source to allow the circulation of the current, and the electrolyte is the internal ionic conductive means.

Three types of water electrolyzers have been developed or are under research. The most-used system is the alkaline electrolyzer. Then there are solid polymer electrolyzers (SPEs), where a proton exchange membrane is used like in a PEM fuel cell. Finally, a younger technique is based on the electrolysis of water vapor at high temperatures utilizing solid oxide electrolytes (SOEs). Figure 9.4 illustrates the three electrolyzers according the principles described in the next section.

9.3.1 Alkaline Electrolysis

The principle of electrolysis in an alkaline medium is based on the circulation of an electric current to produce oxygen and hydrogen. The electrodes are metallic (iron, palladium), and they are separated by a ceramic membrane. The reactions are the following

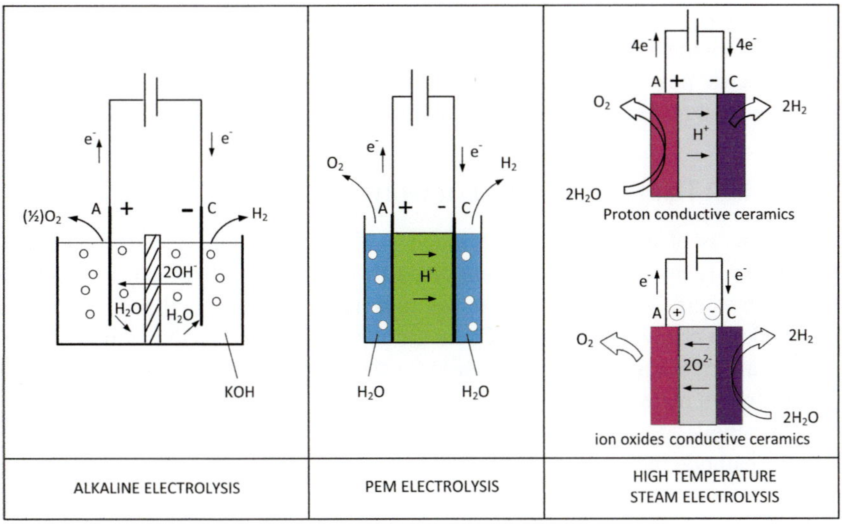

FIGURE 9.4 Three types of electrolyzers.

At the cathode (negative electrode:

$$4H_2O + 4e^- -> \quad 2H_2 + \ 4OH^- \tag{9.1}$$

At the anode (positive electrode):

$$4OH^- \quad -> \qquad O_2 + 2H_2O + 4e^- \tag{9.2}$$

The technology is characterized by relatively high investment costs bound to electricity. Indeed, the alkaline electrolyze demands an high quantity of electricity: around 4 kWh/m^3 of hydrogen.

9.3.2 Polymer Electrolyte Membrane Electrolysis

For polymer electrolyte membrane (PEM) electrolysis, an ion exchange membrane is used that serves simultaneously as electrolyte and as separator between the two electrodes. Ion exchange membranes are composed of organic polymers on which ionogenic groups are transplanted (sulfonic acid $-SO_3H$, carboxylic acid $-$ COOH or ammonium hydroxide $-$ NR_3OH).

At the cathode (negative electrode):

$$4H^+O + 4e^- -> \quad 2H_2 \tag{9.3}$$

At the anode (positive electrode):

$$2H_2O \quad -> \quad 4H^+ + \ 4e^- \tag{9.4}$$

One commonly used proton exchange membrane is Nafion, which shows high chemical and mechanical robustness and high protonic conductivity.

The electric power demand is similar as for alkaline electrolysis. Values depending on the used materials and temperature are given: 4.5–7 kWh/Nm^3 of hydrogen.

9.3.3 High Temperature Steam Electrolysis

A new process of production of hydrogen is high temperature electrolysis (HTE), where steam is injected on the anode or on the cathode, realized on the basis of a solid oxide electrolyte (SOE). Two different structures can be realized, the cathode solid oxide electrolyzer cell (SOEC) or the anode SOEC. For the cathode SOEC, metal-ceramics (cermets) are used, while for the anode SOEC, perovskite oxides are the best-known materials.

For a cathode SOEC, the electrode reactions are:

At the cathode (negative electrode):

$$2H_2O + 4e^- -> 2O^{2-} + 2H^2 \tag{9.5}$$

At the anode (positive electrode):

$$2O^{2-} \quad -> \quad O_2 + 4e^- \tag{9.6}$$

For electric power demand, cathode SOECs are indicated with a consumption of 3.2 kWh/Nm^3 and for anode SOEC 2.6 kWh/Nm^3.

High-temperature electrolysis has the main advantage of benefitting from both thermal and electric action for the dissociation of water molecules. This can be a great advantage for renewable systems based on solar irradiation. Research is ongoing for this promising technique [5, 6].

9.3.4 Advanced Developments in the Domain of Hydrogen Generation

Recent research developments have proposed the principle of indirect water electrolysis as an alternative secondary discharge process for a vanadium-cerium redox flow battery. In such a dual circuit reactor in the charged state, both electrolytes, the V(II) and the Ce(IV), may be directed in an external catalytic bed to be chemically regenerated and then return to the redox flow battery. The catalyzed chemical reactions taking place in the catalytic beds allow the generation of hydrogen from the catholyte (V(II)) and oxygen from the anolyte (Ce(IV)) [7].

9.4 STORAGE OF HYDROGEN

The importance of energy storage has been demonstrated in the viability of any energy system, and several references point out this reality for the choice of a hydrogen-based energy system [8, 9]. The techniques used for industrial and natural gases can be considered for storing hydrogen. These techniques comprise the liquefaction of hydrogen for storage in cryogenic reservoirs, the compression for a storage in pressurized vessels or in a less common form the adsorption by metallic hydrides as solid state.

9.4.1 Liquid Hydrogen Storage

The liquefaction of hydrogen uses the property of the important change of density between the gaseous phase and liquid phase with respect to the adequate temperature for a pressure of the liquid close to atmospheric pressure.

Two main cycles are used for the liquefaction of hydrogen, the Linde cycle and the Joule-Thompson expansion. Other cycles like the Claude Cycle or the Joule–Thompson can be used [3]. Because of the very low liquefaction temperature of hydrogen (around 20K under atmospheric pressure), the liquid storage of hydrogen is not economic.

First, the energy consumed for the liquefaction is equal to 3.228 kWh/kg of hydrogen. Second, high-performance cryogenic reservoirs imply very high manufacturing costs, and a passively isolated reservoir is not sufficient for maintaining stored gas.

An elementary calculation of the efficiency of the liquid storage gives with a heat value of hydrogen of 33 kWh/kg

$$\eta_{liq} = (33\text{kWh} - 3.228\text{kWh})/33\text{kWh} = 0.9 \quad (9.7)$$

9.4.2 Storage of Hydrogen by Compression

Hydrogen storage by compression is the process where the storage is realized by the means of robust high-pressure reservoirs. This process is easier to realize than liquefaction, but the resultant energy density is much lower. For comparison, at atmospheric pressure (1 bar), the volumetric mass of liquid hydrogen (at 20K) is equal to 71.1 kg/m^3. At 293K, the volumetric mass of gaseous hydrogen is 0.0827 kg/m^3 at 1 bar, 14.49 kg/m^3 at 200 bar and 23.66 kg/m^3 at 350 bar. As a consequence, in order to reach important energy densities (energy by volume unit) it is necessary to increase the storage pressure, and as a consequence, the compression work is also increased.

For example, the compression work in isothermal conditions can be evaluated as follows:

One kg of hydrogen at atmospheric pressure occupies a volume of 11 m^3.

The compression from 1 bar to 700 bar leads to a reduced volume of

$$V_1 = 11\ \text{m}^3\ /\ 700 = 0.0157\,\text{m}^3 \quad (9.8)$$

The isothermal work for such a compression can be calculated by

$$W_{comp} = P_1 * V_1 \ln(P_1 / P_2) = 700 * 10^5 * 0.0157 \ln(700/1) = 7.11 * 10^6\,\text{J} \quad (9.9)$$

The same mass of hydrogen (1 kg) has a heat value of 33 kWh, which is equal to 118*10^6 J

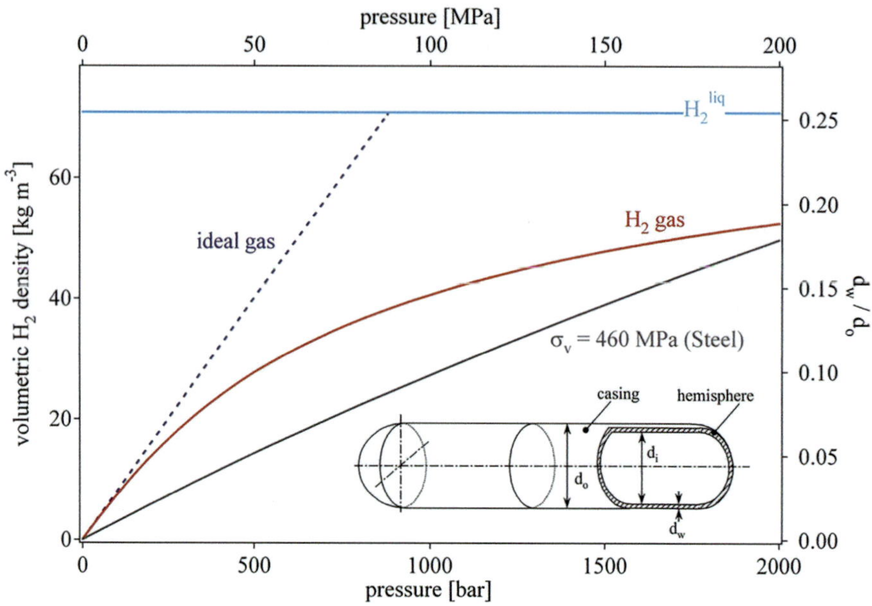

FIGURE 9.5 Volumetric density of compressed hydrogen as a function of gas pressure, compared to an ideal gas and liquid hydrogen. The ratio of the wall thickness to the outer diameter of the cylinder is shown on the right-hand side for steel with a tensile strength of 460 MPa. [2].

Theoretically, the resultant compression loss in percent are:

$$\%\text{loss} = 7.11 * 10^6 / 118 * 10^6 = 0{,}06 \tag{9.10}$$

Taking into account that the compressor and motor have a global energy efficiency of 60%, the resultant efficiency becomes

$$\eta = (1 - (0.06 / 0.6)) = 90\ \% \tag{9.11}$$

For the realization of a high pressure vessel, steel or composite materials can be used.

For the case of a steel vessel, the volumetric density inside a cylinder is represented in Figure 9.5 depending on the pressure, together with the ratio of the wall thickness to the outer diameter of the cylinder.

From Figure 9.2 (bottom left), one can see that the volumetric density increases with the pressure and reaches a maximum above 1000 bar for the steel reservoir. The gravimetric density decreases with increasing pressure.

9.4.3 Storage of Hydrogen by Adsorption

Given materials can adsorb gases in an efficient way. These materials can be carbon structures or metal hydrides. Several metals or alloys have the property of fixing hydrogen in a reversible way by forming solid hydrides. The reaction is exothermic. By pressure and temperature variations, it is possible to change the equilibrium. Some metal hydrides are available from other industrial processes or applications and represent a good solution for hydrogen storage when the weight is not relevant (stationary applications).

As an example, Mg_2Ni allows a storage capacity of 1 kWh/kg but must be operated at 265°C for a pressure of 1 bar. Fe-Ti can achieve 0.5 kWh/kg under a pressure of several bar and at ambient temperature.

9.4.4 Formic Acid for Hydrogen Storage

Recent developments in basic sciences have led to an interesting new method for hydrogen storage, namely formic acid. Formic acid $(HCOOH)$ is produced according well-known methods of catalytic hydrogenation of CO_2 [10, 11] and contains at ambient temperature and pressure 53 g/l hydrogen, equivalent to the double value of the weight energy density of compressed hydrogen at 350 bar.

The new developments concern the production of hydrogen from formic acid with a homogeneous catalytic system based on an aqueous solution of ruthenium catalysts where the formic acid is decomposed in dihydrogen (H_2) and carbon dioxide (CO_2) [12, 13]. The new method does not produce carbon monoxide and solves many problems of classical decomposition methods, such as stability and limited life cycle of used catalysts.

The basic scheme of the transformation of hydrogen into formic acid for storage and further release into hydrogen for users is represented in Figure 9.6.

The return path of carbon dioxide indicates that the CO_2 produced during catalytic decomposition of formic acid can be reused for the catalytic storage process.

The transformation enthalpy for catalytic reactions is given in [14, 15]. The values are represented as

$$H_2 + CO_2 \rightarrow HCOOH \qquad \Delta G = -4kJ/mol \tag{9.12}$$

$$HCOOH \rightarrow H_2 + CO_2 \qquad \Delta G = -32.9kJ/mol \tag{9.13}$$

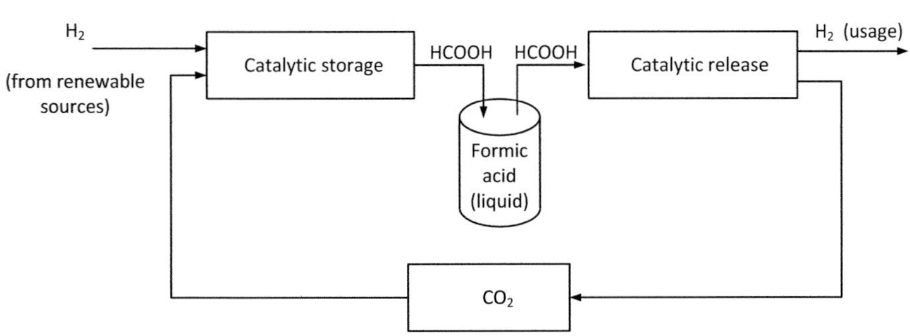

FIGURE 9.6 Catalytic storage and release of hydrogen.

TABLE 9.2
Comparison of the Energy Content of a 50-Liter Reservoir

Energy Carrier	e_m [MJ/kg] (Weight Energy Density)	ρ[kg/l] (Density)	e_v [MJ/l] (Volume Energy Density)	Energy Content (50-Liter Reservoir)
H_2 (700 bar)	118 MJ/kg	0.042 kg/l	4.95 MJ/l	247 MJ
Formic acid (4.4 wt%)	5.19 MJ/kg	1.22 kg/l	6.33 MJ/l	316 MJ
Diesel fuel	46.4 MJ/kg	0.84 kg/l	38.9 MJ/l	1948 MJ

With a molar mass of 46 g/mol

The transformation efficiencies can be evaluated as

$$\eta_{formationFA} = \frac{5.126MJ/kg}{5.126MJ/kg + \frac{4kJ/mol}{46g/mol}} = 0.98 \tag{9.14}$$

$$\eta_{release} = \frac{5.126MJ/kg - \frac{32.9kJ/mol}{46g/mol}}{5.126MJ} = 0.86 \tag{9.15}$$

For an application of formic acid in transportation, the meaning of the value of 53 g/l must, however, be illustrated through a comparison with other carriers, namely pressurized hydrogen and diesel fuel. This is done in Table 9.2; the most representative value is the energy content of a 50-liter reservoir.

9.5 CONVERSION FROM HYDROGEN TO ELECTRICITY

In the structural diagram of Figure 9.2, the output stage of the system that converts the stored hydrogen into electrical power is a fuel cell. A fuel cell allows the direct conversion of hydrogen into current by chemical reaction between the hydrogen and oxygen. In conventional fuel cells, the oxidant is generally taken from the ambient air.

The schematic diagram of Figure 9.7 shows the main components a classical PEM fuel cell with hydrogen and air supply. For such a fuel cell, the energetic balance must be evaluated in detail, taking into account the necessary power for the air compressor when it is operated at partial load [16]. Load current proportional airflow can be a well-adapted control method.

The chemical reactions at the anode and cathode sides are

Anode:

$$H_2 \quad -> \quad 2H^+ + 2e^- \tag{9.16}$$

Cathode:

$$O_2 + 4H^+ + 4e^- \quad -> \quad 2H_2O \tag{9.17}$$

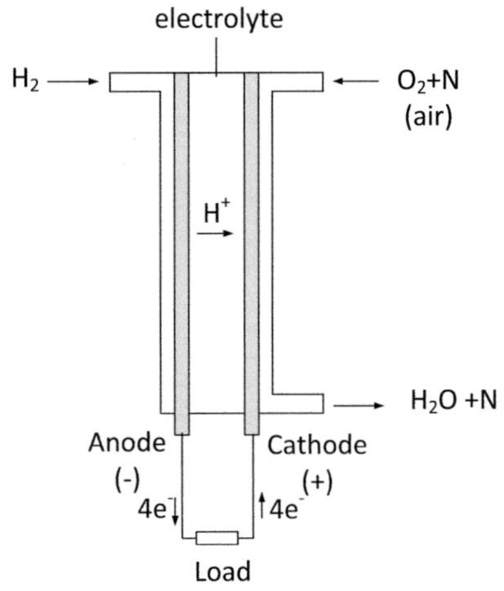

FIGURE 9.7 Standard structure of a PEM fuel cell.

9.6 EFFICIENCY CONSIDERATIONS

The energy efficiency of a hydrogen-based storage system can be calculated by the product of the partial efficiencies of the subcomponents. The first of these components is the electrolyzer, then the hydrogen and eventually the oxygen conditioning devices must be considered, and finally, the back transformation from chemical to electrical power in a fuel cell largely influences the global system efficiency.

9.6.1 Efficiency of the Electrolyze

The energy efficiency of a water electrolyzer is defined depending on the electrical energy consumed per standard volume of hydrogen (in MJ/m^3), assuming standard temperature and pressure. It strongly depends on the size, power and technique used. In Reference [17], the ranges of the real efficiency values of electrolyzers are given as between 43% and 67% for the alkaline electrolyzer and 40% and 67% for the PEM.

9.6.2 Power Needed for Fuel Conditioning

According to the descriptions of Sections 9.4.1 and 9.4.2, the energy needed for hydrogen compression or liquefaction can be evaluated as around 10% of the energy content of the conditioned fuel.

9.6.3 Efficiency of the Fuel Cells

The energy efficiency of the fuel cell is measured by the ratio of the amount of useful energy put out by the system to the total amount of energy that is put in. The useful output energy is measured in electrical energy produced, and the input energy is the energy stored in the fuel. According to Reference [18], fuel cells are generally 40–60% energy efficient. Such values are measured with idealized systems fed by hydrogen and oxygen.

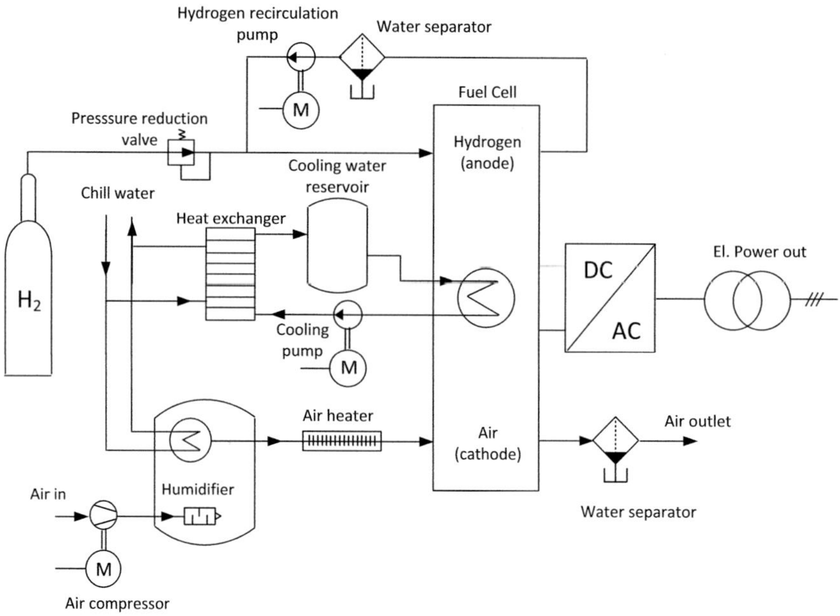

FIGURE 9.8 Fuel cell system. (Adapted from Grasser, F., Rufer, A., PEMFC system efficiency optimization through model based control strategies, *IEEE VPC Vehicular Power and Propulsion Conference*, Windsor, England, September 6–8, 2006.)

In practice, many fuel cells, like most PEM fuel cells, use the oxygen of the ambient air, fed from an air compressor. In these systems, the power used by the air compressor must be taken into account for the evaluation of the efficiency, such as the consumption of many other auxiliary components such as air humidification, hydrogen recirculation or cooling pumps (Figure 9.7).

The power needed for the air compressor can be reduced through specific strategies like electrical output power-dependent control of the air flowrate [16, 19].

9.6.4 Overall Efficiency

The overall efficiency of the hydrogen storage, defined as the ratio of the electrical output energy to the electrical input energy, can be evaluated as

$$\eta_{global} = \eta_{elys} \cdot \eta_{con} \cdot \eta_{fc} = 0.55 \cdot 0.9 \cdot 0.5 = 0.247 \tag{9.18}$$

The use of electrolyzers producing directly hydrogen under pressure, or high temperature electrolyzers which can re-use the thermal dissipation of the fuel cell will lead to higher energy efficiency of the hydrogen storage.

Additionally, the storage of hydrogen or simply the use of fuel cells in the context of combined heat and power (CHP) will increase the benefits of the use of hydrogen in energetic chains based on RES.

9.7 CONVERTING HYDROGEN INTO A SYNTHETIC FUEL—THE EXAMPLE OF E-METHANOL

9.7.1 Principles and Motivations

An alternative to compressed or liquefied pure hydrogen is given by chemical transformations into synthetic fuels. This section will describe the case of so-called e-methanol where green hydrogen produced by PV-fed water electrolysis is combined with CO_2. The main advantage of such a form of storage is that the fuel can be stored in a classical reservoir at atmospheric pressure and temperature. However, the specific energy or weight energy density of methanol is only 22.7 MJ/kg compared to pure hydrogen (120 MJ/kg). Regarding the volume energy density, methanol at atmospheric pressure and temperature has a value of 17.8 MJ/m^3, while hydrogen compressed at 30 MPa (300 bar) has a value of 2700 MJ/m^3. Synthetic fuels like e-methanol are intended to be used in transportation or propulsion with internal combustion engines [20].

9.7.2 Components of the e-Methanol Production Process

The main components of the considered e-methanol production process are represented in Figure 9.9. The output variable $m_{e\text{-meth}}$ on the right side of the figure corresponds to the annual production of e-methanol in mass. This is the output of the process reactor. The input quantities of the reactor are hydrogen and CO_2. CO_2 is captured from the ambient air, and the hydrogen is produced by the electrolyzer, which further produces oxygen. The electric power feeding the electrolyzer is obtained from a large photovoltaic plant. The necessary surface area of this plant is indicated as A_{PV} and will be estimated on the basis of the parameters obtained from an already operating large PV plant of several hundred MW of power.

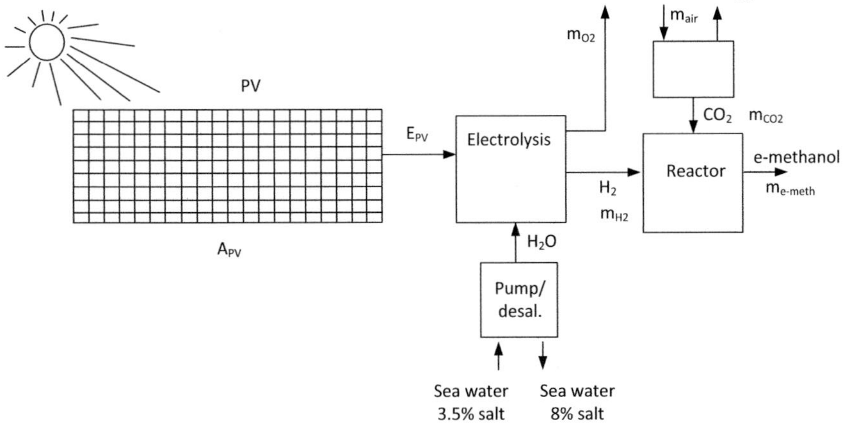

FIGURE 9.9 Schematic representation of the e-methanol production process.

The electrolyzer receives a given amount of desalinized water from the sea. The salinity of the pumped water is 3.5%, and the salinity of the discharged water is 8%.

Near the represented main sub-components, which are used for the main conversions and electrochemical transformations, several auxiliary devices and additional sub-processes are used to complete the system. Between the PV plant and the electrolyzer, diverse power electronic converters and transformers are needed. Especially in the case of a very large PV arrangement, the problem of transporting electrical power over long distances on the order of magnitude of tens of kilometers with an acceptable efficiency demands a dedicated collection network where medium-voltage DC systems or medium- or high-voltage AC systems must be used [21, 22].

At the output of the electrolyzer, the hydrogen generated must be conditioned and stored due to the day/night intermittency of sun irradiation when the CO_2 hydrogenation process is run 24 h per day.

Similar auxiliary conditioning equipment is used at the output of the CO_2 capture system, for the real operation of the reactor or within the seawater desalination system. In a general way, the mass transfer conditions are not considered in this description, nor is the energy demand of auxiliary devices such as water pumps, gas compressors, cooling devices, active heat exchangers or other conditioning equipment.

9.7.3 Hydrogenation of CO_2 (Synthesis of Methanol)

For the synthesis of methanol, different processes have been proposed based on the use of copper/zinc oxide-based catalysts [23, 24]. For the evaluation of the quantities of reactants involved, the process can be described through the three individual reactions illustrated in Table 9.3.

Reaction (1) summarizes the reaction of CO_2 with hydrogen for the production of methanol and water. It is accompanied by Reaction (2) as a secondary reaction, the reaction of the gas with water in the reverse sense (RWGS), which consumes the same reactants but which is not desired in the process (the hydrogenation of CO_2 into CH_3OH_2). The production of CO through RWGS leads to the appearance of an additional path for the synthesis of methanol: by hydrogenation of CO (Reaction (3)). Reactions (1) and (3) are exothermic, whereas RWGS is endothermic.

TABLE 9.3
Reactions Involved in the Hydrogenation of CO_2 into CH_3OH

No.	Reaction	$\Delta_{rx}H_{298\ K}$ [kJ/mol]
1	$CO_2 + 3H_2 \leftrightarrow CH_3OH + H_2O$	−49.51
2	$CO_2 + H_2 \leftrightarrow CO + H_2O$	41.17
3	$CO + 2H_2 \leftrightarrow CH_3OH$	−90.68

Reaction (1) can be rewritten in terms of the different atomic masses of the involved chemical species

as $$CO_2 + 3H_2 \leftrightarrow CH_3OH + H_2O \tag{9.19}$$

or $$12 + 2\times 16 + 3\times 2 \leftrightarrow 12 + 3 + 16 + 1 + 2 + 16 \tag{9.20}$$

or $$44 + 6 \leftrightarrow 32 + 18 \tag{9.21}$$

Related to the masses of matter involved in the process, namely m_{H2} for the mass of hydrogen, m_{CO2} for the carbon dioxide and m_{H2O} for the produced mass of water, the following relations can be written depending on the produced mass of methanol m_{CH3OH}

$$m_{H2} = \frac{6}{32}\cdot m_{CH_3OH} \tag{9.22}$$

$$m_{CO2} = \frac{44}{32}\cdot m_{CH_3OH} \tag{9.23}$$

$$m_{H2O} = \frac{18}{32}\cdot m_{CH_3OH} \tag{9.24}$$

9.7.4 Quantitative Design

The evaluation of the feasibility of a methanol synthesis plant, as described in Figure 9.1, needs an estimation of realistic quantitative conditions. An assumption is made that 500,000 tons of methanol should be produced per year. This value corresponds approximately to the annual fuel consumption of 300,000 cars, each car consuming 10 L per 100 km and driving 20,000 km per year, based on an ICE engine needing around 1.5 times the consumption of classic fuel. The density of methanol is 0.83.

The indicated number of cars that could be supplied from a methanol plant of 500,000 tons should be compared with the real number of cars circulating in a region. For example, Switzerland has more than 4 million cars; while in Germany, the total number of passenger vehicles registered for use on public roads increased to over 48.5 million in 2022. In such a context, a conversion to e-methanol of the whole German car fleet would need more than 160 times the described infrastructure.

Thus, introducing the value of 500,000 t of methanol in Equations 9.22 and 9.23, the following amounts of hydrogen m_{H2} and CO_2 m_{CO2} are needed:

$$m_{H2} = \frac{6}{32}\cdot 500000\text{t} = 93750\text{t} \tag{9.25}$$

$$m_{CO2} = \frac{44}{32}\cdot 500000\text{t} = 687500\text{t} \tag{9.26}$$

According to Equation 9.24, the reactor produces a mass of water m_{H2O}

$$m_{H20} = \frac{18}{32} \cdot 500000\text{t} = 281250\text{t} \tag{9.27}$$

Following the analysis of the hydrogenation process regarding the different amounts of reactants and output quantities as described in Table 9.1, the energetic balance can also be evaluated, in the sense of the produced heat due to the exothermic nature of the main reaction. For Reaction (1), the CO_2 reaction which produces methanol and water, the released thermal energy is given as 49.5 kJ/mol.

The number of moles of methanol generated is calculated with the value of its molar mass $(M(CH_3OH) = 32.04\,\text{g/mol})$; thus, the number of moles becomes:

$$\text{N}\left(\text{CH}_3\text{OH}\right) = 500 \times 10^9 \text{g}/\left(32.04 \text{ g/mol}\right) = 15.6 \times 10^9 \text{ mol} \tag{9.28}$$

The released thermal energy during the reaction for the 500,000 tons of methanol becomes:

$$E_{th\_react} = 15.6 \cdot 10^9 \text{mol} \cdot 49.5\text{kJ / mol} = 773 \cdot 10^9 \text{kJ} \tag{9.29}$$

$$\text{or } 773 \cdot 10^9 \text{kJ / (3600s / h)} = 214.6 \cdot 10^3 \text{MWh} \tag{9.30}$$

For this energy, which is dissipated over one year, if the reactor is run 24 h a day, the instantaneous power becomes:

$$P_{th\_react} = 214.6 \cdot 10^3 \text{MWh / } (24 \cdot 365) = 24.5\text{MW} \tag{9.31}$$

This power must be continuously evacuated from the reactor, requiring auxiliary equipment. It could be reused elsewhere in the process, for example, for a specific type of electrolyzer based on high-temperature steam electrolysis [12, 15]. Otherwise, the heat must be discharged, most probably into a water body. This adds to the complexity of the implementation of the facility and its negative impact on the local environment.

The next sections will be dedicated to the evaluation of the needed infrastructure, from the photovoltaic plant to the CO_2 capture system and the needed amount of seawater to be desalinized.

9.7.5 Hydrogen from Water Electrolysis

The production of hydrogen from water electrolysis is shown in Figure 9.10. The amount of hydrogen to be produced annually is defined in Equation 9.27. This section calculates the electrical energy to be provided for the electrolysis on the basis of the efficiency of the electrochemical converter. This efficiency is defined as the ratio of the heat value of the produced hydrogen E_{H2} to the electrical energy fed to the electrolyzer.

$$E_{H_2} = 33 \cdot 10^3 \text{ Wh / kg} \cdot 93750 \text{ t} = 33\text{MWh / t} \cdot 93750 \text{ t} = 3.1 \cdot 10^6 \text{MWh} \tag{9.32}$$

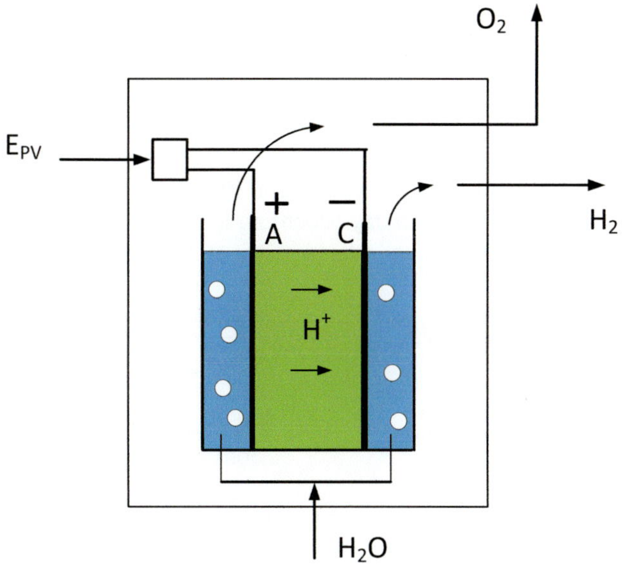

FIGURE 9.10 Hydrogen production from water electrolysis.

The needed electric power feeding the electrolyzer is calculated with its energy efficiency:

$$\eta_{electrolyser} = 0.6$$

$$E_{electric} = \frac{3.1 \cdot 10^6 \,\text{MWh}}{0.6} = 5.16 \cdot 10^6 \,\text{MWh} \quad (9.33)$$

This amount corresponds to an optimistic value calculated with an ideal value of the efficiency and does not consider any need for power for pressurization or purification of the hydrogen [25].

This electric power can be produced by a large photovoltaic plant, the surface area of which is calculated on the basis of the example of the Agua Caliente PV plant (Arizona) [26]. This plant has an area of 971 ha and produces an annual net output of 727 GWh. The annual production per ha of this real plant is then:

$$Yprod_{ha} = \frac{727\text{GWh}}{971\text{ha}} = 748\text{MWh / ha} \quad (9.34)$$

Thus, the surface area of the PV plant in Figure 9.1 for a production of 500,000 t of methanol becomes:

$$A_{PV} = \frac{5.16 \cdot 10^6 \,\text{MWh}}{748\text{MWh / ha}} = 6891 \text{ ha} \quad (9.35)$$

Knowing that $1\ \text{km}^2 = 10^6\,\text{m}^2$ and $1\ \text{ha} = 10^4\,\text{m}^2$, $1\ \text{ha} = 10^4 / 10^6\,\text{km}^2 = 10^{-2}\,\text{km}^2$.

The needed surface area is finally calculated as:

$$A_{PV} = 68.91\ \text{km}^2 \tag{9.36}$$

The required power for the PV plant shown in Figure 9.1 is calculated based on a possible daily production of 6.8 h/day according to the average of daily hours from the Agua Caliente plant, which has a rated power of 290 MW

$$YearHours_{AguaCaliente} = \frac{727\text{GWh}}{290\text{MW}} = 2506\text{h} \tag{9.37}$$

$$DayHours_{AguaCaliente} = \frac{2506\text{h}}{365} = 6.86\text{h} \tag{9.38}$$

The needed power of the methanol plant electrolyzer becomes:

$$P_{PV} = \frac{5.16 \cdot 10^6\,\text{MWh}}{6.8\text{h} \cdot 365} = 2078\text{MW} \tag{9.39}$$

9.7.6 Hydrogen/Oxygen Production and Water Consumed

The electrolyzer simultaneously produces the required hydrogen and an additional amount of oxygen, which is calculated on the basis of the atomic masses of the reactants in the water equation:

$$2H_2O \rightarrow 2H_2 + O_2$$

The relation with the corresponding mass units can be written as:

$$2(2 \cdot 1 + 16) \rightarrow 2 \cdot 2 + 2 \cdot 16 \tag{9.40}$$

Water hydrogen

One can define the two mass unit ratios as:

$$K_{mO_2/mH_2} = \frac{32}{4} = 8 \tag{9.41}$$

$$K_{mH_2O/mH_2} = \frac{36}{4} = 9 \tag{9.42}$$

From Equations 9.25 and 9.41, the corresponding mass of oxygen produced by the electrolyzer is calculated as:

$$m_{O2} = 8 \cdot 93'750\text{t} = 750'000\ \text{t}$$

From Equations 9.25 and 9.42, the corresponding mass of water needed is calculated as:

$$m_{H_2O} = 9 \cdot 93'750 \text{ t} = 843'750 \text{ t} \tag{9.43}$$

9.7.7 Feeding the Electrolyze with Water

Equation 9.43 gives a value of 843,750 tons for the water needed by the electrolyzer. This large amount is obtained from two sources; one is water fed back from the output of the methanol reactor, amounting to a mass of 281,250 tons, as was calculated in Equation 9.27; the other is water pumped from the sea and must be desalinized before use.

Desalination technologies are generally classified as either thermal or membrane-based technologies. Although membrane-based technologies, particularly reverse osmosis (RO), constitute the majority of installed desalination capacity worldwide, thermal desalination remains the dominant technology in the regions where desalination has been used for many years (the Middle East). The two main thermal desalination technologies used are multiple evaporative distillation (MED) and multistage flash (MSF). However, the bulk of newly installed desalination capacity around the world runs on reverse osmosis membranes, which are not only more energy efficient than thermal desalination processes but also offer modularity and ease of installation [19, 20]. In the design of the process described in this study, the reverse osmosis technique is retained, even though the discharge of water with a high salt content can locally create additional nuisances.

The amount of desalinized seawater becomes:

$$m_{H_2Odes} = 843750 \text{ t} - 281250\text{t} = 562500\text{t} \tag{9.44}$$

One can suppose the salinity of the pumped water to be 3.5% and the discharged water to be 8%. With the definitions of the pumped water amount m_p and the discharged water amount m_r, the condition that all salt of the pumped water is returned through the discharged mass to the sea can be written as:

$$m_{salt} = m_p \cdot 0.035 = m_r \cdot 0.08 \tag{9.45}$$

A second relation specifies that the difference between the desalinized pumped water and the discharged water corresponds to the amount of water fed to the electrolyzer

$$m_p \cdot (1 - 0.035) - m_r \cdot (1 - 0.08) = m_{H2O\_dess} \tag{9.46}$$

Combining the two relations, the mass of pumped sea water becomes:

$$m_p = \frac{m_{H2O\_dess}}{1 - \dfrac{0.035}{0.08} + 0.035} \tag{9.47}$$

The amount of discharged water is calculated from Equation 9.45 as:

$$m_r = m_p \frac{0.035}{0.08} \tag{9.48}$$

Numerically, $m_p = 941{,}422$ t and $m_r = 411{,}872$ t.

9.7.8 CO_2 Capture from the Ambient Air

Direct air capture (DAC) of CO_2 from the ambient air has become an important subject in scientific investigation and experimental engineering [27]. DAC is currently categorized as being at "technology readiness level" 6 (on a scale of 1 to 9), meaning it is still in the large-scale and prototype phase and not yet ready for full commercial deployment. However, this also means that there is ample opportunity to improve performance and reduce costs through learning from early iterations of the technology.

In the context of the present investigations regarding the quantitative impacts on used materials, process quantities and involved industrial equipment, one important question is the amount of air that needs to be taken from the surroundings from which the needed amount of CO_2 is extracted, especially considering the low concentration of CO_2 in the air of 414 ppm or 414 g/t.

$$\text{Mass\_}air = \frac{687500 \text{ t}}{0.000414 \text{ t/t}} = 1.66 \cdot 10^9 \text{ t} \tag{9.49}$$

This mass should be increased, as the full content of CO_2 is not extracted in the process. Considering a yield of 80%, the final mass of air becomes:

$$\text{Mass\_}air\_final = \frac{1.66 \cdot 10^9 \text{ t}}{0.8} = 2.08 \cdot 10^9 \text{ t} \tag{9.50}$$

The CO_2 extraction can be run all day, all year round, leading to a flow of air of:

$$Air\_flow = \frac{2.08 \cdot 10^9 \text{ t}}{365 \cdot 24 \text{ h}} = 237560 \text{ t/h} \tag{9.51}$$

or:

$$Air\_flow = \frac{237560 \text{ t/h}}{3600 \text{ s/h}} = 65.9 \text{ t/s} \tag{9.52}$$

with a density of air of 1.3 kg/m^3:

$$Air\_flow\_vol = \frac{65900 \text{ kg/s}}{1.3 \text{ kg/m}^3} = 50760 \text{ m}^3\text{/s} \tag{9.53}$$

This amount of air can be provided using large fans, for example, 30-kW blowers each delivering 27 m^3/s. For the 50,760 m^3/s for the DAC, 1880 blowers would be needed, the resultant electric power for the DAC being

$$P_{el\_DAC} = 1880 \cdot 30kW = 56'400kW = 56.4MW \quad (9.54)$$

9.7.9 Efficiency Considerations

The process of production of e-methanol can be characterized by equivalent energy efficiency. The goal of such an evaluation is to be able to compare two different alternative solutions for future automotive systems, the first being the use of green e-methanol in a conventional ICE-driven vehicle, while the second is an electric vehicle (EV) powered by a battery when this battery is recharged from the same renewable source (PV).

9.7.10 Efficiency of e-Methanol Production and Use

The efficiency of e-methanol production based on a renewable source is here defined as the ratio of the heat value of the produced fuel divided by the renewable energy fed into the process. Two components for the energy input are considered in this evaluation, the electric energy fed to the electrolyzer (5.16×10^6 MWh from Equation 9.33) and the needed power for the CO_2 capture (56.4 MW from Equation 9.54), the corresponding energy of which is:

$$\text{E\_fans} = 56.4 \text{ MW} \times 24 \text{ h} \times 365 = 494{,}064 \text{ MWh} \quad (9.55)$$

The sum of the provided renewable energy from the PV becomes:

$$\text{E}_{-\text{PV}} = 0.494 \times 10^6 \text{MWh} + 5.6 \times 10^6 \text{MWh} = 6.094 \times 10^6 \text{MWh} \quad (9.56)$$

The heat value of the 500,000 tons of e-methanol is calculated as:

$$\text{E}_{-\text{meth}} = 500{,}000{,}000 \times 22.7 \text{ MJ/kg} = 1.13 \times 10^{10} \text{MJ} \quad (9.57)$$

or:

$$1.13 \times 10^{10} \text{MJ} / 3600 = 3.152 \times 10^6 \text{MWh} \quad (9.58)$$

Finally, the efficiency of e-methanol production can be given as:

$$\eta_{\text{emeth\_PV}} = 3.152 \times 10^6 \text{MWh}/6.094 \times 10^6 \text{MWh} = 0.517 \quad (9.59)$$

In Figure 9.11, the whole chain of e-methanol production and use is presented. In this presentation, the efficiency of the propulsion system based on the ICE engine (tank-to-wheels) is given as 0.15. The cascade of the fuel production efficiency (0.51) and the propulsion gives an extremely low value of 0.077.

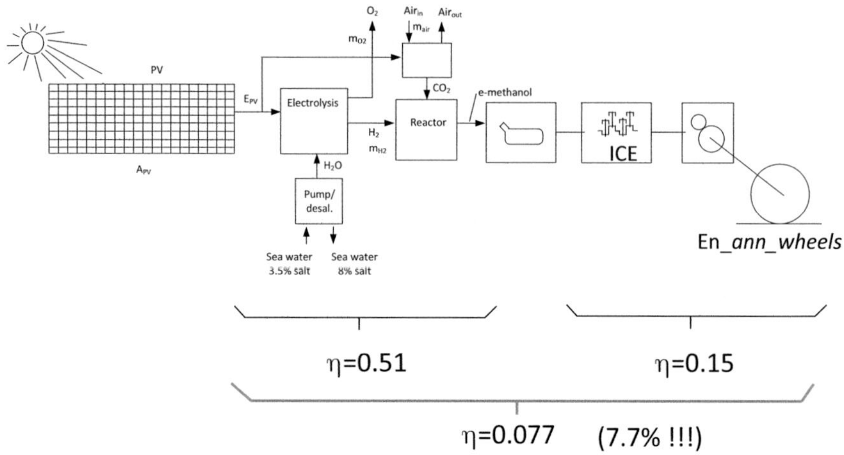

FIGURE 9.11 Efficiency of the e-methanol-powered ICE vehicle.

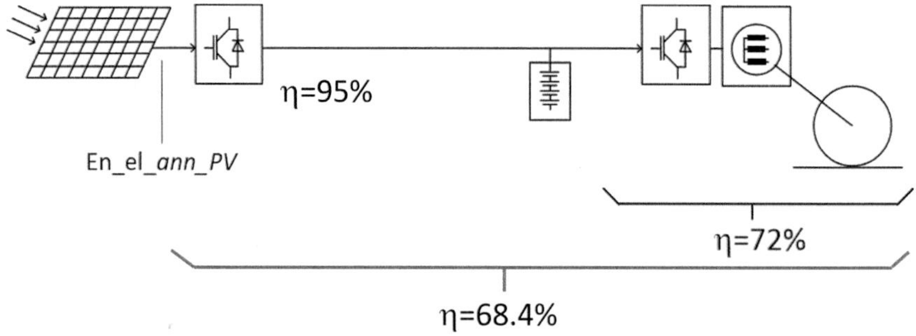

FIGURE 9.12 Efficiency of a PV-powered EV.

9.7.11 Efficiency of an Electric Vehicle Powered from PV

Figure 9.12 shows the complete chain of an EV powered from photovoltaic panels. The indicated value of the efficiency of the power train ($\eta = 0.72$) corresponds to the cascade of an electric motor ($\eta = 0.9$) and a power electronic converter ($\eta = 0.95$) and with 8% standby/idle losses and 6% additional losses for the driveline and auxiliaries.

The resultant efficiency becomes 68.4% if the battery is charged via a power electronic charger of 95% efficiency.

9.7.12 Summary of the Design Parameters

The e-methanol production process as described in this chapter corresponds to a very large industrial production site, involving huge quantities of the different reactants

TABLE 9.4
Summary of the System Parameters

Main Process		
e-methanol production per year	500,000	tons
Hydrogen (from electrolysis)	93,750	tons
CO_2 captured from ambient air	687,500	tons
H_2O from the reactor	281,250	tons
Electrolysis		
Hydrogen	93,750	tons
Oxygen	750,000	tons
Water (desalinized)	843,750	tons
PV plant		
Ground surface	68.9	km^2
PV power	2078	MW
Energy produced (yearly)	5.16×10^6	MWh
Water (sea)		
Pumped water (3.5% salt)	941,422	tons
Discharged water (8% salt)	411,872	tons
CO_2 capture		
Mass of air needed (80% yield in capture)	2.08×10^9	tons
Nb of fans (100,000 m^3/h type)	1880	units
Power of the fan system	56.4	MW

involved. The characteristic parameters of the main process as well as the parameters of the subcomponents are summarized in Table 9.4.

9.8 LIQUID AMMONIA AS GREEN FUEL FOR THE TRANSPORTATION SECTOR

9.8.1 Principles

Hydrogen can be stored in the form of a liquid fuel when combined with nitrogen, forming the well-known and widespread ammonia. Ammonia (NH_3) is one of the most important basic chemicals in today's chemical industry. Industrially, ammonia is synthesized by the Haber–Bosch process at high temperature and pressure (typically 400°C and 200 atm) [28]. In nature, the enzyme nitrogenase can convert N_2 to NH_3 under ambient conditions, motivating the search for similar sustainable technologies for industrial NH_3 production. In recent years, photocatalytic ammonia production using sunlight and photocatalysts has attracted considerable interest, enabling the reduction of N_2 to NH_3 under very mild reaction conditions [29, 30].

9.8.2 Properties and Structure of the System

Ammonia has a volume energy density of 11.5 MJ/l, while hydrogen compressed at 30 MPa (300 bar) has a value of 2.7 MJ/l. Depending on the technology used for the reservoir, the ratio of that densities is between 3 and 4. A decarbonized ammonia production can be realized on the basis of green hydrogen produced by water electrolysis powered from photovoltaic generators. The structure of a system using the Haber–Bosch process is represented in Figure 9.13a. A system based on a photocatalytic reactor is represented in Figure 9.13b.

The reduction of N_2 to NH_3 is described through the chemical equation

$$3H_2 + N_2 \rightleftharpoons 2NH_3 + \Delta H(\Delta H_{300} = -46.35 kJ/mol) \tag{9.60}$$

9.8.3 A Portable Ammonia Cracking System

Dedicated to powering an electric propulsion system in transportation, an ammonia cracking device could be used to feed a fuel cell with hydrogen. A portable cracking system for heavy transportation on road for ships is described in [31]. The cracking technology uses a ruthenium-based catalyst, while the purification is done from a low concentration palladium-based mix. The structural diagram of the thermochemical hydrogen generator is given in Figure 9.14.

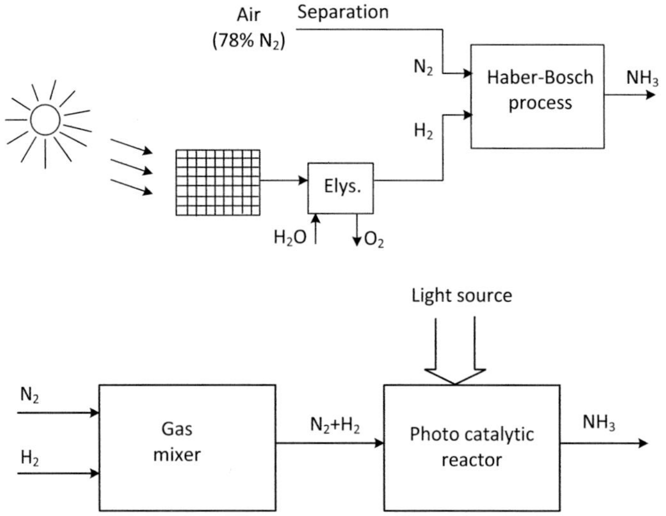

FIGURE 9.13 Green ammonia production process: (a) usage of the Haber–Bosch process, (b) usage of a photocatalytic reactor.

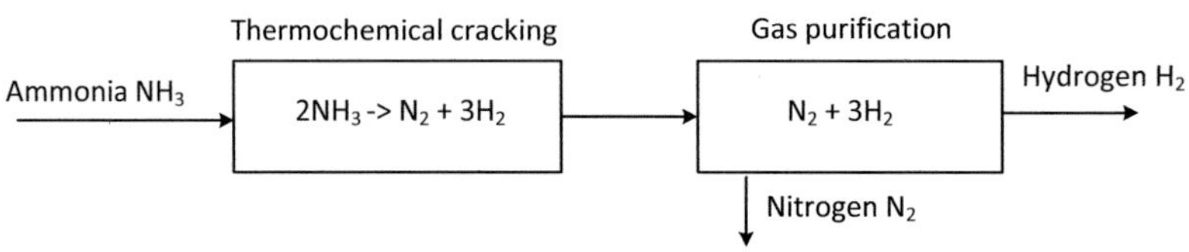

FIGURE 9.14 Ammonia cracking for the recovery of hydrogen.

Exercise 7: Conversion from Electrical Power to Hydrogen

Larger amounts of energy can be stored in a chemical form through the conversion from electric power into pressurized hydrogen. In Figure 9.15, an electrolyzer (1) converts electricity into hydrogen, which is stored in a final reservoir at a pressure of 200 bar (2). The filling of the final reservoir is achieved using a so-called booster station (3). This device is connected to an intermediate buffer-reservoir (4) fed directly from the electrolyzer.

The electrolyzer is fed with electric current from a dedicated rectifier (5), itself connected to a local grid with renewable energy sources.

The booster station is composed of a reciprocating compressor, where the movement is produced with an air drive system (Figure 9.16). The air drive system is supplied by low-pressure air (industrial air) provided by an air compressor and reservoir (Figure 9.15, (7) and (8). The goal of the exercise is to study such a system and to evaluate the energy efficiency of such a transformation.

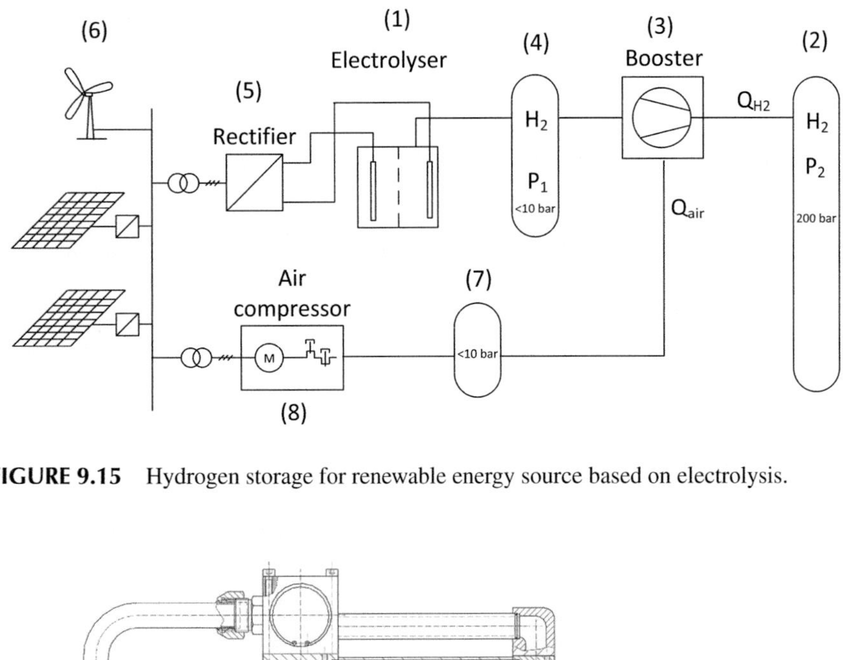

FIGURE 9.15 Hydrogen storage for renewable energy source based on electrolysis.

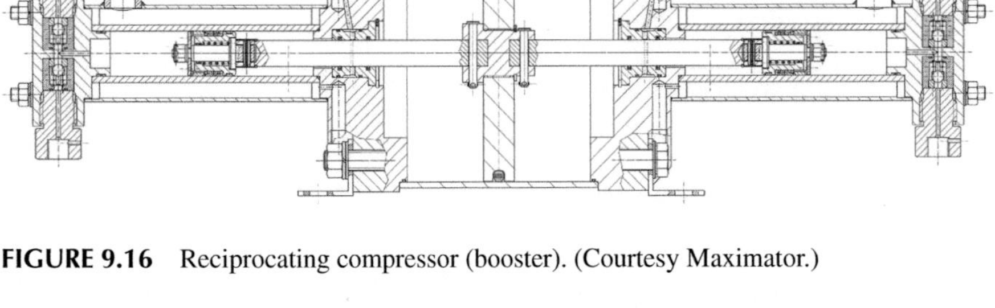

FIGURE 9.16 Reciprocating compressor (booster). (Courtesy Maximator.)

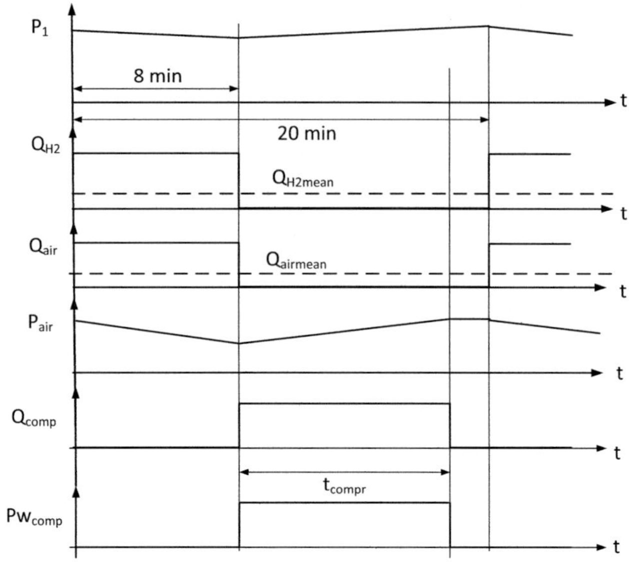

FIGURE 9.17 Time schedule of the hydrogen compression system.

The operation conditions are illustrated by the time diagram of Figure 9.17. The first line shows the variation of pressure in the intermediate hydrogen reservoir. The discharge is caused by the activation of the booster station during 8 minutes. This discharge is followed by a phase of re-pressurization of 12 minutes due to the operation of the electrolyzer. The total cycle time is 20 minutes.

In the same figure, the hydrogen flowrate of the booster stage is also shown (Q_{H2}), as well as its mean value (Q_{H2mean}). According to this hydrogen flow, the consumption of the air drive is also shown (Q_{air}, $Q_{airmean}$). Finally, the airflow provided by the compressor itself to recharge its local air reservoir is also represented (Q_{comp}).

The last line of Figure 9.17 shows the electrical power consumed by the air compressor.

TECHNICAL DATA OF THE HYDROGEN COMPRESSION SYSTEM

BOOSTER STAGE:

Flow capacity at an inlet pressure of 10 bar:	70 l_N/min
Air consumption:	1980 l_N/min

AIR COMPRESSOR:

Rated power (el):	9.6 kW
Flow capacity (10 bar):	1600 l_N/min

ELECTROLYZER:

The specific electrical power consumption per unit of volume of the electrolyzer is (see Section 9.3.2)

$$e_{el\_v\_elyz} = 5000\frac{Wh}{m^3}$$

SOLUTION TO EXERCISE 7

As indicated in Figure 9.17, the booster stage is activated during 8 min and stopped for the rest of the period (12 min). The average flowrate of hydrogen to the high-pressure reservoir is therefore:

$$Q_{H2mean} = Q_{H2}\cdot\frac{4\min}{20\min} = \frac{70\cdot l}{\min}\cdot\frac{2}{5}\cdot\frac{\min}{60s} = \frac{1.16\cdot 2}{5}\cdot\frac{l}{s} = 0.464\frac{l}{s}$$

1 kg H_2 at 1 bar occupies a volume of 11 m^3
So the corresponding density is

$$\rho_{H2} = \frac{1kg}{11m^3} = 0.09\frac{kg}{\mathrm{m}^3} = 0.09\cdot 10^{-3}\cdot\frac{kg}{l}$$

The weight energy density e_m of H_2 is

$$e_{mH2} = 33\mathrm{kWh/kg}$$

The equivalent power or energy flowrate of the hydrogen transferred to the reservoir is:

$$P_{H2mean} = Q_{H2mean}\cdot\rho_{H2}\cdot e_{mH2} = 0.464\frac{l}{s}\cdot 0.09\cdot 10^{-3}\frac{kg}{l}\cdot 33\cdot 10^3\frac{kW\cdot 3600s}{kg} = 4960W$$

The electrical power at the input of the electrolyzer is

$$P_{elyz} = p_{v\_elyz}\cdot Q_{H2} = 5000\frac{W\cdot 3600s}{m^3}\cdot 0.46410^3\frac{m^3}{s} = 8352W$$

EFFICIENCY OF THE ELECTROLYZER

From the specific energy of hydrogen and the specific power consumption of the electrolyzer, the energetic performance can be evaluated

Volumetric energy density of hydrogen:

$$e_{vH2} = e_{mH2}\cdot\rho_{H2} = 33\cdot 10^3\frac{kWh}{kg}\cdot 0.09\cdot 10^{-3}\frac{kg}{m^3} = 2970\frac{Wh}{m^3}$$

The volumetric power consumption of the electrolyzer:

$$e_{el\_v\_elyz} = 5000\frac{Wh}{m^3}$$

Finally the energetic performance of the electrolyzer becomes

$$\eta_{elyz} = \frac{2970\frac{Wh}{m^3}}{5000\frac{Wh}{m^3}} = 0.59$$

GLOBAL EFFICIENCY OF HYDROGEN PRODUCTION AND STORAGE

First the electrical energy consumed by the air compressor must be calculated (average)

Characteristics of the compressor:

Flow capacity (10 bar):	1600 l_N/min
Rated power (el):	9.6 kW

The average air consumed by the booster:

$$Q_{air\_mean} = 1980\frac{l}{\min}\frac{8\min}{20\min} = 792\frac{l}{\min}$$

For the compensation of this air, the compressor must be turned on for the time t_{comp}

$$t_{comp} = \frac{1980\frac{l}{\min}\cdot 8\min}{1600\frac{l}{\min}} = 9.9\min$$

The average power consumed by the compressor is

$$P_{comp\_mean} = \frac{792\frac{l}{\min}}{1600\frac{l}{\min}}\cdot 9600W = 4752W$$

The global efficiency of the conversion of the electrical power into compressed hydrogen becomes

$$\eta_{el-to-press-H2} = \frac{P_{H2mean}}{P_{elyz} + P_{comp\_mean}} = \frac{4960W}{8352W + 4752W} = 0.3785$$

This relatively poor value is due to the limited efficiency of the electrolyzer and to the very poor efficiency of the air compressor.

EFFICIENCY OF THE COMPRESSOR

This value can be easily evaluated as follows

The air flowrate is

$$Q_{air} = 1600\frac{l_N}{\min} = \frac{1600}{60}\cdot\frac{l_N}{s} = 26.66\frac{l_N}{s}$$

At the output side (10 bar)

$$Q_{air10} = 2.666\frac{l_N}{s}$$

The corresponding mechanical (pneumatic) power at the output of the compressor becomes

$$P_{air10} = Press_{air}\cdot Q_{air10} = 10\cdot 10^5\frac{N}{m^2}\cdot 2.666\cdot 10^{-3}\frac{m^3}{s} = 2666W$$

and the efficiency

$$\eta_{compressor} = \frac{2666W}{9600W} = 0.27$$

A more efficient air compressor must be used for the supply of the booster stage.

ADDITIONAL REMARKS

The losses of the power converter (power electronics) at the input of the electrolyzer have been neglected.

The mechanical losses in the booster stage have been neglected.

The efficiency of the electrical motor driving the air compressor is included in the efficiency of the compressor.

Thermodynamic effects like heating by compression have also been neglected.

REFERENCES

[1] Derbal, H., Belhamed, M., M'Raoui, A., L'hydrogène, vecteur énergétique solaire, *Revue des Energies Renouvelables, ICRESD-07*, Tlecem, 2007, pp. 235–247.

[2] Zuttel, A., Materials for hydrogen storage, *Materials Today*, 6(9), 24–33, September 2003, Elsevier Ltd.

[3] Gicquel, R., *Energy Systems: A New Approach to Engineering Thermodynamics*, CRC Press/Taylor & Francis Group, London, UK, 2012.

[4] Gahleitner, G., Hydrogen from renewable electricity: An international review of power-to-gas pilot plants for stationary applications, *International Journal of Hydrogen Energy*, 38, 2039–2061, 2013.
[5] Wendt, H., Kreysa, G., *Génie électrochimique Principes et propriétés*, Editions Dunod, Paris, France, 2001.
[6] Schouler, E., Fernandez, E., Bernard, H., *Electrolyseur de la vapeur d'eau à haute température, Principe et Etat de développement*, Revue Générale de l'Electricité, No. 3, Mars 1982.
[7] Amstutz, V., Toghill, K. E., Powesland, F., Vrubel, H., Comninellis, C., Hu, X., Girault, H., Renewable hydrogen generation from a dual-circuit redox flow battery, *Energy & Environmental Science*, 00, 1–3, 2014, https://doi.org/10.1039/C4EE00098F.
[8] Fazle-Kibria, A. K. M, Mo, Y. H., Nahm, K. S., Yun, M. H., Electrochemical hydrogen storage behaviors of CVD, AD, and LA grown carbon nanotubes in KOH, medium, *International Journal of Hydrogen Energy*, 26(8), 823–829, 2001.
[9] Poirier, E. et al., Hydrogen adsorption in carbon nanostructures, *International Journal of Hydrogen Energy*, 26(8), 831–835, 2001.
[10] Jessop, P. G., Homogeneous hydrogenation of carbon dioxide, in *Handbook of Homogeneous Hydrogenation* (edited by J. G. de Vries, C. J. Elsevier), Wiley-VCH, Weinheim, Germany, 2007, pp. 489–511.
[11] Jessop, P. G., Joo, F., Tai, C.-C., Recent advances in the homogeneous hydrogenation of carbon dioxide, *Coordination Chemistry Reviews*, 248, 2425–2442, 2004.
[12] Fellay, C., Dyson, P. J., Laurenczy, G., A viable hydrogen-storage system based on selective formic acid decomposition with a ruthenium catalyst, *Angewandte Chemie International Edition*, 47, 3966–3970, 2008.
[13] Joo, F., Breakthroughs in hydrogen storage: Formic acid as a sustainable storage material for hydrogen, *ChemSusChem*, 1, 805–808, 2008.
[14] Sordakis, K., Tsurusaki, A., Iguchi, M., Kawanami, H., Himeda, Y., Laurenczy, G., Carbon dioxide to methanol: The aqueous catalytic way at room temperature chem, *Chemistry: An European Journal*, 22, 15605–15608, 2016.
[15] Grasemann, M., Laurenczy, G., Formic acid as a hydrogen source: Recent developments and future trends, *Energy & Environmental Science*, 5, 8171, 2012.
[16] Grasser, F., Rufer, A., PEMFC system efficiency optimization through model based control strategies, *IEEE Vehicular Power and Propulsion Conference (VPPC)*, Windsor, England, September 6–8, 2006.
[17] Bertuccioli, L. et al., *Development of Water Electrolysis in the European Union, Fuel Cells and Hydrogen Joint Undertaking, Final Report*, February 2014 [Online], www.fch.europa.eu/sites/default/files/study%20electrolyser_0-Logos_0_0.pdf. Accessed on August 7, 2017.
[18] U.S. Department of Energy, *Energy Efficiency and Fuel Cell Technologies Program, Comparison of Fuel Cell Technologies*, DOE, Washington, DC, February 2011.
[19] Grasser, F., Rufer, A. C., An analytical, control-oriented state space model for a PEM fuel cell system, *Power Conversion Conference 2007 (PCC '07)*, Nagoya, Japan, 2007, pp. 441–447, https://doi.org/10.1109/PCCON.2007.373005.
[20] MAN Energy Solutions to Supply Methanol Reactor for Porsche AG eFuels Pilot Plant in Chile—World's First Integrated, Commercial, Industrial-Scale Synthetic Fuels Plant to Produce 550 Million Litres of eFuels per Year by 2026 [Online], www.man-es.com/company/press-releases/press-details/2021/07/07/man-energy-solutions-to-supply-methanol-reactor-for-porsche-ag-efuels-pilot-plant-in-chile. Accessed on November 15, 2022.
[21] Steinke, G. K., Rufer, A., Use of a DC-DC step up converter in photovoltaic plants for increased electrical energy production and better utilization of covered surface area, *Proceedings of PCIM Europe 2015; International Exhibition and Conference for Power*

Electronics, Intelligent Motion, Renewable Energy and Energy Management, Nuremberg, Germany, May 19–20, 2015.

[22] Sayed, S., Elmenshawy, M., Ben-Brahim, L., Massoud, A. design and analysis of high-gain medium-voltage DC-DC converters for high-power PV applications, *Conference Paper, Proceedings of the 2018 IEEE 12th International Conference on Compatibility, Power Electronics and Power Engineering (CPE-POWERENG 2018)*, Doha, Qatar, April 10–12, 2018. IEEE, Piscataway, NY, USA, 2018.

[23] Quezada Feliz, M. J., Hydrogénation catalytique de CO_2 en méthanol en lit fixe sous chauffage conventionnel et sous plasma à DBD, PhD Thesis, l'Institut National des Sciences Appliquées de Rouen Normandie, Saint-Étienne-du-Rouvray, France, 2020, www.google.com/url?sa=t&rct=j&q=&esrc=s&source=web&cd=&ved=2ahUKEwiGo6zkhL_6AhX2_rsIHde0B2wQFnoECBEQAQ&url=https%3A%2F%2Fwww.theses.fr%2F2020NORMIR12.pdf&usg=AOvVaw0BICPnoh0TBsBbtYCwofLB. Accessed on November 15, 2022.

[24] Zabidi, N. A. M., Tasfy, S., Shaharun, M. S., Hydrogenation of CO_2 to methanol using copper/zinc oxide-based catalyst: Effect of active metal ratio, *AIP Conference Proceedings*, *1787*, *030009*, 2016, https://doi.org/10.1063/1.4968074.

[25] Ursua, A., Gandia, L. M., Sanchis, P., Hydrogen production from water electrolysis: Current status and future trends. *Proceedings of the IEEE*, 100, 410–426, 2012.

[26] Agua Caliente Solar Project, https://en.wikipedia.org/wiki/Agua_Caliente_Solar_Project. Accessed on November 15, 2022.

[27] Liebling, K., Leslie-Bole, H., Byrum, Z., Bridgwater, L., *6 Things to Know About Direct Air Capture*, www.wri.org/insights/direct-air-capture-resource-considerations-and-costs-carbon-removal. Accessed on November 15, 2022.

[28] Appl, M., *"The Haber–Bosch Process and the Development of Chemical Engineering". A Century of Chemical Engineering*, Plenum Press, New York, 1982, pp. 29–54, ISBN 978-0-306-40895-3.

[29] Zhang, S., Zhao, Y., Shi, R., Waterhouse, G. I. N., Zhang, T., Photocatalytic ammonia synthesis: Recent progress and future, *EnergyChem*, 1, 100013, 2019, https://doi.org/10.1016/j.enchem.2019.100013; www.journals.elsevier.com/energychem.

[30] Zheng, J., Zhou, H., Wang, C.-G., Ye, E., Xu, J. W., Loh, X. J., Li, Z., Current research progress and perspectives on liquid hydrogen rich molecules in sustainable hydrogen storage, *Energy Storage Materials*, 35, 695–722, 2021, Elsevier.

[31] Maroonian, A., *Ammonia for Clean Energy—Driving Clean, Sustainable Off-Grid Power Solution, Neology*, https://neology.ch/.

10 System Arrangements and Applications

10.1 INTRODUCTION

Energy storage systems, mainly based on electrochemical accumulators, appear today as complete arrangements dedicated to specific industrial and infrastructure storage solutions [1]. Such systems can be easily designed and parametrized thanks to dedicated software tools able to simulate the specific application and to facilitate the design. The integration of renewable energy sources together with the storage of energy can be optimized according to real or simulated load profiles and production schedules.

Four main applications are described in Reference [1]: storage as grid component, storage with PV systems, hybrid power plants and quick-charging facilities for electrical vehicles (EVs).

10.2 STORAGE AS GRID COMPONENT

The application consists of the connection of storage capacity to the private grid. The storage means serves to stabilize the load profiles, to cover power peaks and to limit the power consumption from the public grid.

The system offers grid services as primary and secondary control power.

The system structure is illustrated in Figure 10.1.

The connection to the public grid (1) needs a bidirectional counter (2) where not only the fluctuating power consumption (3) is measured but also the possible power injection from the battery (4). The conversion from DC current to AC is achieved through a voltage source converter (5). Such a device makes it possible to impose positive and negative current injection, as well as capacitive or inductive reactive current.

In Figure 10.2, the results that can be achieved with storage as grid component are illustrated. By profiling the power demand, it is possible to manage peak demand using power stored during low utilization periods.

The values indicated with blue bars represent the difference power provided by the storage equipment in addition to the constrained utility value (black dotted line). The shaded light blue values correspond to the additional power taken from the external grid for the compensation of the energy extracted from the battery during the peaks.

10.3 STORAGE WITH PV SYSTEMS

The storage system is integrated within the private grid at the level of a DC link, where the energy produced by the PV panels ((6) in Figure 10.3) is also injected. The

DOI: 10.1201/9781003736486-10

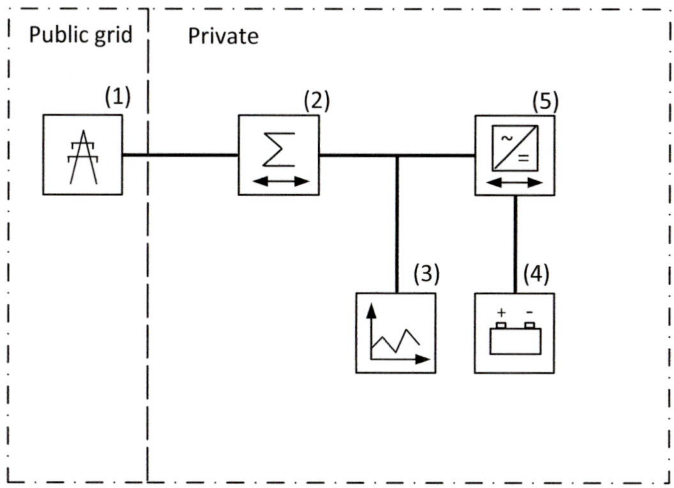

FIGURE 10.1 Storage as grid component. (Adapted from StoraXe, Industrial and infrastructure scalable large-scale storage solutions, ADS-TEC, GmbH, D-72622 Nürtingen, Germany, 2014 [Online], www.ads-tec.de/fileadmin/download/doc/brochure/Brochure_Energy_Industrial_EN.pdf. Accessed on August 7, 2017.)

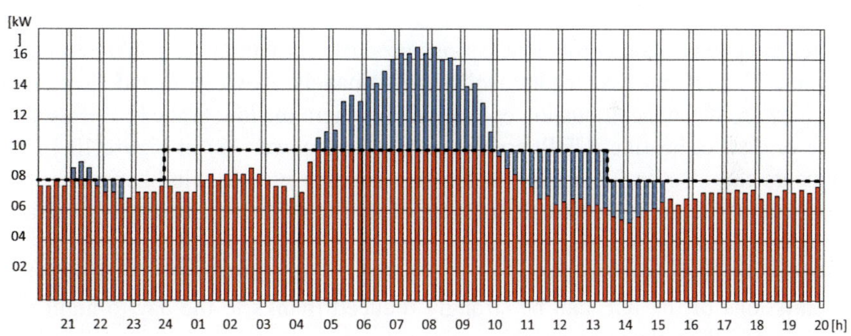

FIGURE 10.2 Profiled power demand using storage. (Adapted from Adams, M., *Solving Power Capacity Challenges with Software Defined Power*, Bodo's Power Systems, Laboe, Germany, pp. 24–28, September 2016.)

interface between the panels and the DC grid is achieved by the DC-DC converter (7) assuming the maximum power point tracking (MPPT) function.

Storage serves to stabilize the load profiles, to cover peaks, to limit the power consumption from the public grid or to limit the feed-in power. One main function covered by the storage is to make the so-called day-to-night shift of the power supply and to optimize the personal consumption. The system structure is illustrated in Figure 10.3.

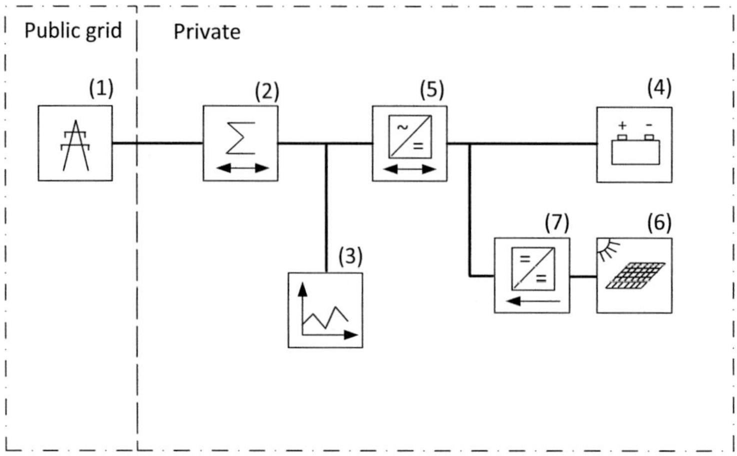

FIGURE 10.3 Storage with PV system. (Adapted from StoraXe, Industrial and infrastructure scalable large-scale storage solutions, ADS-TEC, GmbH, D-72622 Nürtingen, Germany, 2014, [Online], www.ads-tec.de/fileadmin/download/doc/brochure/Brochure_Energy_Industrial_EN.pdf. Accessed on August 7, 2017.)

10.4 HYBRID POWER PLANT

The hybrid power plant consists of various producer structures on a shared medium voltage (MV) DC bus. Near photovoltaic (PV) panels, they can include generators driven by an internal combustion engine (ICE), where the low-grade heat is used for house or district heating (CHP: combined heat and power; (8) in Figure 10.4). The producers can also be fuel cells (10) or other sources such as wind generators (12). The system structure is illustrated in Figure 10.4. The different sources are interfaced through dedicated power electronic converters (9), (11) and (13).

There are more and more examples of hybrid plants with energy storage where the emissions of the ICE can be strongly reduced thanks to the day-to-night shift principle based on electrochemical storage [2].

10.5 FAST CHARGING STATIONS

Fast charging of EV batteries is directly related to a high power availability. This is not always possible due to limited network capacity. Intermediate storage can be filled with cheap energy with low capacity and provide high capacity for fast charging (Figure 10.5).

In addition, the high-power reversible AC-DC converter can provide system-level services to the grid, reactive power compensation [2], active damping and frequency stabilization [3].

Vehicle-to-grid (V2G) concepts can be transferred to a new high power charger–to-grid (HPC2G) concept where the power level and the storage capacity are more significant [1].

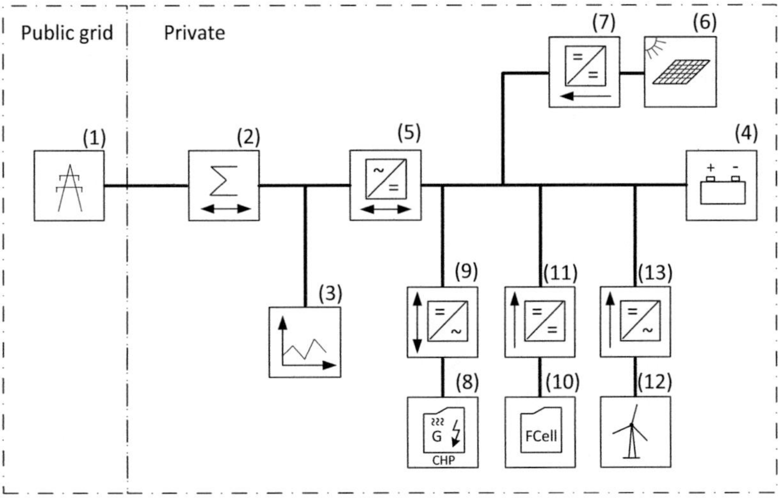

FIGURE10.4 Hybrid power plant. (Adapted from StoraXe, Industrial and infrastructure scalable large-scale storage solutions, ADS-TEC, GmbH, D-72622 Nürtingen, Germany, 2014 [Online], www.ads-tec.de/fileadmin/download/doc/brochure/Brochure_Energy_Industrial_EN.pdf. Accessed on August 7, 2017.)

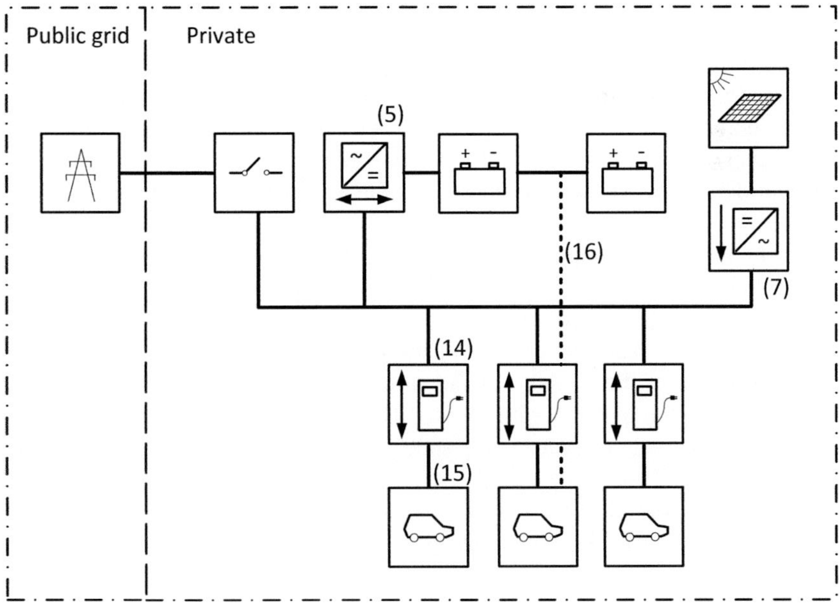

FIGURE 10.5 Fast charging concepts with intermediary buffer. (Adapted from StoraXe, Industrial and infrastructure scalable large-scale storage solutions, ADS-TEC, GmbH, D-72622 Nürtingen, Germany, 2014 [Online], www.ads-tec.de/fileadmin/download/doc/brochure/Brochure_Energy_Industrial_EN.pdf. Accessed on August 7, 2017.)

In the multiport charging station represented in Figure 10.5, a renewable energy source (RES) has been added to the system. In this specific case, the interconnection is done at the AC level (7), which has the consequence of multiple conversions and leads to a multiplication of partial efficiencies. A direct interconnection at the DC level is possible with a DC–DC converter between the PV panels and the battery.

In Figure 10.5, the fast charge of the car battery (15) is done from a charging station (14) connected to the local AC grid.

One possible way would be to integrate fast charge standards for a DC connection (16).

10.6 ADVANCED SYSTEM ARCHITECTURES

Fast charging stations and especially charging stations with multiple ports represent a significant level of the power at the connection point to the grid. Usually, electrical systems of several megawatts are interconnected to an MV grid. The adaptation of the low voltage of the charging devices to the MV level of the distribution grid has to be done with a dedicated power transformer.

Modern power electronics converters with integrated medium frequency transformers for galvanic insulation have been proposed for the realization of modular multilevel converter interfaces (MMCs) [4–6]. Such systems can be directly interfaced to the MV system without using classical low-frequency transformers. The flexibility in the control of the individual modules, together with the possibility to add so-called configuration switches, has led to the proposals of realizing configurable systems [7].

In relation to high power charging stations connected to a MV distribution system, advanced system architectures have been proposed [8, 9]. In such systems, the power interface from and to the MV grid is realized with a series connection of modular converters, allowing it to match the corresponding voltage level. The multiple modules and their dedicated controls allow it to reach a high quality of the exchanged line current, characterized by its very low harmonic distortion.

The principal scheme of the proposed system architecture is given in Figure 10.6.

In such a system, the principle of the use of an intermediary buffer is used. Such a buffer battery system can be used for different purposes, as already described in Section 10.5.

In the configurable multilevel system architecture represented in Figurel0.6, the intermediary buffer battery is split and integrated into the modules, more exactly at the DC output of the modules of the cascaded H-bridges. The interfaces to the vehicle batteries are realized with isolated DC-DC converters. These modules can be operated separately or according a specific parallel connection for an increased charging power if demanded.

The complex control of the system and its performant load balancing algorithm for all phases is described in Reference [9].

10.7 UNINTERRUPTIBLE POWER SUPPLY

An uninterruptible power supply (UPS) is an apparatus that provides emergency power to a load when the input power source or input mains power fails. A UPS provides near-instantaneous protection from input power interruptions by supplying energy

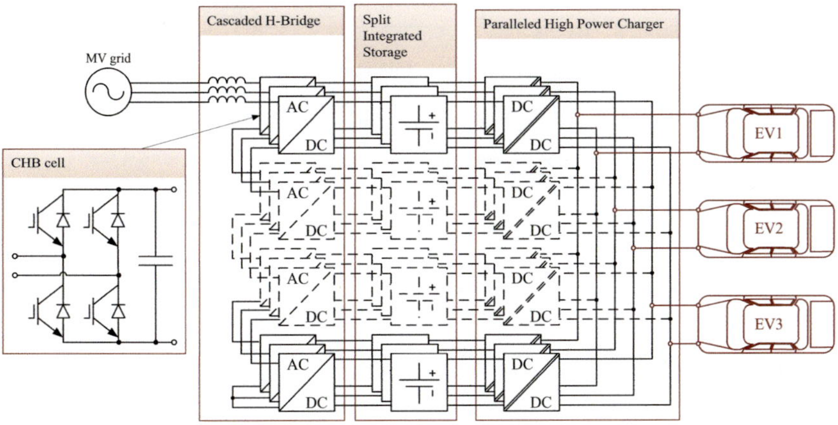

FIGURE 10.6 Configurable multiport charging system based on a modular multilevel converter. (Adapted from Vasiladiotis, M., Modular multilevel converters with integrated split battery energy storage, PhD Thesis 6406, EPFL, Ecole Polytechnique Fédérale de Lausanne, Lausanne, Switzerland, 2014.)

stored in a battery, in supercapacitors or in a flywheel. The on-battery runtime of most uninterruptible power sources is relatively short (a few minutes) and is sufficient to start a standby power generator or to shut down the protected equipment properly [10].

Typical application of a UPS is to protect computers, data centers, telecommunication equipment or other installations where an unexpected power disruption could cause injuries, fatalities, serious business disruption or data loss. Important start-up times of complex computer networks can also be a motivation for the use of UPS even if all individual components are protected against data loss.

10.7.1 UPS Technologies

Various types of UPS systems exist on the market and are presented under different names. In general, there are three different types of arrangements.

10.7.1.1 Offline

When the input mains supply is available, the load is connected to it through a commutation device or switch. The output switch commutates the load to the output of the inverter when the input voltage disappears, and the power is provided by the batteries. The system configuration of an offline UPS is shown in Figure 10.7.

10.7.1.2 Line-Interactive

When the input voltage is available, the load is connected to it through an automatic voltage regulator (AVR). This circuit filters and stabilizes the input voltage using passive components and in many cases a controlled multi-tap variable-voltage autotransformer.

Similarly to the offline circuit, the load is connected to a battery powered inverter when the input supply fails. The circuit is represented in Figure 10.8.

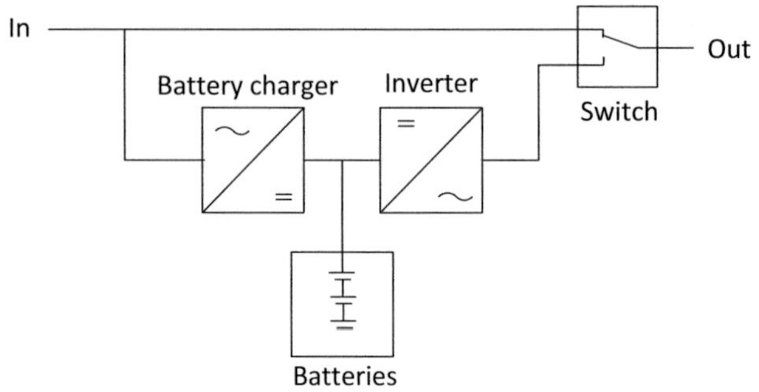

FIGURE 10.7 Offline UPS.

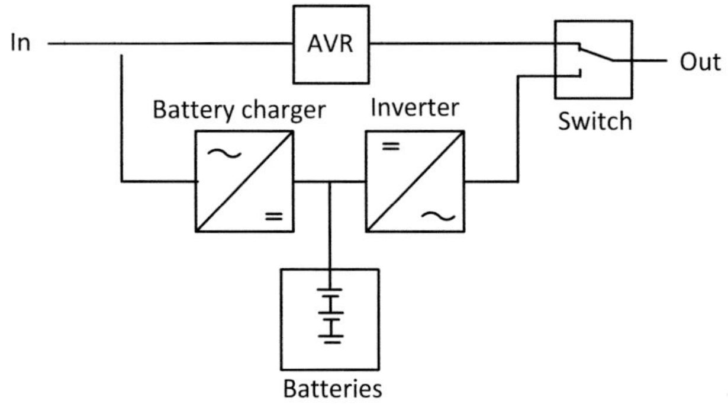

FIGURE 10.8 Line-interactive UPS.

10.7.1.3 Online Double Conversion

The input voltage is rectified and re-converted into alternating current. This way, the output voltage waveform is totally independent from the input. All mains disturbances are eliminated, and there is no transient time switching from the mains to the battery as the output is always powered by the inverter.

In the event of overloads or other eventual problems, this type of UPS has an automatic bypass that ensures the load is powered by switching it directly to the input.

The online double conversion UPS has the disadvantage the power is continuously affected by rectifier and inverter losses, resulting in a lower energy efficiency.

The circuit of the online double conversion UPS is represented in Figure 10.9.

For off-line and line-interactive systems, the battery charger is generally realized with simpler diode rectifiers. The distorted input current of this rectifier circuit is only of a limited impact due to the fact that for normal operating conditions the input current corresponds to the load current, and the fully charged battery imposes a negligible rectifier current.

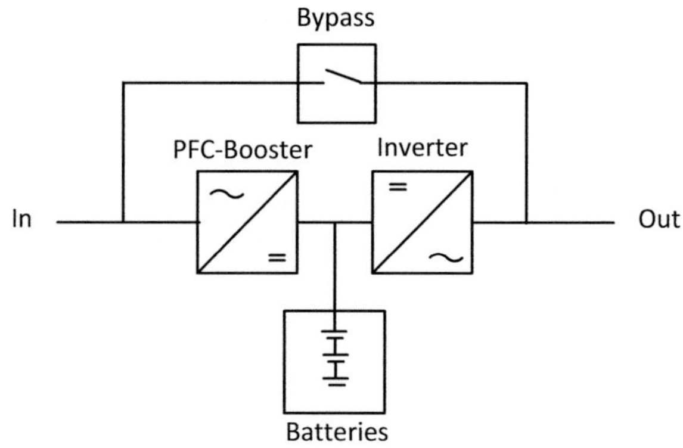

FIGURE 10.9 Online double conversion UPS

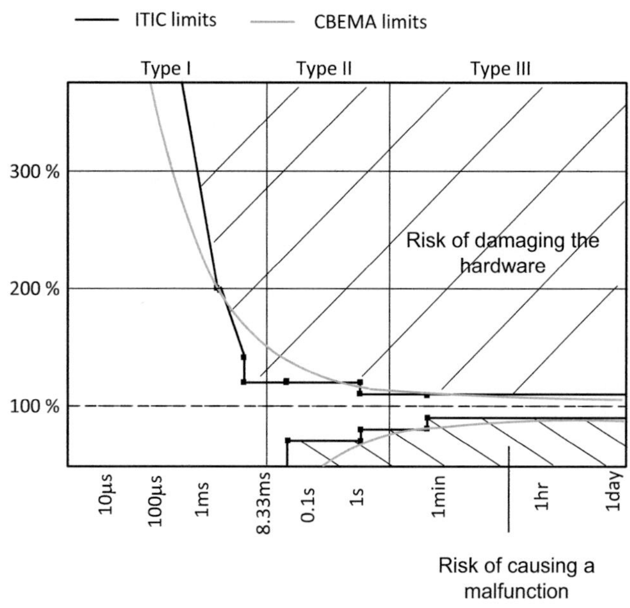

FIGURE 10.10 ITIC immunity curve. (Adapted from Elphick, S., Smith, V., The 230 V CBEMA curve – preliminary studies, *2010 20th Australasian Universities Power Engineering Conference*, Christchurch, New Zealand, 2010.)

For the online double conversion UPS, more sophisticated rectifier circuits are used, such as so-called power factor corrected (PFC) rectifiers or active rectifiers with pulse width–modulated (PWM) conversion circuits with high power factors or low total harmonic distortion (THD) factors.

10.7.2 Performances and Recommendations for UPS

Performances of UPS are defined through classification standards EN 62049–3 [11], and the selection of a solution in a given context can be made according to different recommendations, such as the Information Technology Industry Council (ITIC) immunity curve (Figure 10.10), the Computer Business Electronic Manufacturer's Association (CBEMA) note, and the ANSI/IEEE Standard 446–1995 "IEEE Recommendation Practice for Emergency and Stand-By Power for Industrial and Commercial Applications".

Many recommendation and design notes are available from practically all equipment manufacturers, such as the examples of [11, 12].

10.8 GRID-SCALE CO_2 BATTERY

A new large scale and expandable grid compatible storage system has been recently proposed [13] and is based on high-pressure liquefied CO_2. The CO_2 Battery technology transforms CO_2 from its gaseous to liquid state and vice-versa. When operating in charge mode, the CO_2 is withdrawn from an atmospheric gasholder, the dome, and compressed into an inter-refrigerated compressor driven by a motor. The heat generated from the compression is stored in a thermal energy storage system, whereas the CO_2 is liquefied and stored in vessels under pressure at ambient temperature with zero atmospheric emissions. This process allows for high-density energy storage without using extreme cryogenic temperatures. In discharge mode-whenever energy is needed-the CO_2 is heated, evaporated and sent into an expander before flowing back into

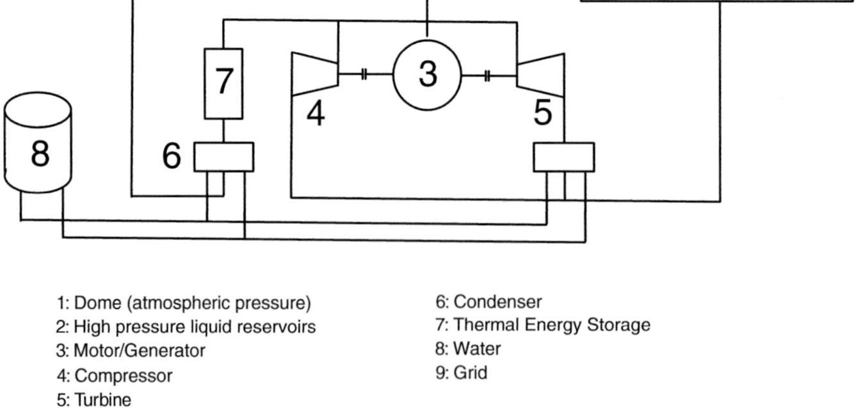

FIGURE 10.11 Structural diagram of the CO_2 Battery System

the dome. In doing so, the expander drives the generator to feed carbon-neutral electricity to the grid.

The CO_2 battery has high round-trip efficiency, meaning it returns 75% of the energy it absorbes during its charge. There is no degradation during its prolonged life (30+years), and its performance is third-party verified. Furthermore, the system relies on off-the-shelf equipment from established global supply chains, ensuring high scalability, other than ease of operation and maintenance.

Figure 10.11 gives a structural diagram of a CO_2 battery system.

REFERENCES

[1] StoraXe, Industrial and infrastructure scalable large-scale storage solutions, *ADS-TEC, GmbH, D-72622*, Nürtingen, Germany, 2014 [Online], www.ads-tec.de/fileadmin/download/doc/brochure/Brochure_Energy_Industrial_EN.pdf. Accessed on August 7, 2017.

[2] Theodorescu, R., Liserre, M., Rodriguez, P., *Grid Converters for Photovoltaic and Wind Power Systems*, John Wiley & Sons, 2011.

[3] John, M. et al., Design and evaluation of a test rig for full-size grid frequency stabilizer modules with ultracapacitors, *EPE-ECCE Conference*, Karlsruhe, Germany, September 5–9, 2016.

[4] Rufer, A., Schibli, N., Briguet, C., A direct coupled 4-quadrant multilevel converter for 16 2/3 Hz traction systems. *Proceedings of the IEEE Conference on Power Electronics and Variable Speed Drives (PEVD)*, Nottingham, UK, September 1996.

[5] Dujic, D. et al., Power electronic traction transformer-low voltage prototype, *IEEE Transactions on Power Electronics*, 28(12), 5522–5534, December 2013.

[6] Hõimoja, H., Vasiladiotis, M., Grioni, S., Capezzali, M., Rufer, A., Püttgen, H. B., Towards ultrafast charging solutions of electric vehicles, *Cigré Session 44*, Paris, France, August 26–31, 2012.

[7] Rufer, A., Schibli, N., Chabert, C., Zimmermann, C., Configurable front-end converters for multicurrent locomotives operated on 16 2/3 Hz AC and 3 kV DC systems, *IEEE Transactions on Power Electronics*, 18(5), 1186–1193, 2003.

[8] Vasiladiotis, M., Rufer, A., A modular multiport power electronic transformer with integrated split battery energy storage for versatile ultrafast EV charging stations, *IEEE Transactions on Industrial Electronics*, 62(5), 3213–3222, May 2015.

[9] Vasiladiotis, M., Modular multilevel converters with integrated split battery energy storage, PhD Thesis 6406, Ecole Polytechnique Fédérale de Lausanne, (EPFL), Lausanne, Switzerland, 2014.

[10] Uninterruptible Power Supply (UPS), *Technical Guide*, 2012, http://ups.legrand.com/wp-content/uploads/2011/07/Guida_UPS_EN.pdf. Accessed on August 7, 2017.

[11] Furlong, E. R., UPS topologies for large critical power systems (>500 kVA), *13th Annual Power Quality Exhibition & Conference*, Atlanta, GA, October 30, 2002.

[12] Elphick, S., Smith, V., The 230 V CBEMA curve—preliminary studies, *2010 20th Australasian Universities Power Engineering Conference*, Christchurch, New Zealand, 2010.

[13] Waltz, E., Grid-scale CO_2 Batteries will take off in 2026, Spectrum, IEEE.org, January 2026, pp. 24–27.

Appendix A: Power Electronics: Basic Circuits and Principles

A.1 INTRODUCTION

In Section 1.2.3, the advantages of the distribution of energy through an electrical vector were shown, and a general system structure representation was given in Figure 1.5. In this representation, an important subcomponent is the power electronics converter. Such a converter is an essential component of any energy storage system and was mentioned in Chapter 2 as well for the feeding of an electrochemical battery or a supercapacitor with a constant current as for the frequency converter used for the variable speed of a flywheel system. This appendix will briefly present the principles and main circuits of the power electronics systems in order to facilitate the understanding of these important interfaces in all presented storage systems of this book.

A.2 VARIATION TECHNIQUE BY SWITCHING

The variation of electrical quantities can be done by the use of a power amplifier or any power converter. Especially in the energy sector, the transformations must be done with very high energy efficiency. Power electronic circuits based on the so-called switching principle are the best candidates for this function and are commonly used in practically all electrical energy applications.

The principle of switching is based on the periodic commutation from one level (for example, a voltage level) to another one and to generate a variation of the average value by controlling the duty cycle of the periodic commutation. Switching from one level to another needs nearly ideal switching devices, characterized by their very low conduction losses. This leads to very low power dissipation during the on-state of the switch when conducting current. Such switching devices are based on fast controllable power semiconductors [1–3].

A.2.1 Ideal DC-DC Conversion Principle

The principle of the ideal DC-DC conversion is represented in Figure A.1. In such a conversion, a variable output quantity (U_{med}) is obtained by switching periodically from one constant level (U_1) to another (U_2).

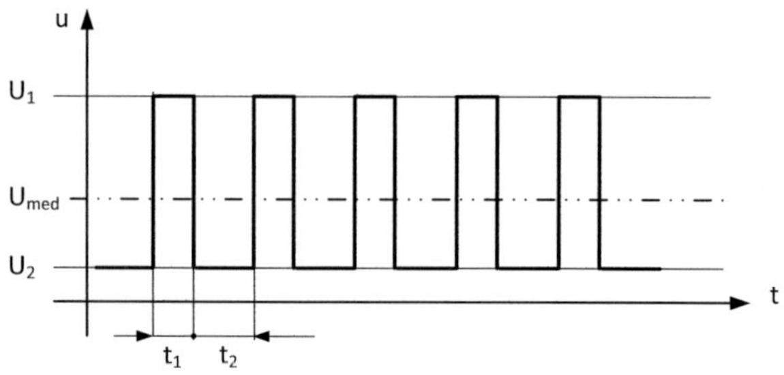

FIGURE A.1 The ideal DC-DC conversion principle.

The instantaneous output quantity shows an important ripple but acts energetically as a controllable average value, generally filtered by smoothing components of the fed system, like inductors, or by specially added filter elements.

For the average value of the output variable, the following relationship is valid:

$$U_{med} = \frac{U_1 \cdot t_1 + U_2 \cdot t_2}{t_1 + t_2} \tag{A.1}$$

The variation of U_{med} is reached by varying the duty cycle $t_1/(t_1 + t_2)$. The switching frequency is equal to $f_{\text{p}} = 1/\left(t_1 + t_2\right)$.

DC-DC switching converters can be realized with the step-down converter (buck converter) described in Section A.3.1. Step-up converters (boost converters) are described in Section A.3.2.

A.2.2 Ideal DC-AC Conversion Principle

The DC-AC conversion principle is an extension of the DC-DC principle where the DC voltage is connected to the output in the positive direction for the generation of a positive half wave and in the negative direction for a negative half wave. The principle is illustrated by Figure A.2.

The duty cycle of the periodic switching between the DC voltage U and zero is modulated with a sine wave, generating a fundamental wave ${}^1\hat{U}$.

A.3 BASIC POWER ELECTRONICS CONVERTERS

A.3.1 Step-Down Converter (Buck Converter)

The step-down converter is used to transfer electrical energy from one DC source to another of lower voltage value. Typical application is the feeding of a DC load. The scheme is represented in Figure A.3, together with the output voltage waveform U_{oi}.

The switching transistor T_1 is periodically switched on and off through the signal S. The inductor current I_{L} flows alternately from the input circuit or through the free-wheeling path of the diode D.

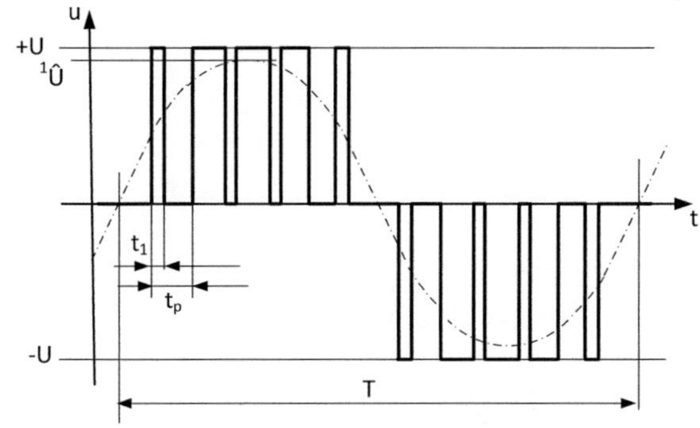

FIGURE A.2 The ideal DC-AC conversion principle.

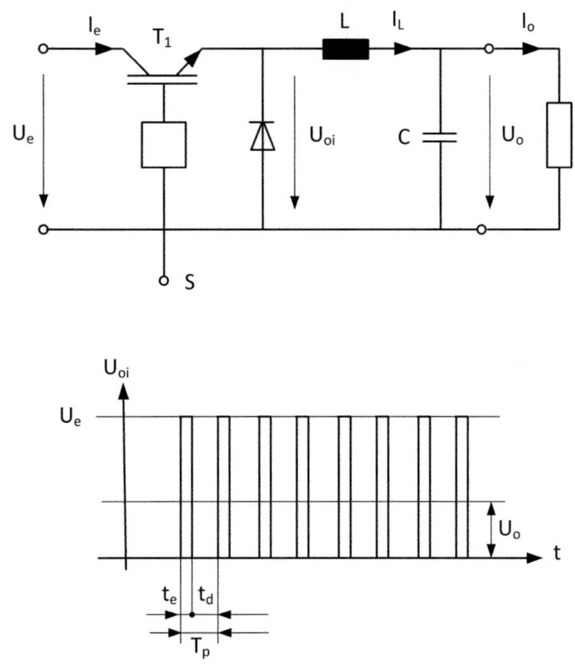

FIGURE A.3 Step-down converter and output voltage waveform.

The output voltage U_o is filtered by the filter capacitor C. Its value is calculated from Equation A1.2:

$$U_o = D \cdot U_e \tag{A.2}$$

where D is the duty cycle and is given by

$$D = \frac{t_e}{T_p} \tag{A.3}$$

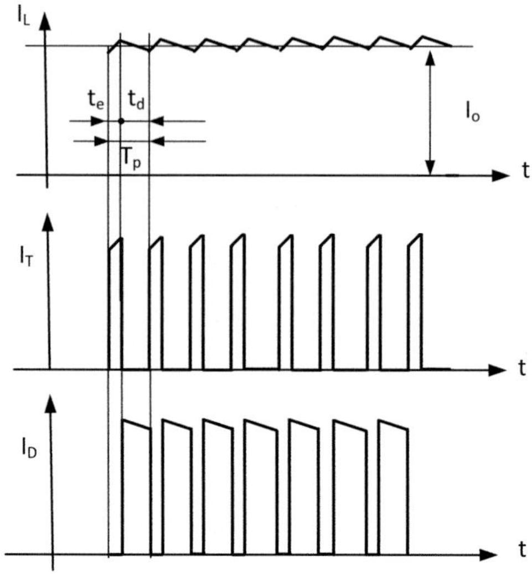

FIGURE A.4 Currents in the step-down converter.

For an ideal converter (without losses), the input power is identical to the output power:

$$U_o \cdot I_o = U_e \cdot I_e \tag{A.4}$$

giving for the input current

$$I_e = (U_o / U_e) \cdot I_o = D \cdot I_o \tag{A.5}$$

The waveforms of the currents in the inductor (I_L), in the transistor (I_T) and in the diode (I_D) are represented in Figure A.4. The input current (I_e) is identical to the transistor current (I_T).

A.3.1.1 Loses in a Switching Converter

In a real converter, energy losses are produced mainly in the used semiconductor devices. From the example of the step-down converter represented in Figure A.3, it is easy to understand that there are losses while the current is flowing through the transistor and through the diode. These losses are called conduction losses [3]. Additionally, the commutation from the transistor to the diode and reciprocally is not infinitely fast, and during a short time, there can be an important dissipation due to a simultaneous current conduction in the switching device and a voltage at its main terminals (delayed fall of the voltage by the turn-on phenomenon or tail current after turn-off).

Simplified curves of the switch voltage and current during turn-on and turn-off are given in Figure A.5. These losses are called switching losses, and they depend on the switching frequency.

A.3.1.1.1 Conduction Losses

For the converter represented in Figure A.3, there are two main states of conduction

1. Transistor ON Diode OFF
2. Transistor OFF Diode ON

Table A.1 gives the corresponding states of the variables associated with the switches for the two states.

In Table A.1, there are two situations where conduction losses occur, namely when the on-state voltage of the switch is not negligible (U_T, $U_D \sim 0$). The notation ~ 0 indicates that the switches are no longer blocking but that their on-state voltage still produces losses.

The on-state losses of a transistor can be calculated through

$$P_{onT} = U_{CEon} \cdot I_T \quad \text{(A.6)}$$

U_{CEon} is the collector-emitter on-state voltage of the transistor

$$P_{onD} = U_{Don} \cdot I_D \quad \text{(A.7)}$$

U_{Don} is the on-state voltage of the diode

The total conduction losses must be calculated according to the time of conduction of each element, namely t_e and $T_p - t_e$

$$P_{cond} = P_{onT} \frac{t_e}{T_p} + P_{onD} \frac{T_p - t_e}{T_p} \quad \text{(A.8)}$$

TABLE A.1
States of Conduction of the Step-Down Converter

State of Conduction	1.	2.
Transistor	ON	OFF
U_T	~0	U_e
I_T	I_L	0
P_T (transistor dissipation)	~0	0
Diode	OFF	ON
$I_e = I_T$	U_e	~0
I_D	0	I_L
P_D (diode dissipation)	0	~0

A.3.1.1.2 Switching Losses

When turned on, a power semiconductor switch effectively turns the current on, while the voltage across its main terminals falls down to the characteristic on-state value. However, the voltage fall takes some non-negligible time (a few hundred of nanoseconds to 1 microsecond). Additionally, when a power transistor is turned on in a so-called hard switching situation where an anti-series diode has to be blocked, the reverse recovery current in the diode makes it unable to overtake voltage during a recovery time (hundreds of nanoseconds).

By turning the switch off, often the voltage across the switch rises while the element is still conducting some current or shows a typical current tail.

The consequence is a kind of overlap of voltage and current, which causes an important but short power dissipation of the active element.

In the framework of this contribution, that phenomenon is strongly simplified, as represented in the schematic transient curves of a turn-on and a turn-off of a power semiconductor switch (Figure A.5).

In practice, it is easy to take into account the indicated switching energies by turning on and off E_{on}, E_{off} from the datasheet of the component manufacturer.

The mean value of the switching losses is calculated as a "switching power" according to

$$P_{sw} = (E_{on} + E_{off}) \cdot f_p \tag{A.9}$$

where f_p corresponds to the switching frequency of the converter.

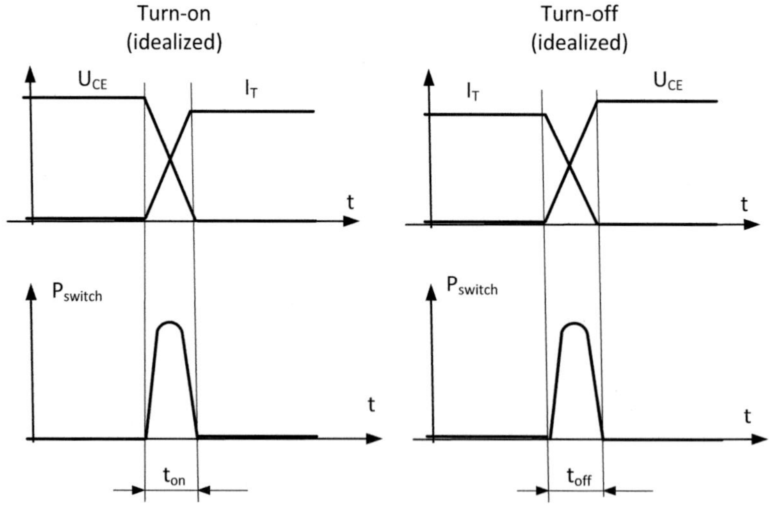

FIGURE A.5 Idealized switching losses.

A.3.2 Step-Up Converter (Boost Converter)

The step-up converter is used to transfer electrical energy from one DC source to another of a higher value. A typical application is the feeding into a DC circuit from a strongly variable source like a photovoltaic panel. The scheme is represented in Figure A.6, together with the input voltage U_e, the output voltage U_o and the voltage across the inductor U_L.

The current waveforms in the elements of the step-up converter are represented in Figure A.7. One can note the continuous input current of the converter, in opposition to the discontinuous input current of the step-down converter (Figure A.4, $I_e = I_T$).

The relationship between the input and the output voltages is given by

$$\frac{U_o}{U_e} = \frac{T_p}{t_d} = \frac{T_p}{T_p - t_e} = \frac{1}{1-D} \quad \text{(A.10)}$$

The relationship for the currents is

$$\frac{I_o}{I_e} = 1 - D \quad \text{(A.11)}$$

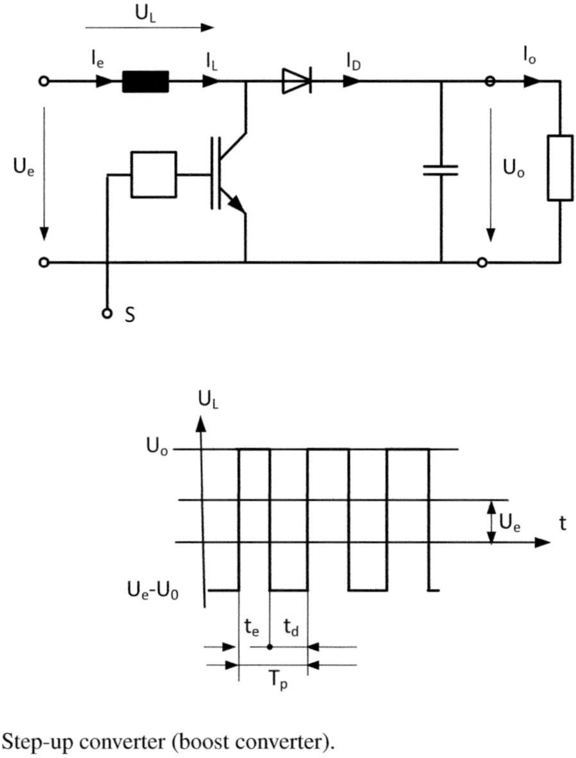

FIGURE A.6 Step-up converter (boost converter).

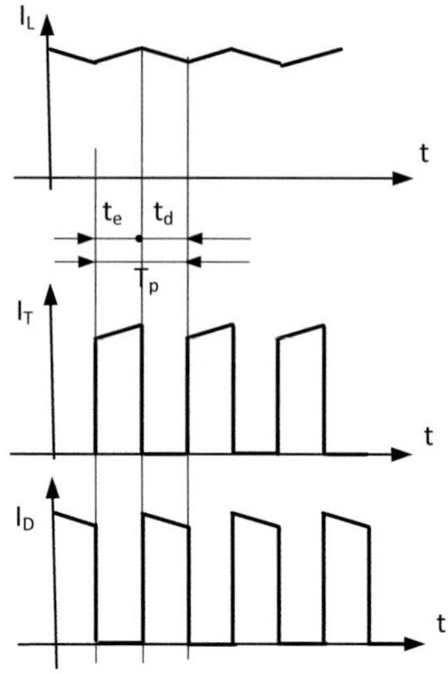

FIGURE A.7 Step-up converter (boost converter), current waveforms.

The definition of the duty cycle D is still valid

$$D = \frac{t_e}{T_p}$$

The calculation of the conduction losses and switching losses is based on similar relationships and definitions as for the step-down converter.

A.3.3 Phase-Leg or Half-Bridge

The phase-leg circuit results as a combination of a step-down and a step-up converter. It allows a bidirectional power flow from one DC level to another. According to the representation of Figure A.8, the input voltage U_{10} must be lower than the output voltage U_d. A power flow from left to right corresponds to the operation of a voltage step-up converter, while the power flow from right to left corresponds to a step-down operation. The input voltage can be between zero and U_d and is strictly positive. The input current I_e, however, can be positive or negative.

A simple comparison with Figures A.3 and A.6 shows the combination of a step-down converter (T and D') and a step-up converter (T' and D). The inductor L is common to both step-up and step-down.

The control of the two power switches T and T' must be strictly alternating, avoiding short circuit situations (T simultaneously ON with T'). This is guaranteed by the logic circuit L (Figure A.8). The control of both switches T and T' from the d signal is represented in Figure A.9.

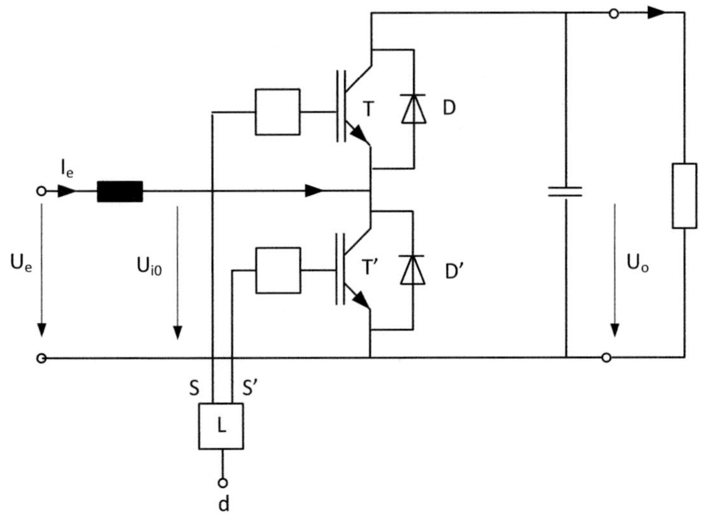

FIGURE A.8 The phase-leg or half-bridge converter circuit.

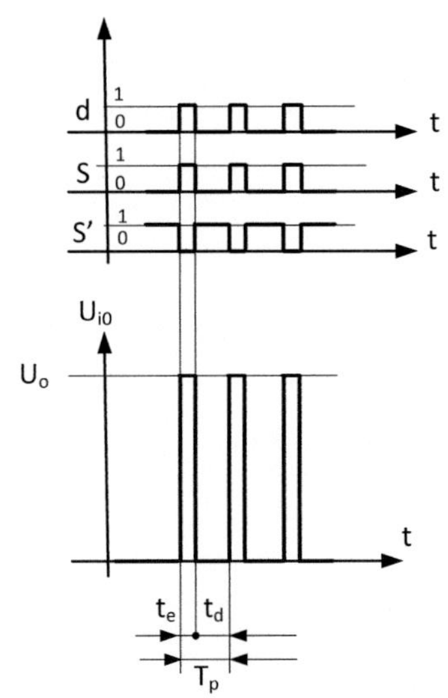

FIGURE A.9 Switching signals and leg voltage of the half-bridge converter.

In Figure A.9, the control signal d is represented, together with the control signals S and S' of the power switches T and T'. The resultant leg voltage U_{i0} of the converter is also represented. Its average value is controlled through the duty cycle t_e / T_p.

A.3.4 Full-Bridge Converter

The full-bridge converter, also called an H-bridge converter, is a combination of two phase-leg converter circuits. The phase-legs are connected together at the level of the constant DC voltage with a common capacitor. The scheme of the full-bridge converter is represented in Figure A.10.

The input voltage U_{12} of a full-bridge converter is equal to the difference of the two leg voltages U_{10} and U_{20}.

$$U_{12} = U_{10} - U_{20} \tag{A.12}$$

This allows generation of instant values of the input voltage that are positive or negative. As a consequence, the full-bridge converter can be used as a reversible DC-DC converter (input voltage average positive or negative) or as an AC-DC or a DC-AC converter, called an inverter for the generation of an AC voltage from a DC level or an active rectifier for feeding into a DC link from an AC source.

The different voltage waveforms of the full-bridge converter are represented in Figure A.11. At the left side of the figure, the so-called bipolar voltage switching mode is represented. The leg-voltages U_{10} and U_{20} are represented, as well as the input voltage U_{12}. In the middle of the figure, the unipolar voltage switching is represented, where the legs are controlled in order to produce a positive average voltage at the input (U_{12}). At the right side of the figure, another unipolar voltage switch is represented, where the produced voltage U_{12} is negative. A combination of positive unipolar switching and negative unipolar switching results in the generation of AC voltage.

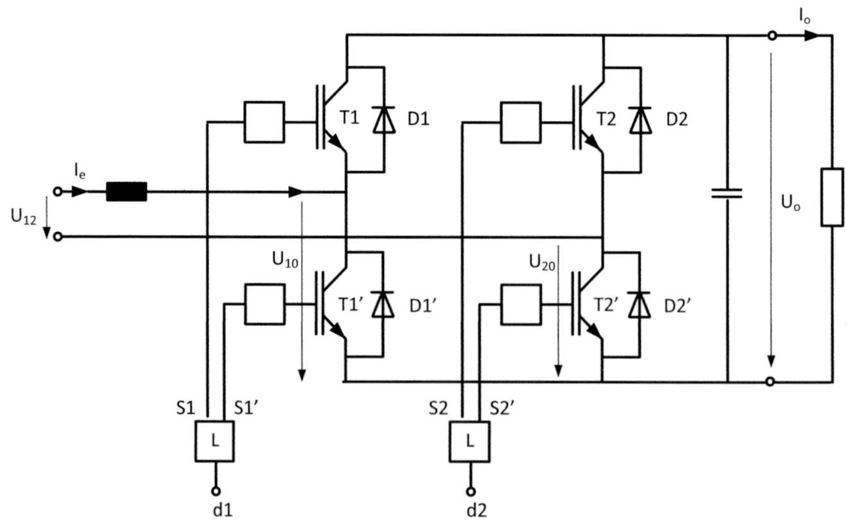

FIGURE A.10 Full-bridge converter.

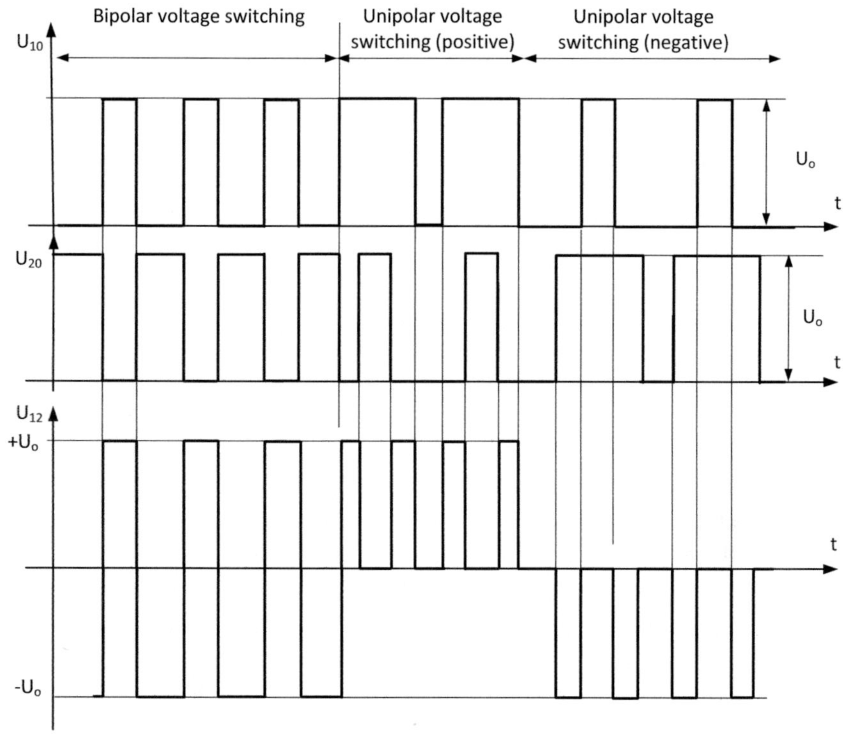

FIGURE A.11 Bipolar and unipolar voltage switching (full-bridge).

A1.3.5 Three-Phase PWM Converter

By adding one leg to the full-bridge converter (Figure A.10), a three-phase converter is obtained. With this converter, power can be injected (inverter mode) or extracted (rectifier mode) from a three-phase AC network. This converter can provide active power (positive and negative), as well as capacitive or inductive reactive power. The scheme of the three-phase PWM converter is represented in Figure A.12.

The different waveforms of the three-phase PWM converter are represented in Figure A.13. The first line shows the curves used for the PWM modulation, namely one triangular carrier signal U_{tr} and three modulation signals for each channel of the modulator U_{ref1}, U_{ref2}, U_{ref3}.

In the modulator, the reference signals are compared with the carrier signal for the generation of the corresponding phase control signals. The leg voltages related to phases one and two U_{10} and U_{20} are represented in the second and third lines of the figure.

Finally, the line-to-line voltage of the converter (U_{12}) is represented in the last line of the figure. The signals related to the other phases of the converter show identical waveforms but are shifted 120°.

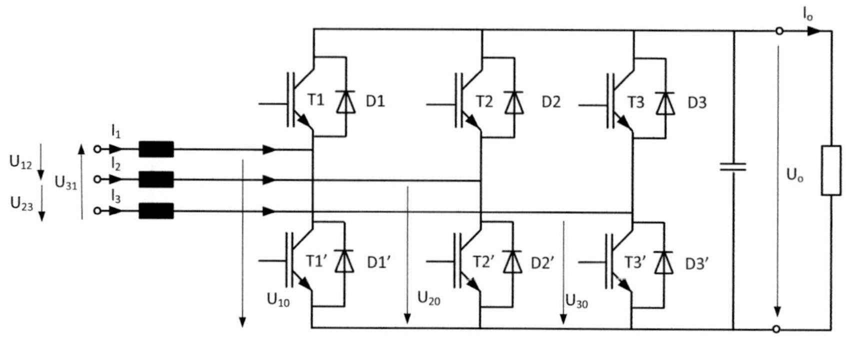

FIGURE A.12 Three-phase PWM converter.

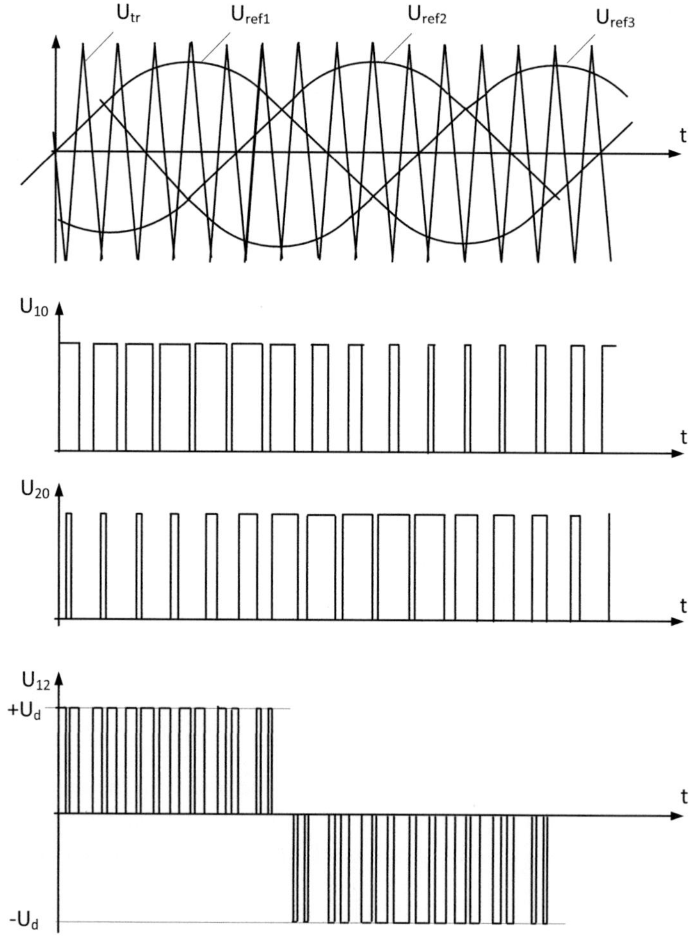

FIGURE A.13 Modulation and voltage waveforms of the three-phase PWM converter.

The control of the three-phase PWM converter is realized on the basis of a so-called vector control for the AC currents. This control imposes the magnitude of the current as well as its phase [4]. The vector control of the current allows to impose the magnitude of the active and reactive power exchanged with the AC network [4, 5].

REFERENCES

[1] Barrade, P., *Electronique de puissance, méthodologies et convertisseurs élémentaires*, PUR, Presses Polytechniques et Universitaires Romandes, Lausanne, Switzerland, 2006.

[2] Bühler, H., *Convertisseurs statiques*, PPUR, Presses Polytechniques et Universitaires Romandes, Lausanne, Switzerland, 2015.

[3] Mohan, N., Undeland, T., Robbins, W., *Power Electronics, Converters, Applications and Design*, John Wiley & Sons, Hoboken, NJ, 2002.

[4] Bühler, H., Rufer, A., *Control of Power Electronics Systems*, PPUR, Presses Polytechniques et Universitaires Romandes, Lausanne, Switzerland, 2017.

[5] Teodorescu, R., Liserre, M., Rodriguez, P., *Grid Converters for Photovoltaic and Wind Power Systems*, John Wiley & Sons, Hoboken, NJ, 2011.

Appendix B: Energy and the Integration of Power

B.1 INTRODUCTION

Several physical components have the ability to accumulate some energy related to the power flow they are subject to. The accumulated energy corresponds to the time integral of the power according to the definition of Section 1.1.2, Equation 1.1.

The most important components belong to the category of the electrical, magnetic or even mechanical sector.

The calculation of the accumulated power is done first on the basis of the example of the electric inductor, and the demonstration of these typical properties will be extended to other devices. Capacitors are concerned but also the moving or rotating mass or inertia.

B.2 THE EXAMPLE OF THE ELECTRICAL INDUCTOR

The inductive power of a magnet or inductor is calculated as the product of its voltage multiplied by the circulating current.

The voltage across an inductor being

$$U_L = L\frac{dI}{dt} \tag{B.1}$$

The inductive power becomes

$$P_L = L\frac{dI}{dt}\cdot I \tag{B.2}$$

which corresponds to the instantaneous power of an ideal inductor and is defined in the general case of AC or DC feeding.

For a feeding with a non-DC-component free voltage, there is an energy accumulated over the time that is calculated through

$$W_L = \int_0^t L\frac{dI(t)}{dt}\cdot I(t)\cdot dt \tag{B.3}$$

The quite strange form of a function of time multiplied by its derivative can be found for different energy accumulating processes as

$$\text{Inductor}: P_L = L\frac{dI_L(t)}{dt}\cdot I_L(t) \tag{B.4}$$

$$\text{Capacitor}: P_C = C\frac{dU_C(t)}{dt}\cdot U_C(t) \tag{B.5}$$

$$\text{Moving mass}: P_m = m\frac{dV_m(t)}{dt}\cdot V_m(t) \tag{B.6}$$

$$\text{Rotating mass}: P_{mrot} = J\frac{d\omega(t)}{dt}\cdot \omega(t) \tag{B.7}$$

For all these systems, the calculation of the accumulated energy over the time is calculated through the integration an expression of the form

$K \cdot f(t)\cdot f'(t)$, $f'(t)$ being the time derivative of the function $f(t)$

So for the accumulated energy

$$W = K\int_0^t f(t)\cdot f'(t)\cdot dt \tag{B.8}$$

the integration is done on the basis of the relation describing the derivation of a product of two functions, namely

$$\left(u\cdot v\right)' = u'\cdot v + u\cdot v' \tag{B.9}$$

by integration

$$u\cdot v = \int u'\cdot v\cdot dt + \int u\cdot v'\cdot \mathrm{dt} \tag{B.10}$$

with $u = v$ it becomes

$$u^2 = \int u'\cdot u\cdot dt + \int u\cdot u'\cdot \mathrm{dt} \tag{B.11}$$

or

$$\int u'\cdot u\cdot dt = \frac{u^2}{2} \tag{B.12}$$

Finally, the energy accumulated in the different components can be calculated for

$$\text{Inductor}: W_L = L\int_0^t \frac{dI_L(t)}{dt}\cdot I_L(t)\cdot dt = L\frac{I^2(t)}{2} \tag{B.13}$$

$$\text{Capacitor}: W_C = C\int_0^t \frac{dU_C(t)}{dt} \cdot U_C(t) \cdot dt = C\frac{U_C^2(t)}{2} \tag{B.14}$$

$$\text{Moving mass}: W_m = m\int_0^t \frac{dV_m(t)}{dt} \cdot V_m(t) \cdot dt = m\frac{V^2(t)}{2} \tag{B.15}$$

$$\text{Rotating mass}: W_{mrot} = J\int_0^t \frac{d\omega(t)}{dt} \cdot \omega(t) \cdot \mathrm{dt} = J\frac{\omega^2(t)}{2} \tag{B.16}$$

The conditions to respect are

- $U_c(t_0) = 0,\ I(t_0) = 0,\ V(t_0) = 0,\ \omega(t_0) = 0$

The accumulated energy depends

- on a constant (C, L, m, J)
- on the square of the value of the variable reached at the considered time

The accumulated energy is independent from the profile or form of the function $f(t)$ with which the final value was reached.

Index

N

O

P

R

S

For Product Safety Concerns and Information please contact our
EU representative GPSR@taylorandfrancis.com Taylor & Francis
Verlag GmbH, Kaufingerstraße 24, 80331 München, Germany